Praise for *The Life Cycle*

'Kate Rawles is quite simply the best travelling companion you could dream up. Her conversational style, ear for an anecdote and searing observations focused on our need to protect biodiversity are a tour de force … I finished the book with a sense of regret that the adventure was over, inspired by the awesome and deeply melancholy at the hells she visited along the way. Welcome to the complexity of the real world.'

Sir Tim Smit OBE,
co-founder of the Eden Project

'Profound and funny, philosophical and gritty, this book shares both the pain of an incredibly brave woman traveller and the enchantment as she meets the pioneers of lifestyles that seek to restore biodiversity rather than exploit it. A gripping read for anyone who cares about what we're doing to the planet and how we can change it.'

David Shukman, former BBC News science editor and visiting professor in practice at the LSE's Grantham Research Institute.

'Rawles built a bamboo bike for one, but with this book she takes each reader on her heart-wrenching and heart-warming ride through South America and into the pounding soul of the vibrant biodiversity we have ignored for way too long.'

Christiana Figueres, co-host of the *Outrage and Optimism* podcast and former head of the UN Framework Convention on Climate Change

'*The Life Cycle*'s pace is brisk, the vistas magnificent, the many characters encountered along the way compellingly and entertainingly brought to life. Even the all-important diversions … leave one feeling stronger, more resolute than ever to support the causes and organisations she champions. This is such an informative, uplifting and truly important book, making all the right connections across many different areas of concern.'

JONATHON PORRITT, author and campaigner

'Kate's epic 8,000-mile journey on a bamboo bicycle was a fabulous adventure, but she also harnesses the power of adventure to inspire environmental action by bringing to life the tragedy of biodiversity loss that requires profound systemic change to tackle.'

ALASTAIR HUMPHREYS, author, adventurer and host of the *Living Adventurously* podcast

'This fabulous book will make you want to live more fully, buy less junk and appreciate our world more. It will also make you want to rewire the whole economy and scream about the mess we are making. And it will make you want to jump on your bike.'

MIKE BERNERS-LEE, author of *There is No Planet B*

'Riveting, poignant and laugh-out-loud funny. From the "heart of the world" in Colombia to the devastating lead mines in Peru and from the coloured lakes of Bolivia to the final breathless dash for Ushuaia, *The Life Cycle* is un-put-downable. Its imagery will stay with you long after the last page is turned. From her own extraordinary endurance – and the stories of those she met along the way – Rawles has conjured up a kaleidoscopic "cosmovision" for our times: a passionate call to fight for the soul of the natural world – and, in doing so, to rescue our own.'

TIM JACKSON, author of *Post Growth: Life After Capitalism*

‘*The Life Cycle* will change your life. Or it should. Here is one of those rare flowers of a story whose message is as powerful, and urgent, as the beautiful writing used to tell it. It will move you, as it did me. Open it, but don’t just read it. Savour it.’

Carlos ZORRILLA, environmental activist,
writer and photographer

‘Kate Rawles is an extraordinary woman – keen adventuress, intrepid cyclist, curious thinker, passionate environmentalist and a fabulous storyteller. Riding with her along high Andean roads but also through terrifying traffic, we get fascinating insights into people, environmental projects and the threat to biodiversity and our beautiful planet. I loved this book.’

ANDREA WULF, author of
The Invention of Nature

‘An epic tale, passionately and powerfully told, which is less a simple travelogue and more a call to arms for urgent action to save our planet’s precious biodiversity ... Rawles is an authentic, compelling narrator who acts as a living epitome of the eco-values she espouses ... A deeply thought-provoking and essential read.’

REBECCA LOWE, author of
The Slow Road to Tehran

‘A beautifully written story of eco-adventure and eco-pilgrimage... It is as much an inspiring travelogue as it is a plea to care for the diversity of life on our precious planet ... an enchanting narrative told passionately by an eco-warrior. Reading this book is an immensely engaging and entertaining as well as heartbreaking experience. Read this book, you might become an eco-activist!’

SATISH KUMAR, editor emeritus
Resurgence and Ecologist and
founder of Schumacher College

'In this remarkable journey, Kate Rawles brings the biodiversity crisis to vivid life. And she does it in a way that is at once thrillingly gripping, intimately heartbreaking, touchingly funny and full of fierce hope. This is a book about perseverance and determination ... Few books have illuminated so clearly and honestly what is at stake. A magnificent, inspiring and unforgettable ride.'

JULIAN HOFFMAN, author of *Irreplaceable*

'I was captivated by Kate's unique ability to take such complex and paramount matters and craft them into a thrilling, meaningful and accessible story ... *The Life Cycle* will be taking pride of place on my bookshelf'.

JENNY GRAHAM, world record-breaking endurance cyclist, presenter and author

'A call to arms to protect what's left of our precious natural world. Kate's explorations open up new perspectives, helping us understand how our daily choices impact on people and species that may be far away, but with whom we are intimately linked and co-dependent.'

HELEN BROWNING OBE, organic farmer, author and CEO of the Soil Association

THE LIFE CYCLE

To environmental and community activists everywhere –
and especially in South America

THE LIFE CYCLE

8,000 MILES IN THE ANDES BY BAMBOO BIKE

KATE RAWLES

First published 2023
Icon Books Ltd, Omnibus Business Centre,
39–41 North Road, London N7 9DP

www.iconbooks.com

British Library Cataloguing in Publication Data.
A catalogue record for this book is available from the British Library.

ISBN 978-178578-787-4 (hardback)
ISBN 978-178578-788-1 (ebook)

1 2 3 4 5 6 7 8 9 10

Typesetting by SJmagic DESIGN SERVICES, India.

Printed in the UK.

CONTENTS

ABOUT THE AUTHOR

Kate Rawles is a writer, cyclist and former university lecturer in environmental philosophy who uses adventurous journeys to raise awareness about environmental challenges. She writes for a range of publications, is a mountain and sea-kayak leader and a fellow of the Royal Geographical Society. She lives in Cumbria.

SOUTH AMERICA
CARIBBEAN SEA
Cartagena
VENEZUELA
GUYANA
SURINAME
FRENCH GUIANA
COLOMBIA
ECUADOR
PERU
NORTH ATLANTIC OCEAN
BRAZIL
BOLIVIA
PACIFIC OCEAN
PARAGUAY
SOUTH ATLANTIC OCEAN
CHILE
ARGENTINA
URUGUAY
Islas Malvinas
Ushuaia

INTRODUCTION

The Fish Story and Other Cameos

I am young, perhaps ten. A small scrap of a kid, mess of pale blonde hair, walking alone along Aberdeen beach in Scotland. Where my parents are or why I am alone is not clear. The beach is clear, though – that long, long stretch of dark wet sand, a thick grey/dun wedge between the road and the receding tide. In a line along the length of the beach are occasional structures that look like goal posts; a wooden frame with a net across the back, designed to catch fish as the tide goes out. Between them drives a man with a tractor and trailer. He plucks thrashing fish from a net and throws them onto the trailer. Then he drives across the sand to the next one to do the same.

I've walked up to one of the nets he's already been to and can see that he's left some of the fish behind. They are still alive. I lift one – cold, wet, grey, pulsing – and carry it carefully back to the sea where it swims fast and downwards, disappearing in a cloud of sand and saltwater. Then I walk back for the next. I'm completely absorbed in this task of fish rescue.

Suddenly the man and the tractor are right alongside me. There is a smell of diesel and hot machinery. The man is angry. Very angry. These fish are the most valuable, he shouts, and he was leaving them until last so they were as fresh as possible when they reached the fish market. I've been liberating his best catch.

The memory fades with me standing there, small and scared next to the big man and the big tractor and the trailer-load of writhing animals. What did I think I was doing? I hadn't given it much thought. I remember how obvious it seemed – in the way things appear clear-cut when you are ten – that I should take the fish back to the sea. And I remember how right it felt as each one disappeared into the shallow waves; the knowledge that I had saved at least a few of these creatures from a slow, suffocating death gratifying in a way that left a mark.

I've often wondered since then. Do most people feel as I do about other living things? Or like the man, who, after all, was only doing his job?

December 2016. I am standing with a heavily loaded bicycle on a cold Calais dockside, beside a massive blue hull. The ship is so huge I can't take her in all at once. Towards the bow, a towering slice of off-white, multi-storey accommodation block with windows and railings stretches skywards, an external stairway zigzagging its way up each level. At the stern, the words 'CMA CGM Fort St Pierre, Marseille' are painted in white on the half-moon-shaped rear; thick, rust-coloured stains running in frozen motion down through the words from the railings above. The ship is in turn dwarfed by a gigantic gantry, the crane arm temporarily suspended above us.

A man in black and white checked trousers and an immaculately white T-shirt has come down a precarious-looking stepladder to find me. He introduces himself as Christian, from the Philippines, the passenger steward. He's joined by a man in navy overalls and a hi-vis vest. I start to take the panniers off the bike, but they stop me with hand gestures and wave me to stand aside. Christian holds the bike while the other man attaches a thick, grubby loop of grey webbing to the frame, front and rear. I watch in astonishment as the bike – complete with handlebar bag, map case, two front panniers, two bulging back panniers and several bungeed-on dry bags – is hauled up parallel to the side of the ship, dangling unceremoniously at a mildly alarming angle. I wait for the panniers to tumble off. They don't. Instead, there's a shout.

'Take it in!'

An arm appears and my bike disappears over a railing and into the belly of the boat.

I made the bike, Woody, myself, from bamboo. Or at any rate, the frame is bamboo.* I learned how to build it at a workshop in London, run by the Bamboo Bicycle Club. Most of the bamboo they use comes from China – I wanted to find a more local source. We did: a thick cluster of tall gold and green plants growing behind one of the massive domes at the Eden Project in Cornwall.

It took about five days to build the bike, a fascinating, occasionally alarming process, starting, quite literally, with a pile of bamboo canes. The first cuts were hair-raising; the unnerving sound of metal on cellulose seared in my memory, alongside the pungent odour of vegetable-based glue. Much cutting, scraping and sandpapering was involved, and a lot of learning how to use equipment I'd not encountered previously – jigs, vertical cutting machines, high-powered sanders. There was a lot of dust and disorganisation. By the end of it, though, I was riding around the club's car park on the UK's first 'home-grown' bicycle. Somewhat eccentric in appearance, admittedly, but apparently functional.

I'm not sure where I got the idea that I might be able to build my own bike, never having done so before. I had basic if rusty bike mechanic skills and remember thinking I'd be better able to maintain it on the road if I'd constructed it in the first place. Using bamboo would surely involve some interesting learning curves. Most of all, though, there was an irresistible

* Bamboo is, of course, a grass: but 'Grassy' didn't cut it as a name.

fit with what the journey ahead – *The Life Cycle* – was about. I wanted to celebrate and explore the importance of biodiversity, by riding across one of the most biodiverse continents on the planet, causing as little environmental harm as possible in the process. Woody's joints look like fibreglass but were made from tightly wrapped strips of hemp soaked in vegetable resin, without the chemical pollutants of fibreglass or epoxy. And, while all bikes are pretty good on the carbon footprint front, a bamboo frame has an even lower footprint than steel or aluminium.[1] What could be more in keeping than a biodiversity bike ride on a bike that used to be a plant? It made me grin.

As for *Fort St Pierre*, she is a cargo ship, and I'm about to cross the Atlantic on her. In comparison with flying, this reduces my carbon footprint much more than the bamboo bike does.[†] I'm heading for Colombia. My plan is to cycle the length of South America, from Colombia to Cape Horn (or as close as you can get to it by bike). I'll be riding most of it solo, following the spine of the Andes, the longest mountain chain in the world. The journey will take me through an astonishing variety of landscapes and ecosystems; from Caribbean coastline to high-altitude Andean páramo; from cloud and rainforests to the Bolivian salt flats, the Atacama Desert and the classic, spiky, white mountain peaks of the Peruvian Cordillera Blanca. And finally, of course, to the Patagonian Steppe, known for its brutal, relentless, cyclist-soul-destroying winds.

I grew up wanting to be 'an adventurer'. But I was a weedy, unathletic kid and my image of an adventurer was overly shaped by reading too many books by Wilfred Thesiger walking across deserts on a handful of rice. And by seeing those photographs of Edmund Hillary and Tenzing Norgay heavily kitted up, brandishing flags and ice axes, about to 'conquer'

[†] According to Mike Berners-Lee, while a second-class return transatlantic flight would account for approximately 2 tonnes of CO2e, travelling by cargo ship reduces it to around 50kg.

Everest – the ultimately daunting ice-clad death-zone mountain in the backdrop. In other words, utterly out of reach. There was something else that troubled me, too; something I struggled to articulate about the relationship between adventurers and the chunks of nature they were conquering. Was it possible to adventure in a different way?

It was the bicycle that cracked it for me. A bike is a magician, transforming a journey that would be utterly mundane in a car or on a bus into a mini adventure. On a bike you are really *in* the landscape you are travelling through. You are not conquering nature so much as being absorbed by it. It's fabulously low-impact – I once read that a cyclist on a flat road with no wind (if such a thing exists) can do ten miles per peanut, and that bicycles are the most efficient way of moving humans around ever invented. And people treat you differently when you turn up on a bike. You are perceived as vulnerable, a touch eccentric, perhaps, but probably harmless. Uniquely free. With a bike – any bike – the world opens out. An unfit person, as I was as a teenager, can cycle ten, twenty, even thirty miles with minimal training and not too much pain. Over the years, I've gone further and further from a very unathletic starting point. If I can do such trips, so could pretty much anyone if they wanted to.[‡]

The idea behind *The Life Cycle* is one I've started to call 'Adventure Plus': using adventurous journeys to raise awareness and inspire action on some of our most urgent environmental challenges. Or trying to. My previous attempts were *The Carbon Cycle* ride from Texas to Alaska, exploring

‡ Well, anyone who enjoys the privilege of having been born into a relatively 'rich' society, which grants passports and travel freedom to its citizens.

climate change, and a voyage on Pangaea Exploration's sailing yacht, *Sea Dragon*, investigating ocean plastic pollution. For *The Life Cycle*, my focus is biodiversity loss, and it's been inspired in part by a startling article I read while searching out material for an environmental degree I taught when I used to be a lecturer in environmental ethics.

The article was based on work lead by the Swedish scientist and sustainability expert Johan Rockström, and his team at the Stockholm Resilience Centre. The starting point for the article was the often-made observation that we've enjoyed a uniquely stable era in the history of our planet for the past 12,000 years or so. It has been this stability – in relation to climate and sea level, predictable seasons, a thick cloak of forest across the globe reliably performing various ecological functions – that has allowed humanity and the phenomena we call 'civilisation' to flourish. Rockström and his team argued that we owed this stable era, the Holocene, to a series of interlocking, self-regulating natural systems such as the climate, carbon cycles, water cycles and so on. Where 'the safe operating space for humanity' might lie in relation to these systems was the subject of their research. How far can we push, for example, climate change or chemical pollution or overuse of global fresh water supplies before we get into the danger zone in relation to human life and well-being?[§] Where are the tipping points that will likely take us over the edge into dangerous – and irreversible – change?

When I first saw Rockström's work presented visually, I was dumbfounded. The basic model was a circle overlaid on a globe. The circle was divided into ten segments, each representing a major environmental issue. At the centre of the

[§] These have already been pushed far enough for geologists to argue that the Holocene is sliding into a new era, shaped and dominated by the influence of humans to such an extent it should be called the Anthropocene. This is not a compliment.

circle, crossing each segment, was a smaller green circle. The green area represented the 'safe operating space' – where things are still OK, at least from a human perspective. Red was used in relation to each segment/issue to indicate where we'd transgressed. And the segment that was further out into the red zone than any others, further even than climate change, was biodiversity loss (see plate section).

I couldn't shake the image of that brutal chunk of red bursting out of the safe zone from my head. Could biodiversity loss really present an even greater threat than climate change? And what must that mean, in terms of actual, living animals and plants?

We are currently losing species of both at such a rate it has been called the 'sixth great extinction' caused, for the first time in earth's history, by a single species – us. As someone who would undoubtedly be diagnosed with biophilia,[¶] it felt right on a personal level to make this the focus of my journey. I have always been an animal kid, the toddler drawn like a magnet to animals of any size, tame or wild, friendly or fierce; the exasperating child who kept snails in jars and constantly pleaded to have a dog and tried to rescue fish. I've become increasingly enamoured of plants too. But if the analysis that Rockström offers is accurate, then this issue of 'biodiversity loss'[**] is one that *everyone* should care about – because of the implications it has for the stability of earth's life support systems. *Our* life support systems.[††] Yet this cataclysmic draining away of life has had relatively little press, eclipsed by climate change (immensely important though that is).

¶ Love of living things, the term coined and explored by the biologist Edward O. Wilson.

** The bland and boring nature of this phrase is dangerously misleading.

†† I don't think this is the only reason to care about biodiversity loss by any means – but it's a powerful one.

And so, as I pedal south on the longest bike ride I've ever attempted, I will be looking for answers to some of the questions that Rockström's diagram raises. I'll be visiting a wide variety of conservation-related projects and people, from grassroots, community-based forest regeneration to a school whose entire curriculum is based on turtles; from local advocates of sustainable fishing to a former radical priest, now a member of the Peruvian Senate. I'll be cycling through some of the most biodiverse landscapes on earth, and some of the most ecologically degraded. And I'll be exploring, not just mountains – much as I love mountains – but biodiversity. What is it? What's happening to it? Why does it matter? Does its loss really present us with as great a threat than climate change? And if so, what can be done to protect it?

Climbing the steeply angled stepladder that leads up beyond *Fort St Pierre*'s hull, I realise I have no idea what my time on board will be like. I am reassured by the friendliness of the crew – or at least the Filipino crew. There are big, smiley hellos and lots of 'welcome ma'ams'. The officers are all French and, when I see them, are much more reserved.

My accommodation is generous. An en-suite cabin, with a comfortable bed, a small desk and a view out across the vast deck. This becomes a view of orange metal as the ship is loaded and the stacks of containers mount up to my level. There are just two other passengers on board – Alicia and Olivia from Austria. We bond over our first meal together, which, like all our meals, must be eaten in the vicinity of the officers' table, though we are separated from them by a screen (and from the rest of the crew altogether – they eat in a different location we never find). I arrive exhausted from the inevitable pre-big-trip preparations, and spend a lot of the voyage sleeping, catching up with the many jobs you can do even without Wi-Fi, and while sitting on deck watching flying fish. As we steam west, each day is warmer, and Alicia, Olivia and I become fast friends, allies in exploring the ship and in the aim of securing permission to descend to the cargo decks and sway down the long, narrow walkway to the bow for sunset wave-watching. The crew are about 30 in total, living on board for months at a time. On Sundays we are invited for a lunchtime glass of champagne with the senior officers. On Friday nights, it is karaoke with our Filipino shipmates. Apart from the cargo deck, we have the freedom of the vessel, including the bridge, provided we do not speak to the captain during manoeuvres.

The journey up to that magical point of departure had been turbulent. To turn *The Life Cycle* from an idea into reality, I had to take a large chunk of time off work. This didn't seem unreasonable. It had been ten years since I had done something similar and, upon returning, I could feed my adventures and what I'd learned into my teaching, like a minor Indiana Jones.

My head of department hadn't quite seen it that way. Instead of offering me unpaid leave, she suggested I resign.

'Kate,' she'd said, visibly supressing a smile, 'if we can manage without you for a year, how can I justify your salary?'

'Kate,' a friend echoed later, his face cracking into a grin as I recounted this tragedy, 'the system just spat you out. And maybe that's a good thing.'

Chris, my ever-supportive partner, was prosaic. Over a particular Scottish liquid especially crafted for moments of crisis and inspiration, he pointed out that I didn't earn much anyway and that he could cover my share of the bills while I was away – and lend me what I needed for the journey until the bestseller came out.

Having always been keen on independence, this offer felt problematic as well as fabulously generous. The clincher came by cliché: a book that changed my life. It was written by Bill Plotkin, a North American writer and environmental activist, and, as I read it in my van on wet days on a camping/cycling holiday, a cluster of words leapt out from the page:

> If you can find the intersection between something you are passionate about, and something the world actually needs, that sweet spot will be where you are happiest AND most effective.

I wanted to be back on the road, back in the mountains, back on my bike. And I wanted to be a sort of investigative activist, not an academic. The chance to do this, to live my best life while making some kind of contribution, was surely more precious than my financial so-called independence. Plotkin's words completed the justification jigsaw. I resigned.

On the deck of *Fort St Pierre*. It is night. We've just left Calais. There is a line of ships on the near horizon, sparkling like long jewels between the dark sea and the dark sky. It is beautiful, just beautiful. I feel a huge uprush of sheer excitement. Underway! Heading for Colombia! I have over a year ahead of me. I've been set free to do what I most want to do. Woody and I are steaming into the sweet spot.

PART 1

Colombia

1. NORTH IS NOT THE DIRECTION OF CAPE HORN

COLOMBIA

Happy Planet Index score: 60.2
*Rank: 3rd out of 152**

It's typical. You set out with a linear idea of what your journey should be and soon discover the world has other plans.
STEFFEN KIRCHHOFF, SHAMANIC MUSICIAN

Cycling out of the coastal city of Cartagena was probably one of the most dangerous things I've ever done – right there on the border between brave and stupid.

Before I got going, swaying out of my quiet, temporary haven of a street on the overloaded, unstable-feeling bike, I'd mostly been worried about taxis.

The taxis, flocking the city in yellow, hooting multitudes, turned out to be just fine. The buses, though, were demonic. They were all in some rival warfare with each other, and

* The Happy Planet Index measures the quality of life a country provides for its citizens against its environmental footprint. See: https://happyplanetindex.org/.

they blasted past, multicoloured and blaring, cutting up anyone on their inside, focused on outmanoeuvring other buses rather than on the road. I inched out into traffic, stopped at the slightest doubt and pushed the bike across every busy junction. It was still hair-raising. And you can't focus solely on the buses because the road is prolific with potholes, motorcycles in all lanes and all directions and constantly switching, people pushing carts, and sudden piles of gravel. Out on the coast road, things calmed down slightly, but only slightly. A huge, hot wind was blowing – against me, of course – initially with added sand and grit.

Having changed course to avoid a road that led to a tunnel clearly signed 'No Cyclists', I came to a halt at a complicated flyover and asked a man in old jeans and a blue T-shirt, walking in my direction along the gutter, which of the confusing array of roads on offer was the one to Barranquilla. He pointed and then said:

'*Pero es muy peligroso. ¡Peligroso!*' 'But it's very dangerous. Dangerous!'

I think I had figured that out by then, but I thanked him anyway.

'*Amen*,' he said, seriously, and shook my hand.

Colombia. Such an evocative word. A country laden with stereotypic associations. Violence, drugs, guerrilla warfare. Pablo Escobar. Football. I had cycled across it on a journey from Venezuela to Ecuador in 1992, when the complex civil war between the Colombian government, crime syndicates, right-wing paramilitary groups and left-leaning guerrilla movements was ongoing. I had returned with very different stories. Me and the friend I was cycling with were both underinformed and full of an idealistic, half-naïve, half-youthful, untainted optimism. To us, the terrible, life-decimating conflict was largely invisible.

What we mainly encountered was huge friendliness and astonishingly beautiful landscapes. Enormous mountains that were green all the way to their summits. Music, everywhere: salsa and merengue rhythms ricocheting through villages and towns, exploding from houses and cafés and bars. Fabulous fruit.

What would I find now, 25 years later and less than a year after the official declaration of peace? A country still dancing that strange and dangerous dance between violence that's long been, as the writer Tom Feiling puts it in *Short Walks from Bogotá*, part of the fabric; and overwhelming warmth, humanity, humour and hospitality? A country recalibrating itself after almost topping the charts of countries with the highest numbers of internally displaced people (IDP) – over 5 million – and where extreme wealth coexists with the more than 25 per cent who live below the poverty line. A country whose politics, landscapes, conflicts and peoples have been made vivid through literature; homeland of Gabriel García Márquez and a host of modern writers. A country still overflowing with music. A vibrant cycling culture, thanks not least to brilliant Grand Tour riders like Nairo Quintana and Egan Bernal. A country bursting with life, hosting a dazzling variety of species and habitats: the country with the second-highest biodiversity in the world, second only to Brazil.

I was delighted to be back.

If your starting point is Cartagena and you are heading for Cape Horn, Santa Marta is not on the way. Santa Marta lies about 150 miles up the coast to the north-east. But, with invites to visit a cluster of environmental projects in that direction, it seemed like a small and worthy detour. One invitation – to meet a woman called Manuela Perez – had especially caught my imagination. Manuela worked with local farmers on forest regeneration and rural sustainability in the mountainous Sierra Nevada area. Visiting her offered the prospect of diving straight into some big questions about how biodiversity conservation can be achieved in a way that also enhances our quality of life: a great way to start the journey. Which is why Woody and I were launching ourselves into the Cartagena traffic and heading north instead of south.

Ferocious winds dominated the rest of that day. They were relentless. We slogged along in the blistering heat, the slowness not only wind-related. Maintaining cycling fitness on a cargo ship is not straightforward.[†] As for Woody, he was almost completely untested. Aside from his eccentric appearance, there was a question mark over whether he would hold together when faced with the open road.

The road ran alongside the sea. The sea looked unfriendly, a dark khaki colour, littered with off-white waves. After a while we left the coast for what was, to me, a strange new landscape. I'd never seen mangroves and cacti in the same place before. Road signs promised anteaters, snakes and geckos. In between roadworks and stretches of low, tangled woodland, hotel developments were sprouting up alongside the occasional golf course. Sometimes there were cattle and white cattle egrets and once a cluster of black vultures, hopping around a roadkill. Not me, at least not yet.

Finally, after the hardest 38 flat miles I'd ever ridden, I turned off for Galerazamba, where I knew there was a hostel. I felt both intensely alive and alarmingly tired. *Let's*

[†] Woody and I left Cartagena on 31 January 2017, after a six-week detour by boat and bus to Panama and Costa Rica. Between that and the eleven-day voyage on the cargo ship, I'd done zero cycling for two months by then.

not even think about the mountains at this point. You don't have to climb the Andes today. The small town at the junction – a cluster of roughly built houses – unexpectedly unnerved me. There was visible poverty, a lot of people milling about and a sudden outbreak of shouts directed at me, most of which I didn't understand. I felt elsewhere, a bit out of my depth and a little ashamed of the feeling; the elegant and easy-going Cartagena tourist quarter already far away. After the town, the road ran along sandy, wooded strips of land with rivers full of people swimming and washing motorbikes. Then no one.

I was at crawling pace by the time I saw a sign saying 'Hostel'. The sign pointed in two different directions. One led to a corner of a building with no door. The other to a huge iron gate, behind which were two loudly barking, lightly slavering Dobermans. No bell. No sign of humans. A woman walking by confirmed this was the hostel. The only hostel. She took out a mobile and rang someone. Then explained, patiently, in basic Spanish, that the *chico* running the place was having dinner and would be back soon. I should wait. I waited. Eventually the *chico* turned up, jeans, black jacket, black hair and a big smile, went through the huge gates and caught the dogs, depositing them where we could still hear their outraged cries at confinement. Inside the gates, a big open space, various buildings, a few short trees in the dwindling light. I was led to a large room full of bunk beds. Electricity? No. Drinking water? No. Two other travellers arrived, and we walked along sandy tracks, hopping with small frogs, through the darkness to a small store, candlelight flickering across faces, fruit and shelves of tins.

Later, sitting in a saggy chair, eating bread, cheese and apples and slowly coming to, I found myself grinning. It was good to be here.

Day two turned out to be an intensive refresher course on bicycle travel for a rusty, unfit *gringa*. Having survived it, I wrote the following in my journal:

> *Lesson One: No matter the distance, you will always arrive late.*
>
> *At times so strong I have to get off and cling to the bike to stay upright, the headwinds slow me down more than a little. It takes me all day to ride relatively few miles and I have the unnerving sense that I am encountering a law of (my) cycling that will likely apply across the entire trip. End result: loss of daylight. And so:*
>
> *Lesson Two: Cycling in the Dark. As a general rule, avoid it.*
>
> *Cycling in the dark is a wonderful thing in many circumstances, but not as an introduction to the city of Barranquilla.*
>
> *I assume that finding somewhere to stay in populous Barranquilla will be a breeze after Galerazamba. But I end up slogging uphill on a never-ending main street in fast-fading light with horrible traffic – then weaving about looking for cheap hotels that appear on the maps.me app but not on the ground. After the third one fails to materialise and it is pitch-dark but still traffic-crazy, I bump into a 'Park Inn' complete with liveried doorman. Deciding my life is at stake, I tentatively enquire about the price of a room. Expensive. Though not the multiple $100s expensive I am expecting. I put frugality on hold, pull out the emergency credit card and am abruptly transported into off-the-road, air-conditioned, double-bed luxury, Woody leaning against the mini-bar.*

The hotel was posh in a very Latin way: on my way down from my twelfth-floor room (the lift wasn't working), I found several floors still under construction. Steel poked out of concrete and glassless window-holes gave out to dizzying squares of cityscape below.

The next day, I'd come up with a super-smart plan to outwit the traffic by taking a minor road out of town close to the sea. I found myself cycling with a razor-wired industrial site on one side and acres of grim-looking favelas separated from the road by a thick sludge of water on the other.

A woman on a scooter overtook and flagged me down.

'*¿Adónde vas?*' Where are you going?

'Santa Marta.'

The route from Barranquilla towards Santa Marta involves crossing a road bridge, some distance ahead. She shook her head in horror.

'*¡Muchos rateros!*'

The literal translation is 'many pickpockets' but her throat-slashing sign language suggested something rather more unpleasant. In the decision to escape the traffic it hadn't occurred to me that I might be venturing into perils of another type, the stereotype of dangerous Colombia long displaced in my head by that of friendly, helpful and beautiful Colombia.

My scooter guardian insisted on escorting me through the industrial site on the 'quiet coast road' all the way to the bridge, stopping at checkpoints to inform the officers there of the stupid *gringa* on a bicycle with panniers full of presumably valuable stuff. Finally, we reached the bridge junction where she pulled up at a parked police van and explained the situation. The policeman grinned, shook his head and then my hand and waved me onto the bridge the wrong way up a slip road and into a gap in the traffic. Taking a cue from a motorcyclist, I heaved Woody up a steep curb onto the pavement through a miniscule pause in the blistering truck conveys and caught my breath in the now truly sweltering heat. The pavement was bumpy and potholed but at least it was out of reach of the HGVs.

Not far on the other side, in an area lined with shacks, small houses and stalls, I met a touring cyclist! First of the trip. I was to meet many more but, at that point, Tom-from-Australia seemed downright astonishing. He had just turned south from Santa Marta, having cycled there from Ushuaia in Patagonia – pretty much as close as you can get to Cape Horn on a bike, and hence, my destination. We chatted at the hot, noisy roadside. Barranquilla had left me resolved to avoid cities altogether as my best chance of survival, fast followed by the realisation that this was not a feasible option. Talking with Tom, I was unnerved to notice myself feeling something close to incredulity that a cyclist had journeyed the entire length of South America and survived. I told him about my encounter with the 'quiet road' and he grinned.

'Yes,' he said, 'I always take the traffic over people with guns. In cities, you should stay on the busiest roads you can find.'‡

For the rest of the day, the surrounding lands stayed flat. There were intermittent stretches of sand and the occasional cactus, merging into mangroves and river-fed swamps. Pelicans. And beautiful brown and cream hawks, scavengers, presumably, as I often saw them on the road. Occasional explosions of raucous, emerald-green parakeets, their harsh cries ricocheting through the ferocious sidewind.

The sidewind veered to a headwind. Then, quite suddenly, it was getting dark. Having cycled through Barranquilla with no front light, I now had a strong urge to have one. I stopped near a toll station and, something missing from the front-light attachment, taped it onto the handlebar. Then it was off into Colombian small-town chaos: traffic, pedestrians, motorbikes and scooters, cyclists, tuk-tuks, the occasional horse and cart, all moving erratically at different speeds and changing course without warning. I was super-focused on staying upright, not hitting anything, not being hit; on staying alive, essentially. I couldn't move very

‡ In hindsight, I think his advice was dubious.

fast because of the people wandering in the hard shoulder, making me feel suddenly visible and a bit vulnerable to the wolf whistles – something I had hoped my advancing years would inoculate me against.

Just as I was thinking, *OK, this really is not so smart*, a sign appeared by a petrol station. 'Popeye's: Hotel Restaurant Playa' A hotel! I pulled over. It was not clear where the hotel actually was, so I asked a pump assistant, a young woman who dealt with my terrible Spanish by miming, 'Do you want to sleep?' and laughing a lot. She called someone on a walkie-talkie. Several policemen appeared and then some truck drivers, and eventually, I figured out that I was to leave Woody with them while I went to check in. I was given a key and relieved of about £7, for which I was suddenly safe and off the dark, noisy, traffic-ridden road. Someone insisted on carrying the bike fully loaded down some steps to Room 42, which was small and super-hot. I didn't care. I felt only relief. After I'd eaten, I lay under a sheet directly below the fan and slept for a good ten hours.

Somewhere in those first few crazy days on the road, I had picked up an email from Manuela Perez. Not only did her work sound fascinating and relevant but she was fluent in English, a significant bonus. As part of my pre-trip preparation, I had diligently attended language classes, to virtually no effect. My Spanish was nowhere near adequate to understand discussions about biodiversity conservation. (Now I was immersed in Spanish-speaking countries, it was bound to get better, right?) Meanwhile, I arranged to meet Manuela in a town called Palomino, an easy bus-ride from the town of Santa Marta, where I had somehow arrived in one piece.

Positioned on the Caribbean coast, Santa Marta is probably where the Europeans first 'discovered' what we call South

America. Founded in 1525, it is one of the oldest-surviving Spanish towns on the continent. Símon Bolívar retired there having liberated Gran Colombia – now Colombia, Ecuador, Panama, Venezuela, parts of northern Peru and north-west Brazil – from the Spanish in the early 19th century.

The hostel I'd been recommended, Casa del Ritmo, was a tonic: a wonderful place, run by musicians, full of music, colour and friendliness. On top of that, it was – perhaps uniquely for Colombia – vegetarian.[§] A hand-painted sign on the kitchen wall read: 'Animal Friendly Area. Don't cook them.'

Next day, I sought out the bus. The route took me east along the coastal road, though I saw little of it thanks to the local custom of travelling with curtains drawn across all windows.

I liked Manuela from the start. A friend of a friend, her work with local farmers on forest regeneration was done under the auspices of an international organisation called Environomica, for which Manuela was the Colombian lead. She was slim, with dark hair and warm smile, a beguiling mix of elegance and sheer toughness, as I soon found out as we set off for one of the farms she worked with.

It took us all day to get there, initially on moto-taxis. We stopped for lunch in the neat, colourful, hamlet of Marney, one of the first towns on the well-known trek to *Ciudad Perdida*, 'The Lost City'.

After the meal, we would be heading further inland for the farm, a small, family affair, high in the mountains. It promised to be a dramatic journey. In the space of only 26 miles, Manuela told me, scooping omelette from her plate, the eco-systems change from ocean, to coast, to desert, to tropical dry

---

[§] According to Mike Berners-Lee, that my cycling was powered by vegetarian food rather than a meaty diet saved me even more CO2e than I saved by not flying.

forest, to tropical humid forest, to páramo – grass and shrub ecosystems above the treeline but below the snowline – to the snow-capped Sierra Nevada de Santa Marta, the highest coastal mountain in the world, with twin peaks rising to nearly 19,000 feet above sea level. Such diversity of habitat across such a small distance is very unusual, possibly unique.

'There's extraordinary diversity in the plants and animals here because of it,' Manuela said, as we finished eating and settled the small bill. 'And intensity. You'll *feel* it.'

We'd been expecting to travel onwards with the moto-taxis, but they hadn't returned. The drivers had been called to a meeting, apparently with the paramilitary, who, I learned later, controlled the area in the absence of any kind of government presence. They were said to extort a cut from all tourist-related revenues and to keep a tight rein on, for example, moto-taxi safety, and anything else that could compromise such useful income streams. As a result, the tourists here were among the safest in the world (not a little ironically given the violent, gun-running, drug-running reputation the paramilitary had earned during the civil war), though most never knew why.

We left our luggage at the café – 'they'll bring it later,' Manuela said – and walked. The trail took us up through the rainforest, into the Sierra Nevada. Vivid lime-coloured lizards scuttled across the hot sand-coloured path towards the densely forested, angular, green mountains. As we walked, a machete swinging from Manuela's waist, we talked about Environomica. Part of the organisation's work involved job re-creation for cattle farmers.

'Cattle farming here takes huge amounts of land,' Manuela said, 'about ten acres per animal. Every acre cleared of trees.'

If the farmers could make a better living by growing organic cacao or adding value to other products by producing them organically, without the need to fell trees, that was a win–win for the forest ecosystem and the farmers.

Those two interconnected aims – conservation of forest ecosystems and sustainable rural development – wound through everything Environomica did. As we climbed slowly through the lush complexity of foliage, Manuela set out layer after layer of their work. Tree planting, community development, education, research, local engagement … She fizzed with ideas and energy, undiminished by the ever-present heat.

One of the biggest challenges was explaining – to farmers, to businesses, to potential funders, to anyone – why biodiversity matters.

'People think it's a luxury, something that may be nice to have, but not essential,' Manuela said. 'Couldn't be further from the truth.'

'Do you think biodiversity loss is as great a threat as climate change?' I asked, thinking of Rockström.

There was no hesitation. 'Yes, definitely,' she said.

'Why?' I asked, and not just because I was short of breath. I really wanted to know what she thought.

'Well, in a nutshell, biodiversity is our life support system,' she said. 'It's fundamental to our well-being and indeed, survival, in all sorts of ways. We literally cannot live without it.'

'Literally?' I asked.

'Literally,' she said.

One way Manuela liked to illustrate this involved sunshine.

'We hear "solar power" and tend to think of panels on rooves,' she said. 'But *we* are solar-powered.'

The sun, Manuela reminded me, is where we get our food energy from. But we can't access it directly. The only living things that can are plants and other organisms with green chlorophyll – algae, some of the seaweeds – who capture that energy and turn it into carbohydrates, via photosynthesis.

'It's an astonishingly important chemical process when you think about it.'

Manuela paused, gesturing towards the lush plant life that bordered the track.

'It's going on all around us, right now.'

We both stared at some mundane-looking grass, suddenly imbued with new significance.

'Everything else alive on earth,' she continued, 'bar a few extremophiles who have figured out how to get energy from sulphur in deep ocean vents,¶ either eat plants – including grass – or things that eat plants. And plants depend on the huge diversity of often tiny organisms that make it possible for them to grow and reproduce. So yes, we literally couldn't live without plants, or without the myriad lives that make plants' lives possible. And that's just one example.'

Manuela's favoured approach to unpacking this was to talk about how different species of plants and animals all play different roles within ecosystems. Ecosystems depend on this diversity to function and flourish, and this is vital in turn because of the so-called ecosystem services that functioning ecosystems deliver.

'Ecosystem services include breaking down waste and recycling nutrients to create fertile soil. And pollination. Then there's fresh air and clean water,' Manuela said. 'Also, flood management. Carbon and oxygen cycles. Far from luxuries.'

'The language of "ecosystem services" is useful, then?' I asked.

'I hate it,' she said. 'It makes it sound as if the rest of nature is there for us, that nature matters only in relation to what *we* get from it. But the point is that ecosystem services make life possible – all life, not just our own – and yes, I think the phrase is useful in communicating that.'

An hour or so later, the light fading from the distant mountains, we stepped off the narrow, rutted path to let a train of mules plod slowly past us. They carried bulky loads – luggage and provisions for tourists hiking into The Lost City, a popular four- or five-day trek.

¶ Not great role models for humans.

Later we would see the mules on the return journey, sacks now stuffed with rubbish, including hundreds of empty bottles and cans. Most seemed to be Coca-Cola bottles, the distinctive red labels visible through the worn hessian.

Finally, the light gone, we pitched up a steep track and into a small crowd. Manuela, incredibly, did the majority of her Environomica work alone, and I had envisaged that it would be just the two of us at the farm. Instead, we walked into semi-organised industriousness in a small, dimly lit courtyard. There were various Colombians, some of whom turned out to be the farm owners, Wilmer and Sandra, and their visiting relatives. A small huddle of European volunteers were chopping vegetables by candlelight in a corner, closely watched by a dog and two puppies, a cat and some chickens, all criss-crossed by heavily laden washing lines above them. A large water trough seemed to be the main source of drinking and cooking water. I sat in a shadowy corner, disoriented by the hot walk, the sudden busyness and the flickering lights, and slowly figured out how life functioned in the place I'd just parachuted into.

The next morning, emerging banana-shaped from a night in a hammock into the briefly chilly air, I found to my delight that Manuela's plan for the day involved riding mules. Mine, Lola, was a dark dun, and bold. Looking out over her long ears to the incredible lush complexity of ferns and trees and foliage towards the angular mountains as we rode further inland, I was shot through with happiness. Some of the slopes were visibly scarred and bare of trees from the cattle grazing. But where the hillsides were still forested, I could feel the sheer presence of diverse life as a sort of vital yet peaceful intensity. And hear it, humming and calling and crying and buzzing into the hot blue sky.

Biodiversity isn't typically understood in terms of the feelings it evokes, though perhaps it should be.

It's common to think of biodiversity mainly at the species level, the undeniably astounding variety of plants and animals that inhabit the earth. Within that, there's often an association of biodiversity with 'charismatic megafauna' or big, sexy animals, such as polar bears and tigers.

But biodiversity is the variety of *all* life, and at multiple levels – from genes to species to ecosystems. It is diversity within a species, the myriad genetic variations present in, say, a single species of starfish. It is the dizzying range of ecosystems on our planet, from deep ocean to high mountain and everything in between. And it is also the diversity and abundance of species themselves, from sandflies to blue whales, from starlings to oak trees, from centipedes to silverweed. It is life in our oceans and rivers as well as on land. And it is the small stuff – the critical minifauna – as well as the big, eye-catching stuff; the many millions of small-to-minute organisms doing their all-important ecosystem work.

Biologist Edward O. Wilson sums it up perfectly. Biodiversity, he says, is:

> All the variation you find in living creatures around the world: from ecosystems to species to the genes that describe the traits of species – nested levels of biodiversity within the biosphere, that thin and fabulous layer of our planet and its atmosphere that supports all life.[1]

Manuela and I stopped at a bend in the trail and tethered the mules to a small tree. Its larger, older neighbour – the trunk too wide to loop the reins around – was hung with beards of lichen. A confusion of lime-green ferns sprouted from the rough, grey branch-junctions. At its foot, a fallen, long-dead branch hosted dark orange fungus, the size of small plates, among the myriad of different shapes and colours in the foliage all around us. There must be hundreds of species just here on this trackside, I thought, before we even start contemplating the multitudes living in the soil underneath. About a billion microorganisms in a teaspoon of good soil, I'd read; 'more bacteria than there are people on earth' in a tablespoon.[2] The thought was dizzying, multitudes in motion beneath our feet, unseen but absolutely vital. Microorganisms are crucial to soil fertility: no microorganisms, no plants.

'Did you know that soil is one of the largest reservoirs of biodiversity on earth?' Manuela asked. 'About 90 per cent of terrestrial organisms spend at least part of their life cycle there. We don't live in it, but we wouldn't last long without it.'

Manuela paused. 'I guess we often think of soil as pretty samey,' she continued. 'But soil habitats are actually very varied. So the variety of species they host are extraordinarily varied, too.'

Our planet has a fantastic variety of species in general. So far, about 1.7 million species of plants, animals and fungi have been identified, and just over 1 million species of insects. The estimates of how many species there are in total range from 8 million, to around 100 million.

'This way,' Manuela called, bringing me back to the here and now.

I followed as she scrambled up the bank to a grassy plateau where we climbed into a small structure, half-full of young trees. Manuela sat beside one and touched a large, bright green, classically leaf-shaped leaf.

'They are called mastre, *Pterygota colombiana*. One of the native species that would be growing here were it not for the cattle,' she said. 'We sell them to the tourists trekking into the Lost City. They take a young tree and plant it in a strategically situated, pre-dug hole, helping forest regeneration. Plus, the income from the trees goes into the local community and supports farmers diversifying away from cattle. And the people trekking learn something about the role and value of forests. Forests provide so much we need – from water, to soil stability, to habitat for thousands of species, to taking carbon out of the atmosphere – but so many of us don't know much about this.'

The Lost City trail tree-planting was only part of Environomica's ambitious forest regeneration plan. They aimed to plant 10,000 trees a year in the Sierra Nevada region, and to join up numerous forest fragments to create a far-reaching conservation corridor.

Manuela eased back out from among the trees and stood up. 'And the trekkers get to give something back,' she added. 'Buying and planting a young tree helps them leave a positive footprint rather than just stomping in and out of a sacred area, leaving their rubbish behind ...'

'Sacred?' I asked.

'Sacred to the Kogi people,' she said.

The Kogi, whom we encountered occasionally on the trail, walking barefoot in long, white garments, have lived in this area for centuries. They are descendants of the Tairona, the civilisation known for their brilliant engineering; the dramatic buildings and extended pathways they constructed in the jungle, their irrigation systems and the spectacular jewellery and ornaments they crafted from gold.

The Tairona culture extended across the Sierra Nevada region, with a complex trading system between societies from the coast to the mountains. When the conquistadors blasted onto the scene in the 15th century, this system, which had been functioning for five centuries, was violently disrupted.

Initially, the Tairona and the Spanish coexisted. But the cultures were deeply incompatible.

'What the Spanish saw, when they looked at the Tairona,' the British historian, writer and broadcaster Alan Ereira tells us, 'was a society of devil-worshippers in a state of untamed nature. And what the Tairona saw when they looked at the Spanish were savage barbarians with outlandish weapons of extraordinary power.'[3]

In 1525, a battle centred on Santa Marta was followed by an extended massacre. The Spanish moved along the coast, setting dogs on people, sacking and burning towns, capturing, killing and looting, stealing gold that was sacred to the Tairona. Those who escaped fled through the jungle and headed upwards.

Many starved, and many more were decimated by European diseases, as they were across South America. In the Sierra Nevada region, the survivors hung on in the mountains. With the areas below them now inaccessible, they couldn't recreate a way of life based on trading. The descendants of the Tairona gradually established a new social order built on simplicity, self-sufficiency and equality. They maintained their core values, including a deep respect for the earth; and they stayed high, surviving the ongoing brutalities of the Spanish invasion by being out of reach.

The Kogi have lived there ever since, doing their best to stay isolated, resist colonisation and the modern world and keep their culture intact. Their greeting to strangers is, 'When are you leaving?' Should you need to know anything about how to farm and live well in mountainous terrain, these are the folk to ask. They have now farmed there successfully for more than 1,000 years. Out of touch with everything we in the West think of as 'progress', what would 'living a good life' mean to them?

It would probably include protecting diversity at the ecosystem level, the diversity the Sierra Nevada is most famed for – that dazzling transition from coast to tropical forest to ice-capped mountains. The Kogi believe that the variety of the Sierra Nevada ecosystems is so comprehensive, in fact, that the region acts as a mirror to the world; that what is happening there, ecologically, reflects what is happening globally.

Despite deliberately eschewing technology and many forms of communication we take for granted in the West, their knowledge of local and global ecosystems, the way they are currently changing and why this matters, largely parallels the same conclusions reached by Western ecological science. Two converging streams of wisdom from utterly different sources. In the Kogi's case, it is a stream that has continued, unbroken, since before the arrival of the Spanish. Almost everywhere else, the European invaders of America 'destroyed the worlds they found'.[4]

Back at the farm, Manuela and I inspected ourselves and each other for ticks and settled in for the evening. Casa del Ritmo had been reasonably basic, but this small farm was much more so. Cold-water bucket showers, no electricity, no telephone and definitely no Wi-Fi. The few days I spent there provided a glimpse into a particular kind of conservation in action, and how people lived in these remote mountain regions. Despite its brevity, it was to prove a deeply affecting visit, a short pause in my new road life that left me feeling alive and appreciative of what each day brought – landscapes, people, conversations, food, water.

For me, a recent arrival from a life of comparative wealth, the absence of Wi-Fi and phone signal and, well, *things*, felt like a sort of holiday, a release, a kind of liberation. How it would be as a normal state of affairs was another question. I tried to envisage Wilmer and Sandra's life, the small farm and everything in it, positioned on a quality-of-life spectrum, a tightrope of possessions. Not enough stuff vs too much stuff. Quality of life is undermined, even made impossible, at both ends. When does real poverty begin at one end, and when do our things become a life-eroding trap at the other end? How much 'stuff' *do* we need to be happy? Given that every item has carbon and other environmental footprints, and that these footprints bear down on the natural world, the possibility of being happy with less clearly has implications beyond the personal ones.

Manuela and I left by moto-taxi the next day. It's a uniquely hair-raising experience, sitting on the back of motorbike as a young man drives you and your luggage over terrain you had no idea a motorbike could tackle, often at high speed, on the wrong side of the track. There is no option but to trust. Leonard Cohen's 'If It Be Your Will' played in my mind the whole way, intercut with the unnerving, yet bizarrely reassuring knowledge that, however dubious their motives, the paramilitary had a sharp eye on moto-taxi safety.

Manuela was right. That landscape was powerful. And my brief experience of living with so little was powerful, too. As we came back down the mountain, I felt both more at peace and more alive than I had done for a long time. But as we finally reached the road and Manuela's car, both were already draining away. What is it about 'civilisation', about roads, traffic, towns and shops, that makes it so hard to sustain that wonderful, energised, peaceful feeling; that saps it out of you so quickly?

Whatever the reason, coming down the mountain was harder than going up.

2. MONKEYS, TURTLES AND FISH

More turtles, less plastic.
School motto, Colegio Los Manglares

Manuela had driven me back to Casa del Ritmo in Santa Marta. From there, I visited a school whose entire curriculum was based on turtles; possibly the only such school in the world.* The school was set up and run by a woman known as Chile. Chile lived high in an apartment block, with fantastic views.

We talked over lunch and all afternoon about the turtle school, which she had set up not least because she wanted a decent education for her adopted daughter. The school's links with Roots & Shoots, an organisation set up in 1991 by the renowned primatologist and environmentalist Jane Goodall, helped support and structure the environmental aspects of their work.

There are thousands of Roots & Shoots projects in 50 countries across the world – Chile was the Colombian coordinator – and every project has three strands: social justice, environment and animal welfare. Beyond that, the particular

* Colegio Los Manglares in Rodadero, near Santa Marta.

focus is chosen by the group. In the case of Chile's school, it was turtles and plastic. The school did turtle-based creative writing: tell the story of a turtle from egg to ocean voyage; and turtle maths: if the turtle is swimming this fast for this long, how far does she travel? The students got to know the young turtles at the aquarium and to learn about turtle biology, how they lived and what they needed. Then, when the time came, they worked with the aquarium staff to release the juvenile turtles into the ocean – across a beach they had cleared of rubbish. It was hard to imagine a more impactful experience.

For Colegio Los Manglares, the ripples included pester power: the children would go home and nag their parents into not dropping the litter in the first place; into using less plastic. Each year, the school built on the knowledge and commitment that the children had gained.

'So, you end up with a group of informed, confident, well-rounded, activist kids,' Chile was grinning, 'whose impact spills out across the community around them.'

By the end of the afternoon, I was fairly convinced that all we needed to save the world was for Chile to take it over.

At Chile's suggestion, I went next to the aquarium. It was a visit full of surprises: I hadn't expected the young turtles to be so beautiful. Their shells gleamed at the water's edge, a rich caramel colour patterned with dark hexagonals, thickly traced as if with kohl eyeliner. They crowded against the pool-sides, alert and inquisitive. Making eye contact with them was unexpectedly moving; a tantalising half-glimpse, an almost-window into the inner life of creatures living out their lives alongside us.

I was shown around the aquarium by Calu Norriega, a marine biologist who worked for a turtle and marine mammals conservation organisation, Programa de Conservación de Tortugas y Mamíferos Marinos. Calu looked after turtles brought to her by fishermen who had caught them accidentally. She and her colleagues also hatched turtles from eggs and released them into the ocean. As well as the loggerheads I'd just been wondering at, they had tanks full of leatherbacks. The largest of our living turtles, with an average curved length the same as a tall man, leatherbacks are unique in having a leathery skin rather than a shell. Impressively hydrodynamic, their teardrop-shaped bodies and huge, flattened front flippers allow them to power through the water. One leatherback was tracked swimming from Indonesia to the USA for 12,000 miles on a foraging journey for jellyfish that lasted nearly two years. They are the fastest moving reptiles in the world. And they are fantastically versatile, found in oceans all over the world.

Despite their versatility, leatherbacks in Colombia, as elsewhere, are endangered. 'Bycatch' – marine creatures caught by accident while fishing for something else – is one of the biggest drivers of marine biodiversity loss worldwide, second only to taking unsustainable quantities of fish and other animals, including turtles.[1] The tremendous increase in plastic in the ocean is another of their challenges. By 2050, it's been estimated, by weight, there will be more plastic than fish in the world's oceans. Plastic is everywhere: the ocean surface, the beaches and shorelines, the deep water. It stays there for hundreds of years, larger pieces breaking down into smaller pieces, with more species able to accidentally ingest them the smaller they get. For the turtles, it's bag-sized plastic that presents the biggest problem as it looks like the jellyfish they eat.

Calu's organisation worked with fishermen and other local people and businesses to try and tackle all of this.

'Many of the people we collaborate with are really poor,' Calu explained. 'They can make reasonable money selling turtles for meat, eggs and shell. Or they eat them themselves when the fishing isn't good.'

As an alternative source of income, Calu was developing turtle tourism.

The tourists pay to meet the turtles. But on Calu's programmes, they also learn about them and get involved with their release into the sea. Sometimes they dive with them. Those taking part, she told me, almost always become passionate, informed advocates for ocean conservation, while the fishermen and other local people earn a decent living.

'And the turtles, most importantly, get to live,' she said.

'*Why* is that important, though, beyond the fact that they are amazing animals and have as much right to live as any other creature?' I asked, thinking about critical mini-fauna, and their supreme ecological importance.

'We keep learning more about the turtles' ecological roles,' Calu said. 'Like many of the so-called charismatic megafauna. We used to think they were supported at the top of a kind of an ecological pyramid by all the plants and animals below. But it turns out that they contribute more than we realised.'

When you look at a turtle, Calu explained, you aren't just looking at one animal but a habitat. Barnacles and algae live on and under turtles' shells, providing food for fish and shrimp. Some species of fish feed exclusively on these 'aquatic hitchhikers', and, because turtles travel such long distances, they help distribute nutrients and diversity from reefs to seagrass beds to open ocean. The turtles themselves are food for many different ocean animals, as are their eggs – most of which are eaten before they hatch – while egg-shells and unhatched eggs fertilise beach vegetation. Turtles graze on sea grass, helping to keep it healthy – which provides an important habitat, especially for juvenile fish. And leatherbacks eat jellyfish in such large quantities it controls their numbers, while loggerheads eat hard-shelled crustacea, increasing the rate at which the shells are broken down and hence the rate of nutrient recycling on the ocean floor.

To listen to Calu talk was to suddenly see interconnections between turtles and other parts of the marine ecosystem, and to understand that turtles play essential roles there.

We may not realise it, but *all* life depends on ocean life, Calu explained – and ocean ecosystems are kept healthy by the diversity and abundance of species that live in them.

Despite the main challenges, Calu was optimistic. The outlook for turtles at least was, she believed, beginning to change for the better. In her view, it wasn't possible to do nature conservation if the local community wasn't involved and didn't benefit. But if they were, then conservation could be really effective.

'Here, the fishermen are gradually coming on side,' she said, 'and others too. Turtle numbers are slowly increasing.'

I left Casa Del Ritmo the next day. For the first time in my journey, I was actually heading south, back towards Cartagena.

I'd planned a short cycling day, with a reservation in an 'eco-hostel', only 30 miles down the road. I was carrying full camping kit, but so far I'd felt uneasy about using it. Campsites were not really a thing in Colombia, and there were frequent towns on my route, making discrete wild camping difficult. I'd decided to use hostels and cheap hotels until I felt more confident.

Today, I had a plan in place and plenty of time. I stopped for a second breakfast, then plunged back into the chaotic traffic, noise and poverty in the town of Ciénaga that I had found so hair-raising in the dark. It was hair-raising in daylight too, though at least second time round I knew what to expect. But there was a deeply unsettling atmosphere to the place that didn't seem fully explained by the traffic mayhem.

I rode on, the mountains fading behind me. Rows of colourful buildings appeared, somewhere between very small houses and sheds, painted yellow, green, turquoise, blue. Social status was clearly closely correlated with sound-system volume and the blasting roar of merengue or cumbia rhythms was often close to ear-splitting as I passed, the music merging with the hooting of cars.

The sheds gave way to more industrial-looking complexes interspersed with stretches of mangrove, sand and cacti. Then an entire field of miserable-looking concrete shacks. Acres of grey. Rubbish everywhere, and stench. A man emptying a bucket of waste into a trench. I couldn't imagine living in one of those hot, buggy, stinky shacks, with no clean water, no toilet, no sewage system. I felt an uprush of anger. Ernesto 'Che' Guevara had written about this same poverty, and felt this same outrage, on his now famous motorbike journey back in 1952. How could this level of hardship still exist? The poverty seemed heightened by the grim reality that little had changed since Che had ridden by 65 years previously.

Suddenly there were stalls lining each side of the road, all selling one thing: fish. Piles and piles of fish, carefully arranged, heads one side and tails the other, baking in the hot sun. I wondered who bought them, and where they had come from. Then, crossing a bridge over a wide estuary, I saw hundreds of small wooden rowing and sailing boats, all putting out nets or fishing from lines. The water was covered in boats and the roadside was covered with fish.

Before leaving Cartagena, I'd gone to visit a university-based project called Pescando Para La Vida – Fishing for Life – working with fishing-dependent communities to manage the fish more sustainably. The lead researcher had explained that it wasn't just industrial-scale fishing that had a devastating impact on fish populations. Poor, local, small-scale fishermen could too. The difference was that the latter usually had no other options.

On that stretch of road, the truth of this was as visible as the dead fish themselves.

Not long after the rows of fish stalls, I turned off down a small sandy road where I thought my hostel should be. There was nothing down there bar a small group of people looking at me as if I were a unique mix of *gringa* and alien.

I hauled back up to the main road. It was now 3.30 in the afternoon and Barranquilla was a good 30 miles down the road. Which would mean, almost inevitably, another city ride in the dark, something I'd only recently vowed never to do again.

I set off anyway into the long, hot, 'uninhabited' (by humans) section of road ahead of me, with not quite enough water, but an uprush of energy and exuberance, the wind behind me. Pushing on past the now familiar cacti and sand and ocean and mangroves, a scatter of lizards shot off the roadside and into the verge. Heading south, I was counting off chunks of five miles instead of one, as I had been when heading north. Working hard, trying to out-pedal the swiftly sinking sun. Herding lizards.

At fifteen miles, I put some music on. I was hot, dry-mouthed and shot through with bursts of happiness, Sheryl Crow singing loud in my ears. Where the ocean lapped right up to the road, a group of motorbikers had stopped to take selfies with the surf behind them. They waved. Briefly cooled by spray, I had one of those flashes of realisation. *I love this. I love riding this eccentric bike on these hot, crazy roads, despite the hard and dangerous bits. I am happy. I feel alive. It is so very good to be here.*

Towards the end of the 'empty' stretch, I half-remembered a hotel, just this side of the bridge, that could save me from a second bout of Barranquilla in the dark. A while later, a bit dazed, I turned off at the toll I remembered as nearby and asked. A young man escorted me to the hotel – a regular truck stop, with rows and rows of multicoloured trucks parked out the back.

'No,' said the receptionist. 'No, full.'

'Really? Any room, small or large. I'm on a bicycle ...'

After being told to wait, I sat on a large leather sofa with a kindly truck driver, engaged in the kind of linguistically challenged conversation I was getting used to. And hoping, as the last of the light faded and it became properly dark, that a room would materialise. Or that I could kip on the sofa. Che Guevara hadn't worried about where he would sleep. And he hadn't paid for hotels. He didn't appear to have worried about his own safety or comfort much at all, in fact. Che had had his eye firmly on the bigger injustices of the world.

'Be more Che,' I thought.

Eventually, I was summonsed back to reception and given a key. Oh, that gratitude-feeling.

Next morning, I lug all my stuff back down the stairs. Truck drivers materialise from nowhere and hold the bike as I load it. Four panniers, the front box, an armful of waterproof bags with my sleeping bag, tent and other bulkier things that I bungee on top of the pannier rack. I am troubled by the stuff. It weighs too much and is a burden – yet I feel I need it all. Ah, the parallels with wider life. How much stuff do we need to lead a good existence? How much is too much and who pays the costs? On a bike, the distance between your stuff and its weight – literal and metaphorical – is radically reduced. Yet I am still finding it hard to contemplate living with less of it.

I pushed out into the road and took it slowly. On the far side of the bridge, I dropped down into Barranquilla and total congestion, stuck behind buses and trucks, cooking in diesel smoke when the lights were red, inching forwards on green through the heat and exhaust fumes.

The main route through Barranquilla is, as the travel writer Michael Jacobs puts it, 'an interminable straight road'.[2] At one point, I could see along it for miles, past the huge 'Barranquilla' sign – the road I had slogged up in the dark a few days before. Now, finally on the way out, I was going down it in hot sun, with a tailwind. Same endless straight road, utterly different experience.

While in Barranquilla the first time around, I'd made contact with Rosamira Guillen, or rather, she'd made contact with me. I'd taken the call in the evening, while standing gormlessly with the bike, hot, tired and dishevelled in the early part of the hostelry hunt.

Next morning, I'd gone to her office in a nice, cool taxi. Rosamira, long dark hair, warm welcome, full of humour and focus, is the executive director of the conservation organisation Fundación Proyecto Tití, or Project Tití, and a winner of the prestigious Whitley Award for Nature. She used to be an architect.

'My dream in those days was to create homes for people,' she'd said. Then, while working for Barranquilla Zoo, she had been shocked to learn not only that there was a monkey found only in her part of north-west Colombia, the diminutive cotton-top tamarin, or Tití, but that it was critically endangered.

'I'd never even heard of these monkeys,' she'd admitted. 'And when I did, they were one step from extinction.'

Rosamira, moved by the monkeys' plight decided that she had to do something. She co-founded Project Tití and completely altered the course of her career. Her work was now entirely focused on conserving pint-sized primates.

'One of our main strategies involves collecting waste plastic,' she'd said. *Plastic?* I was just trying to figure this out when she'd asked the question I'd been hoping for.

'Would you like to see the monkeys?' There was a Project Tití reserve not far off my route.

Two phone calls later it was arranged: on my way back to Cartagena from Santa Marta, I would ride to the reserve to meet some of the staff on the ground – and the monkeys in the trees.

'They spend their entire life in the trees,' Francy Forero, a biologist and one of Project Tití's key educational staff, told me when I arrived. 'They never touch the forest floor.'

Young, slim, super-smart, fluent in Portuguese as well as Spanish and English, she translated for the deeply knowledgeable Felix, who had had been working with cotton-tops for 30 years, as we all walked into the forest, crunching over dry leaves in the immense heat with the extraordinary sounds of howler monkeys ricocheting around us.

Then we waited. Rosamira had been optimistic about my chances of meeting the monkeys. It was a well-studied troupe, and their habits were known. But, as the hours sloped by, I felt increasingly doubtful that I was going to see any monkeys at all. Then, just as I'd resigned myself to their not turning up, both Felix and Francy pointed to the same tree.

'They are here!'

With the agility of squirrels, first one, then several diminutive creatures swung down towards us, their white, tufty mops of mane swept back from dark, skull-like faces. They had white fur on their tummies and legs, long fingers and toes and grey-brown, furry backs.

'And that one', said Francy, pointing to the monkey who appeared boldest, 'is Tamara'.

According to the textbooks, cotton-tops in the wild live to be thirteen at most. Tamara was known to be sixteen. What's more, her mate of many years having died, (they are monogamous), Tamara had not long acquired a new mate, thirteen years her junior. She was clearly quite a character. All the Project Tití staff knew and loved her.

Cotton-top tamarins are endangered not just by lack of trees, but of the right kinds of trees. They need trees to provide food and aerial transport infrastructure, and trees for sleeping in, about which they are particularly choosy.

'They sleep altogether in a circle on one of the branches,' Francy said, 'shoulders touching, kids in the middle.'

For night-time security, they prefer tall trees, ideally with spiky trunks to deter intruders, a generous array of large, high branches and good, thick leaf cover. Such trees are usually a couple of decades old, at least, and since they change their sleeping abode nightly, the monkeys needed a good supply of them.

According to satellite data, only 5 per cent of their habitat remains. Field observation reduces this to 2 per cent.

'That's why forest conservation and regeneration is absolutely critical,' Felix explained.

'Where does waste plastic come into the story?' I asked. Francy grinned.

Forest protection and regeneration were predictable aims for a conservation body focused on animals who utterly depend on trees. Cotton-top tamarins, with their tragic clutch of endangered statuses – Colombia's most endangered native primate, one of the world's 25 most endangered primates, classified on the International Union for Conservation of Nature (IUCN) Red List as 'critically endangered' – are inevitably threatened by the relentless loss of their forests. Research was predictable too, as understanding the kinds of trees they need, and what was behind the trees' demise, would be key to improving their chances. This was where the plastic came in. The main driver of deforestation in this part of Colombia was not multinational beef corporations – often the case in, say, Brazil – but local people cutting trees for firewood or clearing the forest to farm, often with no other way of getting fuel or food. On top of that, there was a lucrative, albeit illegal, trade in wildlife as pets. It was obvious that these cute little mop-haired monkeys would be a good catch. And so, Project Tití also tackled poverty.

'We involve local communities in collecting waste plastic,' Francy explained, 'then we show them how to turn this plastic into a sort of twine, and from twine into fashionable shoulder bags.'

They'd also developed a process for chopping waste plastic into small pieces, from which they could then construct fence posts.

'Much more durable than wooden ones in this climate,' Felix added. 'They sell for a good price. Many of the slightly better-off farmers buy them.'

The income from these products meant that people could afford to buy food, rather than having to cut down forests to grow it; and earn money without catching monkeys.

It was another brilliant win–win–win. I'd only been on the road a few days and had already seen shocking quantities of plastic and other rubbish dumped by the roadsides – unsurprising once you know that parts of northern Colombia have no rubbish collection at all, let alone recycling facilities. Rubbish was either burned or dumped. What else were people to do with it? Project Tití had taken a problem – plastic waste – and turned it into a solution to several other problems: poverty, forest destruction, and the threat they posed to monkeys.

Carlos, a keyring-sized, cotton-top toy that Rosamira gave me, sat in the handlebar box, now invested with a whole lot more meaning. Pedalling out from the forest on hot, plastic-strewn roads back towards the Cartagena coast road, I was thinking about these charismatic little monkeys as biodiversity ambassadors. This had been Francy's answer to my question about whether the focus on another charismatic megafauna – admittedly in this case a relatively small one – was justified. Like the turtles, these monkeys play important ecological roles, including distributing seeds as they move around the forests. Her main point, though, was that cotton-tops are cute and interesting. People will pay to come and see them, unlike your average insect, detritivore or zooplankton. Once seen, almost everyone falls in love with them and wants to see them protected. But that didn't mean that the monkeys were saved at the expense of other animals. On the contrary, as with the turtles, once you cared about the monkeys, you had to care about their habitat. Restore and protect the habitat and you've gone a good way towards protecting everything else that lives there.

Scientific research, habitat protection and restoration, education, community development, community engagement, income generation: the many, interconnected strands of nature conservation. It was already becoming a key theme. It wasn't quite what I'd expected, somehow. The sheer creativity of conservation was taking me by surprise, too.

Above all, I was learning that any attempt to protect biodiversity and particular species had to include sustainable, nature-friendly ways of meeting *people's* basic needs, if it was to work. Conservationists had been saying this for years, but I was experiencing the reality of it much more vividly in Colombia than I had done in the UK. In Colombia, it was fast becoming evident that conserving nature and tackling poverty were inseparable.

As Woody, Carlos and I turned back onto the main road to Cartagena, the sea still whipped white by the wind, I thought that this basic principle must surely be true everywhere. Nature conservation and improving human quality of life had to go together. Neither could be successfully pursued at the expense of the other, or not in the longer term at least – as Rockström and so many others were arguing. In the shorter term, the dilemma was the same everywhere that poverty was high, whether it involved cutting down critical forest habitat in order to grow food in Colombia or creating jobs in a deprived area by proposing a new coal mine in the middle of a climate change emergency in the UK. The solution in all such cases had to be to create *both* rather than *either/or* ways forward.† My first two weeks had been almost entirely populated with brilliant examples of *both* – Environomica, the turtles, Pescando Para La Vida and now Project Tití.

It wasn't always to be quite so positive. I was about to encounter powerful enterprises that put profit at the expense of both environment and local communities – not *both* but *neither*.

† Where people are already relatively wealthy, the challenge is a different one. In that case, what is meant by quality of life needs a radical rethink.

3. INTO THE MOUNTAINS AND MEETING THE SLOTH LADY

Of the world's major food crops, the banana is the most chemically treated.

PETER CHAPMAN, *BANANAS*

Having started cycling from Cartagena, it was some weeks before I pedalled away from it again in a southerly direction. Cartagena is full of ambiguity, and hard to leave. One of the most beautiful cities in Colombia, it's rich in music and galleries and handsome squares and quirky streets where colonial architecture clusters around buskers, sculptures and street food stalls. Beyond the tourist zone, miles of densely packed poverty and, at its heart, ancient, brutal theft. It was in Cartagena that the conquistadors melted down the gold they had stolen and shipped it back to Spain. I cycled untouched through all of it, constantly reminded, with my passport and ability to finance a ride like this, that I was a privileged descendent of a former colonial country that had wreaked similar havoc in other parts of the world.

After Cartagena, I touched the Caribbean coastline. Palm trees. A slight sea breeze and turquoise waves. Then I spent nearly a week cycling inland, a fantastically hot, lush, bird-filled, briefly flat interlude, pedalling towards the mountains with a stomach-swirling mix of excitement and apprehension. Excitement because I love mountains. Apprehension because these mountains were so very big.

It was hot in a way that bore down like a weight: almost impossible to imagine being able to fix a puncture, say, if there were no shade. There were watery, swampy areas, full of squabbling egrets and moorhens and many other birds whose names I did not know. On the road were geckos, some a vivid green and several decked out in grey and brown stripes, their tails particularly striking. Sometimes the geckos were utterly still, soaking up the sun on the sweltering tarmac. Others hurtled away in a frantic panic.

One day I saw a group of people coming up the road. They were waving banners and placards and my brain veered straight into 'Uh oh, I've read about this on the British embassy website' mode. *Stay away from political demos! Keep well clear of marches and other possible trouble!* The banners suggested the march had to do with protecting local farmers and the environment. By the time I realised that this was a small and friendly bunch, they had gone by – with lots of smiling and waving. Instead of turning round, I carried on. What was I thinking? Che Guevara would definitely have stopped to talk.

Not long after, the probable reason for the march became clear. Palm oil plantations. Miles and miles of them. Then a huge mill pouring thick grey smoke into the blue sky. The palm oil plantations were presumably lucrative and meant jobs – for some – but their presence meant the absence of small farms. Palm oil plantations involve heavy use of pesticides and often deforestation, both heavily implicated in biodiversity loss.

None of this was new. Modern palm oil companies can be understood as part of a tradition: a plantation tradition. In the late 19th century, a New Yorker called Minor C. Keith began experimenting with bananas as a cheap source of food fuel for workers on a Costa Rican railway line whose construction he was overseeing. The sale of bananas proved more lucrative than the railroad, and Keith's business came to dominate the banana trade across Central America and Colombia's Caribbean coast. A merger with a rival company created the United Fruit Company, worth a good $11 million in the early days, and vastly more after that. Its headquarters were in Boston, and it was soon wielding immense influence across the United States.

Its power grew across Latin America, too. United Fruit, nicknamed *el pulpo* – the octopus – had tentacles everywhere. It ran railroads and telegraph, radio and postal services. It had its own fleet of steam ships. It owned vast tracts of land and actively prevented others from owning land, including local farmers as well as rival banana companies. Ostensibly contributing to the 'development' of its host countries, most of the financial benefits accrued to United Fruit. On top of that, it was regularly accused of bribing government officials and, in at least one case, after that failed, it arranged to have an elected government (in Guatemala) overthrown.

Environmentally, its approach to banana production was not dissimilar to that of palm oil. It involved draining swamps and clearing forests, destroying entire ecosystems. And, as with pretty much any monoculture, it required the heavy use of ecologically dangerous pesticides and herbicides with terrible effects on insects – including bees and other pollinators – other wildlife and freshwater systems. Overall, it was a disaster for biodiversity (and, eventually, the bananas themselves).

In Colombia, United Fruit grew bananas at the foot of the Sierra Nevada and then sent them out by steam trains to the ports of Barranquilla and Santa Marta. Peasant farmers were thrown off the land and labourers treated badly, something for which the company was famous.

Reading about this later, I discovered quite how badly. Ciénaga, the one place I'd felt truly uneasy so far, had been the site of the *Masacre de las bananeras* in 1928. Workers, striking over their abysmal pay and working conditions, had been murdered en masse when the government sent in the army. The military defended their actions, though later admitted they'd been pressured by the US government, keen to protect the interests of United Fruit. The slaughter was covered up, the inspiration, if that's the right word, for the massacre that Gabriel García Márquez depicts in *One Hundred Years of Solitude*. How much had changed? I wondered. Surely the power of these corporations – the power to destroy and degrade ecosystems, to exploit and even kill people in pursuit of profit – would by now be constrained, tamed and controlled by human rights and environmental legislation.

For a while, a slight breeze made the cycling easier. Flat roads. Beyond the stretch of plantations were big open spaces, scattered with trees. Parched-looking land, dun-coloured grass. A man with a huge yellow backpack walking across a field spraying weedkiller. Cattle ranches, a low line of hills off to one side. Occasionally, I rode through avenues of trees with pink blossom, where large, pendular birds' nests hung down precariously from the branches, right above the traffic. It was a beautiful, albeit damaged, landscape; degraded from a biodiversity point of view by the cattle farming and the pesticides. I saw a few rich-looking farms (*fincas*) and many small hamlets. There were chickens, kids and dogs, all doing their thing a few feet from the road.

Often there were soldiers on the roadside, too. I had no real idea why they were there, but my best guess was that their presence, highly visible and obviously armed, was intended to help maintain the peace. After decades of conflict, years of negotiation and a 'no' vote in a peace referendum, a peace agreement between the Revolutionary Armed Forces of Colombia (FARC) and the Colombian government had finally been signed toward the end of 2016, just months earlier. It was a controversial deal in many ways, including the proposal to resettle FARC members into society by giving them land and jobs, a proposal often resented by civilians who had neither. But most people I talked to, however they felt about FARC or the government (many had suffered violence from both), passionately wanted the peace to endure.

For my part, I was loving cycling all day, through landscapes and communities that were novel to me. Merengue leaking from a roadside café; the smell of hot trees; the rocking bass-beat in a passing car; the sweet, sweet sound of a chick calling to a hen in the dust. My first taste of tree tomato juice, astonishingly delicious, somehow both tangy and sweet. The ordinary extraordinariness of the world.

I loved the absence of constraints and obligations. I loved the sense of purpose that simply cycling gives. And I especially loved the wider purpose that my journey was motivated by; the notion, even if delusional, that my travels had a point to them beyond that of fulfilling my own wish to be on the road. I was feeling invigorated by the whole thing, properly alive.

Yet I was also still vaguely anxious. About finding a place to sleep. About the fact that my Spanish didn't seem to be improving, and what a handicap that was. About finding the stories and whether I was cycling far enough each day. I knew I should let go of these worries. That what was needed was simply to set off: to take that step forward out into the world and trust that things would work out. But that wise head came and went.

On the first day of climbing, I cycled out of the small town of Tarazá not long after nine. Tarazá had a custard-coloured church with short, white spires, on a street off the main thoroughfare, unusually quiet in terms of traffic and the opposite in terms of music. That first moment of motion, having loaded the bike – panniers and dry bags hauled, clipped and bungeed onto front and back racks – was always a good one. Woody took the load without complaint and, pushing back out onto the road each new day, he felt increasingly like a solid and dependable ally. My early worry that he might literally fall apart now felt almost disloyal. Yet our allegiance was not untroubled. I was uncomfortable while riding and would get pins and needles in my hands and shoulders after only a few miles. I was constantly trying new ways to solve this, lifting the saddle or changing its angle, or that of the handlebars.

That morning, I wriggled on the bike and tried to focus on what was around me. I was in lush, forested countryside, running alongside the Río Cauca, wide, coffee-coloured, flowing quite fast in the other direction. For miles and miles there were hosepipes feeding from the river to the roadside, and trucks being washed there. Men, women, boys, all drenched, scrubbing away at wheels and undercarriage with river water. I could smell the washing powder on trucks that had just been cleaned when they came up behind me.

And then, around a corner, suddenly and with no gentle lead on, the road started to rise. I was elated. Despite my lack of athletic prowess, I have always loved cycling in mountains – especially hot, sunny ones. I'm drawn to mountain landscapes like a magnet and when I'm in them, I feel more alive. There is no feeling like it: when you power yourself upwards and gain views of mountains stretching away in front of you, range after range, and your whole body gives a sigh of relief and release.

Now rearing up ahead of me more like an gigantic roadblock than a vista of freedom were the Andes. The longest mountain chain in the world, stretching north to south through seven South American countries. The highest range in the world, outside Asia, with some of the highest road passes. Over the years, I'd cycled in the Scottish Highlands, the French Alps, the Spanish Pyrenees, the North American Rockies. The Andes are in a different league. There's nowhere in Europe or North America that you could cycle upwards for days without running out of mountain. Now I was cycling straight into these giants, into the northern end of the Cordillera Occidental, one of the three immense mountain spines that track roughly north-east to south-west down Colombia.

I made slow progress. After steeper sections, I paused in patches of shade. Sometimes wisdom dictated pulling over for a truck, and if there was one truck there were often ten. On even the briefest straight sections of road, they tried to overtake each other. Some were slatted and packed with livestock. I could see strips of hide and eyes or muzzles through the gaps. A grim journey in this heat. Others carried containers, unloaded from a cargo ship at Cartagena, perhaps. Occasionally, trucks coming from both directions, loud music blasting from houses and a heat trap on the road all collided, and the noise and heat would be almost overwhelming. Sometimes there was no traffic at all for a while, and I could hear birdsong, Woody's tyres on the road and the river.

Around four in the afternoon, with the town of Valdivia somewhere up ahead of me, I stopped to chat with a woman in a white shirt with a net over her dark hair, standing outside a small restaurant. 'Yes, this is also a hotel. Yes, I do food. Good food! It's very hilly ahead.'

The usual inner monologue. It was a couple of hours until darkness and I was nowhere near too tired to carry on. But I was definitely on the hot side and trying to break myself into the mountains gently. I was also trying to figure out a different way of moving through the world, one that didn't simply involve cycling as many hours as possible before falling off my bike into some hostelry and going to sleep.

I stopped. And then sat with the woman, whose name was Elizabeth, drinking a thick, white, guanabana juice, an utterly delicious fruit with a flavour that eludes description. We chatted, or tried to. She had three sons and had always lived in the area. As we talked, a tree in Elizabeth's backyard was visited by a hummingbird, then a black bird with a yellow, oval patch under the tail. Maybe I didn't need to worry about finding ways to visit the national parks, something that was proving unexpectedly difficult. Much of what I wanted to see and learn about was right there by the roadside.

I spent the rest of the afternoon reading about the impact of oil palm plantations on jaguars. The plantations are the primary cause of the jaguar's habitat loss, I read. This forces the animals out onto the plains in search of cattle, where they are regularly killed by ranchers, though this is illegal.

Afternoon drifts into a lovely, laid-back evening. From time to time, Elizabeth's lads come and join me, gracefully manoeuvring around my poor Spanish by showing me pictures on their phones. I have a beer and buy them one too. Dinner is eggs. An early night. I switch from jaguars to The Invention of Nature, *by Andrea Wulf, about the now little known but then wildly famous explorer and naturalist, Alexander von Humboldt. Humboldt travelled from Cartagena to Lima, Peru, in the early 1800s with a train of mules, carrying a crazy quantity of scientific instruments and specimens. I am often on his route. He was excited by the hugely varied habitats he travelled through, and the plants and animals he found there. But Humboldt, like Guevara 150 years later, was also angry, outraged at some of the impacts of a 'developing' society he witnessed. Humboldt talked about colonialism as disastrous for people and environment – exploiting raw materials, destroy-*

ing nature, ensnaring people into mining, cultivating cash crops and poverty. Subsistence farming was, in his view, the route to freedom. The farmers in the march I witnessed earlier would have agreed with him, I think. Finally forcing myself to put the book down, I wake in the night to the sound of rain.

The next morning, I was, unusually, up early. But no one else was. The rain had stopped, and it was cool and slightly overcast. My early getaway thwarted by the need for breakfast, I answered emails instead. Eventually, one of the lads appeared. It was, he said, Elizabeth's day off. She appeared as I was leaving and gave me two large chunks of carefully wrapped pineapple for the journey.

'Please stop here again if you ever come back,' she said.

I suppose we both knew that was unlikely, the bitter in the sweet of these precious moments of connection on the road.

I pushed Woody back out onto the tarmac. The sun came out and mountain vistas unfolded through the slight mist on both sides of the ever-climbing highway. The mountains were green, all the way up, densely forested where the trees hadn't been cut down for cattle.

I cycled past occasional small, multicoloured houses, some right by the road, with lines radiating out at random angles to the electricity wires above. Then a series of hamlets, oozing loud, fast-beat dance tunes that temporarily upped my pace.

Day two of cycling skywards. The road went up and up. And up. It was some hours before I reached Valdivia, and I cycled straight through it. Another road life lesson: yesterday is not always a good predictor of today. After Valdivia, there was little on the highway in terms of food. The occasional stalls sold only pineapple juice, delicious shots of cool sweetness punctuating the climb. I passed a group of school children in uniform, walking home up the hill. Sometimes, a truck would overtake me and I would see a lad or two on tiny BMX bikes hanging off the back of it like cleaner fish on a guppy. Later, I would see them whizzing back down the hill in freewheel, crouched over the handlebars and grinning like mad things.

The road climbed on, relentlessly. I was cycling at walking pace. It was getting cooler. I ate the last few wafer biscuits from a purchase days ago. I had one piece of pineapple left. My feet were cold, not something I'd experienced on this hitherto swelteringly hot ride. Eventually I stopped to take a grip. *Come on, Kate, you know about mountains. They get colder as you go up. This one has a cloud on top of it. Sort out the kit you need now, not when you are already up there, frozen and wet.* I dug out cycling shoes from a pannier – I'd been riding in sandals – and another top. The rush of warmth was wonderful.

Around mid-afternoon, the road flattened out. To my joy, there was a small café. The café had a large dog outside it. I pulled over. The dog was called Caramello and the woman running the café also had a kitten and numerous hamsters. I drank coffee and ate cake as it began to rain. Her partner emerged from the back, admired Woody and offered a swap for a scruffy black mountain bike. I handed him a sample strip of hemp I kept in my handlebar bag.

'You can keep it,' I said, noting that he had already put it in his pocket.

He grinned. 'The road ahead goes up and down a bit from here, but the main climbing is done. And then you are on a plateau.'

I pulled away in waterproofs, with lots of waves, full of coffee/cake/cheerfulness and expecting the road to flatten out. Which it did for about three minutes. And then it started climbing again. *Just be here. Not minding the climb, not wanting it over, not jumping mentally to the future. That's the challenge.*

Dusk fell and then, of course, I was cycling in the dark. Finally, I reached the town of Yarumal, in the middle of which was a triangle of taxis, trucks, petrol stations and lots of apparently drunk, shouting men. The hotel marked on maps.me was a dubious-looking establishment above a string of roadside bars.

Heading out in search of food later, I was pointed up the street to a small canteen-like restaurant. Two women and a man were eating in there. A smiling cook made me eggs, salad and rice. While I ate, I watched *Yo Me Llamo* (My Name Is) on the TV, a sort of Colombian *X Factor*, which seemed to be on constantly everywhere I went, and which was weirdly impossible to ignore.

A few days later, Medellín suddenly appeared below me, the second largest city in Colombia, a vast sprawl of brown filling an entire valley bowl. The bowl-shape helped keep the heavy traffic fumes trapped in the valley, and the city was in the thick of a pollution crisis blamed for 'more deaths than guns'. By the time I arrived at my host Diane's flat, I was close to euphoric to have survived cycling through it.

Next day, I cycled out again, heading for the town of Guatapé and a couple of weeks at a language school. It had dawned on me that cycling alone all day was not conducive to magically 'picking up' Spanish as I'd hoped. Two weeks of focused study would surely accelerate my progress.

There were said to be hippos somewhere near Guatapé, on a vast former estate that had belonged to Pablo Escobar. The hippos were descendants of four animals he imported in the early 1980s for a private zoo at his country hacienda. After his death and the collapse of his drug empire, the hippos escaped. They have been breeding in the 'wild' ever since. Hippos are not, of course, native to Colombia, and whether their presence was positive was the subject of a lively debate. One view was that the hippos' adopted river ecosystems may be benefitting from these enormous herbivores, as they filled ecological niches left vacant by long-extinct species such as the giant llama – similar to the hippo in its size and diet – and a strange, hooved mammal, notoungulata – similar in its size and semi-aquatic habitats. The idea of using substitutes for extinct animals, or those difficult to reintroduce, was enjoying an amount of attention in general, thanks to the exciting upsurge in an approach to nature conservation known as 'rewilding'. If the focus is on the ecological role that a species plays, rather than the species per se, a hippo and a giant llama can start to appear weirdly similar. Others argued that the hippos, who now numbered around 100, were more likely to be destabilising the local ecosystems, with their dung reducing the oxygen in the river water.

Gautapé itself is famous for its *zocalos*, huge painted boards skirting the base of each house, depicting leaping dolphins, farmers in fields, or the Pink Panther, all in vivid colours. It is a place that makes you grin and vow to shake off any lingering shade of beige in your life. It was also a lot more touristy than I'd expected. I would have loved to have seen the hippos, but the crowds were disorienting, and I was happy to retreat to the school, a small eco-hostel/permaculture farm up in the hills. I kept quiet about my 54th birthday, celebrating in secret with new verbs. But then our tutor, Paula, appeared with a surprise cake in the coffee break; my partner Chris having somehow found a way to contact her.

After two weeks at the language school, I cycled back to Medellín and spent a week catching up with emails and research in the relative luxury of Diane's apartment. For respite from the laptop, I visited Medellín's fabulous art galleries, enjoying Fernando Botero's 'fat sculptures' – plump women, birds, children – and his startling painting of a (plump) dishevelled, barefoot Pablo Escobar in black trousers and an open, white shirt being shot as he tries to flee across rooftops, a line of bullets like fat grey bees heading for his bare and already wounded chest.

By the time I left Diane's, I was ready to be back on the road – despite that the air pollution in Medellín was so bad an emergency had been declared. Certain car numberplates were banned from the roads each day, and people were worried about going outside.

I headed out of town, aiming for my first Casa de Ciclistas. The Casas are dotted across South America, offering a welcome to touring cyclists and facilities ranging from a space to pitch your tent to an entire building. They are run on a trust and exchange basis, visitors paying what they can, or helping out with jobs, or buying food. It was a stiff climb to my first one which, combined with air pollution and some navigational confusion, left me dizzy with breathlessness.

When I finally reached it, the hosts welcomed me into a beautiful, eccentric little place, full of bicycles, paintings and illustrated notes left by other cyclists – and cyclists. Tara and Aidan – aka 'Portland to Penguins' on Instagram – two North Americans who had ridden there from Portland, Oregon. A woman from Belgium, who had ridden from Mexico, travelling solo on her first ever cycle-touring journey. A man travelling by tandem and offering lifts to people en route, and his current companion, who was in the kitchen chopping onions. They were all heading south for Patagonia, in much the same time frame as me and Woody. Suddenly, just like that, I knew people, not just in Colombia, but on the road.

We spent the evening looking at routes on apps, an actual paper map of the whole of South America spread across most of the floor. My ride so far, from Cartagena to the Casa, looked dauntingly short compared with the distance ahead – and I'd already been in Colombia for a couple of months. I had initially planned to take six months for this journey, but friends who knew South America had persuaded me to give myself a year instead. I was convinced I had all the time in the world; the others had set off thinking they had more time than they needed to get to their destinations too. Now it was dawning on all of us that we might not have enough.

The next day was a short one, in terms of cycling miles at least, on account of a woman I'd been told I must visit, reputed to rescue sloths. I'd left the Casa de Ciclistas early, with Tara and Aidan, enjoying their company (and their superior urban navigation) back down the hill and in and out of the Medellín outskirts. Then we went our separate ways. I hoped we'd cross tracks again, but for now, my destination was a detour for them. The directions I'd been given to find the sloth lady took me off the main road out of the town of Caldas, down a steep track and along a traffic-free lane to a huge black gate. The gate swung open to reveal large and beautiful grounds. Trees, grass, a handsome timber-clad house, numerous outbuildings and a tall blonde-haired, blue-eyed woman in her sixties.

Tinka Plese had left Croatia several decades earlier to do a PhD in Miami. While there, she'd met many Colombians.

'They all said they were studying in order to take what they learned back to their country,' Tinka said. 'I was intrigued. In those days, nobody who left Croatia ever went back.'

Arriving in the country some 30 years previously with her now ex-husband, Tinka had taken on the Colombian wildlife trade.

'It's totally illegal,' she told me. 'But very widespread. People catch wild animals – parrots, baby sloths – and then sell them, often by the side of the road on main routes between, for example, Medellín and holiday destinations like Cartagena. When the small, cute, clingy young sloth becomes a large and pissed-off adult, they often get dumped.'

Not long after Tinka arrived in Colombia, someone had asked her to look after two of these dumped sloths.

'And I realised', she said, 'that no one here really knew that much about them'.

Tinka was now known as the Jane Goodall of the sloth world. Over the years, her organisation, Fundación Aiunau, had developed a huge body of knowledge about anteaters and armadillos, as well as sloths, all members of the super-order Xenarthra – meaning 'strange joints' – and all of which they looked after when needed.

I'd seen an armadillo close up in Costa Rica and found its appearance captivating. Bands of armor-plating, striped tail, small snout, unexpectedly kangaroo-shaped ears. Armadillo behaviour can be intriguing, too. One strategy for crossing rivers, for example, involves the armadillo filling her lungs with air and swimming for the other side. The other involves emptying her lungs and walking underwater across the riverbed. That armadillos could make this choice implies the ability to assess the width of the river – with stiff consequences for getting it wrong. What would it be like to be an armadillo, making that decision?

Armadillos are threatened across the Americas by illegal hunting and trade, among other things. The scale of the wildlife trade is colossal. Hundreds of millions of wild plants and animals are traded each year. It's worth billions. And vast swathes of it are illegal, which makes it extremely hard to monitor. It can, of course, be conducted both legally and sustainably. But it often isn't. Across the world, overexploitation of species – including by the wildlife trade – as well as overfishing and hunting – is one of the most significant direct threats to species after habitat destruction.

In Colombia, while Tití monkeys are one of the animals targeted, it is sloths that are probably most affected by trafficking. A distressing number of captured individuals – more than 90 per cent – die. Put this together with the deforestation and fragmentation of their habitat, and it's not surprising that they are endangered.

To try and tackle this, Aiunau work on community engagement and education as well as running a nationwide anti-wildlife trade campaign. And they take in any 'strange jointed' creature that needs help, always with the aim of releasing them back into the wild.

'We operate on the principle of "compassionate conservation",' Tinka explained, 'bringing conservation biology and animal welfare together. In other words, every individual animal counts, not just the species as a whole.'

Conservation, she explained, tends to prioritise protecting species – at the expense of individuals if necessary, and often in a way that doesn't acknowledge individual animals and their well-being. Compassionate conservation aims to protect both.

'We have two giant anteaters here at the moment, injured when their forest was burned down,' Tinka said. 'Ironically, the forests were protected by the civil war. The guerrilla lived in them, and everyone else kept away.'

Now that there was peace, and the guerrilla were being integrated back into society, the forests were at risk, Tinka explained. From farmers, loggers, gold-miners; from people who saw the potential to make money in a way that involved felling or burning the trees. It was often poverty-driven and almost always money-driven. Forests were at risk across the whole of Colombia, animals among the many casualties.

Tinka sighed. 'We also have an orphaned, three-month-old anteater who we are teaching how to be independent. Would you like to see him?'

The juvenile anteater, not quite a foot long and most of that nose, was probably the most outrageously cute creature I had ever seen. Because he was being prepared for release, I wasn't allowed to get too close or to handle him, but even at a distance, I was smitten. I could see why people would fall for young anteaters as pets.

'It's basic supply and demand,' Tinka said. 'If you buy one of these creatures, you are creating demand. And more of them will be extracted, usually brutally, from the wild.'

Some of Tinka's rescues came via the police. The police were supposed to act on any report of wildlife being sold. According to Tinka, though, they often didn't; or worse, they took the animals themselves and passed them on to rich friends who would keep the wheels of bribery well-oiled in return.

'It's part of a much wider issue,' Tinka said, with feeling. 'There is corruption at all levels. Local government, national government, the police, the wildlife trade. Colombia has some of the best environmental laws in the world. But these laws are utterly worthless unless they are properly implemented and policed.'

This, she said, was not just about the wildlife trade. I thought about Medellín, where bus companies bribed MOT inspectors so routinely that bus fumes were one of the biggest contributors to the pollution emergency. And about the people who had told me Colombia's national parks were nature reserves only in name. In practice, the tiny number of rangers, and the lack of interest in prosecuting violations, left them unprotected from illegal logging and mining. I knew in theory that 'good governance' was critical to achieving environmental goals: now I was beginning to see in practice how the issue of good governance underpinned everything.

For Tinka, it had a vividly personal dimension. Back in the house, she told me how she had been threatened after complaining about local water pollution.

'There was a factory dumping straight into the river,' she said. 'When I objected, their main response was to try and scare me off with thugs.'

Her next anecdote was even more alarming. After reporting an incident in which the police had failed to deal with a case of wildlife trafficking, three men had turned up at her house with guns. One of the guns had been put to the head of her housekeeper, while the other men ransacked Tinka's office and bedroom.

'I had been on my way home and would have been here,' Tinka said. 'Luckily I'd received a call about a sloth and had turned around and gone to the airport instead.' She paused. 'If it hadn't been for that sloth, I doubt I'd still be alive.'

It was both depressing and inspiring. How good to know that people like Tinka; Manuela at Environomica; Chile at the turtle school; Rosamira, Francy and Felix at Project Tití; and the folk at Pescando Para La Vida were out there working to protect wildlife and biodiversity in so many different ways. The more time I spent in Colombia, the more people like them I was finding. And the more I was experiencing Colombia as a country of contrasts. Immense warmth, friendliness, generosity, diversity, beauty and brilliant environmental projects on the one hand; the ongoing decimation of forests and other ecosystems, the chasmic differences between rich and poor, the disturbing power of the corporations and the still-present undertone of violence on the other.

Next morning, I was on the road again by nine. I'd had an early night, and my first off-asphalt day began well. The beginning of the 'green road' to Fredonia – a small town about twenty miles away – was beautiful. Lush vegetation, a mix of farmland and wilder lands, a big blue sky. A cyclist at the Casa de Ciclistas had said this road was unpaved, and unpaved can mean many things; I was half expecting thick gravel. I was glad to be riding the road first thing, with the whole day ahead, though they'd also said it was 'an easy two-hour ride'. Initially, the surface was fine: hard and not too loose. It was still truck infested, though, meaning multiple stops on steep sections on which it was difficult to get going again.

When I eventually came to a shoulder, I assumed I was at the top of the first main climb. Apart from the trucks, and the occasional piles of roadside rubbish, it was gorgeous. But the top of the first climb it was not. The top was some considerable way later. And the second climb was much tougher than the first. The road surface got looser and rougher. The temperature rose. The scenery became increasingly stunning – big drops down to an angular valley and complicated green mountains all around, some farmed on crazy angles, bananas and palms bursting from the steep slopes and houses on insane overhangs with spectacular views – though my capacity to appreciate it rather came and went.

By the end, I was pushing more than I was cycling. The last couple of miles were the hardest and hottest. After being overtaken by two chatty seven-year-olds with one bicycle between them, I crawled into Fredonia and the most welcome gallon of water, coffee and cake ever. That 'easy two-hour ride' had taken most of the day, one of the hardest twenty miles of my life. Those high-altitude, off-road routes I was planning in Peru? I needed to make some dramatic improvements to my fitness before then.

I left Fredonia to a tremendous thunderclap. Thunder rolled and crashed around the mountains as I raced downhill on lovely smooth asphalt, trying to stay ahead of the storm. There were huge views through the curtains of rain across hills and farmland to another, distant, mountain range. Woody and I plummeted down through the rain, dodging evidence of landslips, into a different ecosystem, patches of forest studded with silver-leaved trees like huge umbrellas. A river valley, trees with salmon-pink flowers, subtly beautiful against all the green. And then lots of roadworks, with groups of orange-clad men every few miles. I put on an album by the Brazilian singer Dona Onete, who kept me swinging along at a cracking pace before a final uphill pull slowed me down and I let the serious dusk catch me.

It was a relief to be working less hard. But I found my mind see-sawing between exuberance and half-formulated questions about how the violent Spanish conquest, the exploitative banana companies and the environmentally disastrous palm plantations might all be linked. I had arrived thinking that colonialism was in the past. Sometimes the thought that a version of it was alive and well would, like an unseen patch of thick gravel, catch me quite off-guard.

4. THE RIVER HAS GOT US

With a shared ethic of respect for the planet, we can create common solutions and lead by example in showing how to coexist with nature.

César A. Franco Laverde,
director of Serraniagua

An avalanche of mud, rocks and water hit the town of Mocoa, in the southern Colombian province of Putamayo, in the early hours of Saturday 1 April 2017. Most people were still asleep. Engulfed by the ferociously swollen and boulder-laden Río Mocoa and two of its tributaries, large swathes of the town and its neighbourhood were 'basically erased' according to Mayor José Antonio Castro. Over 300 people were killed, including 62 children. Nearly 400 people were injured. It was the deadliest of a wave of flood-related disasters that had swept across South America, the warmer seas of El Niño and climate change leading to astonishing quantities of rain. In Mocoa, almost half a normal month's rainfall had fallen in three hours. One of many heartbreaking phone calls to the emergency services that night had been from a woman on the roof of her house, watching the coffee-coloured mass race up towards her: 'The river has got us. Help us, please.'

I heard about the disaster while I was staying with Frances and David, friends of a friend, in Manizales. The press photos showed a town half-submerged in a wide, now stationary, mud river; rooves and half-houses visible; boulders the size of trucks; a white estate car at an impossible angle pointing downwards from the second-storey window of a half-demolished building; homes split open; shards and splinters of colour that were fragments of people's lives in the overwhelming brown; mud-covered people scrabbling and searching in mud; shattered, upended trees; trucks scattered and broken apart like toys among the rocks and sludge.

President Juan Manuel Santos, declaring a state of emergency, attributed the catastrophe to climate change. There were other factors, too. Mocoa sits at the confluence of three rivers, below steep-sided mountains. Multiple smaller landslides in the catchment above had contributed to the massive one that had decimated the town and its people, triggered by the unnaturally heavy rainfall. But the destructive footprints of poor farming practices and the removal of trees were also discernible. Putamayo province had one of the highest rates of deforestation in the country. Not only did this mean vastly fewer trees to help soak up the rainfall, it also contributed to the instability of the precipitously steep slopes. And, on top of some shabby urban planning 500 years previously, Mocoa had expanded in recent years, with many new houses constructed on deposits built up during previous debris flows. This was the archetypal disaster waiting to happen.

As the story of the Mocoa mud avalanche unfolded, I felt something shift in me. Among the complexity of causes, the contribution of biodiversity loss, or at least from deforestation and slope-destabilising habit degradation, was clear. I knew, intellectually, that biodiversity loss could, literally, be life threatening. But this was a gruesomely vivid, real-life demonstration. And in the here and now.

The WWF's 2016 'Living Planet Report' had come out not long before I left. If I'd needed reassurance that the biodiversity focus of my journey was apposite, this was it. The headline: populations of vertebrate wild animals – mammals, birds, fish, amphibians and reptiles – have decreased in abundance by an average of 58 per cent since 1970.* Over half! And in only 40-something years. In less than my lifetime.

Other reports have a species extinction focus, assessing the numbers of species that have been permanently erased from the planet, or those in danger of being so. The United Nations has recently estimated the figure for the latter at a horrifying 1 million. We are losing species at somewhere between 100 and 1,000 faster than the so-called 'background' rate of extinction – the rate species went extinct before human pressure became such a significant driver.

Biodiversity, of course, refers to all life forms, including plants and insects. WWF's report looks at wild vertebrate animals in part because this is where we have most available data.

Recent figures for insects are every bit as alarming.

The first study that hit the press was of insect numbers in German nature reserves: they had fallen by more than 76 per cent in less than 30 years. This was soon followed by attempts to figure out what was happening to insects globally. The results? More than 40 per cent of insect species were found to be on a downward curve around the world,[1] with evidence of 'recent, rapid decline in insect abundance, diversity and biomass'. A third were considered to be endangered. The picture is complex, and based on information from a limited number of countries, most in the northern hemisphere. But, overall, the rate of insect extinction is

* Based on many thousands of monitored populations of almost 4,000 different species.

currently thought to be about eight times more rapid than that of mammals, birds and reptiles. And given that the main drivers of this precipitous decline are primarily the loss of habitat to intensive agriculture and urbanisation, and pollution by pesticides and synthetic fertilisers – trends that are set to increase – the projections are grim. And not just for the insects themselves.

How long do we think we would last without insects? Einstein famously wrote that if bees disappeared, humans would follow four years later. And that's just bees.

Insects are fundamental to the functioning of our ecosystems, playing key roles in pollination – three out of four crops depend at least in part on insect pollinators – food chains, and nutrient cycling. Perhaps their most important role, as Manuela had explained, is in creating and sustaining fertile soil on which more than 90 per cent of the world's (human) food depends, and on which almost all plants depend.[2]

'Many people think of insects as a nuisance,' the introduction to a piece about insect decline on the UN Environment Programme website reads. 'They don't realise that without them we are doomed.'[3]

Strong language for the United Nations.

I had vacillated for ages over the route I should take to Manizales. There were numerous hilly options, and my mountainphilia pulled in that direction. It would be wonderful to spend more time in the high hills and small hamlets; to hear more birdsong, smell more trees and less truck fumes. But it would take longer and be considerably tougher. In the end, I reasoned that the priority was to get to Manizales and meet Frances. Frances was researching some dubious-sounding activities related to the oil industry in southern Colombia, and I wanted to learn more. Feeling oddly guilty, I took the easy option and stayed on the main road. There would be plenty of mountains ahead.

I rolled down the hill to La Pintada and onto the main Manizales road. The road ran alongside the Río Cauca, running fast and tawny-brown in the other direction. There was a lot of lushness. Steep-sided green hills, a mix of trees, farmland, occasional cattle. Many birds, including some small yellow-green finches with orange heads that I began to see everywhere;† the electronic whine of crickets and yellow rectangles warning of 'Animals on the Road', displaying their shapes in black and names in Latin and Spanish.

The traffic was light. I cycled and grinned. My body was tired from the Fredonia ride, and I had to admit a sense of relief not to be facing 50 miles of serious uphill. The main road and its relative flatness suddenly seemed close to idyllic.

As I got nearer to Manizales, the traffic returned with a vengeance, as if someone had turned on a traffic tap. An outbreak of roadworks meant waiting in a sweltering, fumy, noisy chaos of trucks and cars. When the lights finally changed, sometimes after people had unpacked and eaten picnics on the roadside, I would have to sprint madly through the exhaust of charging motorbikes to get to the other side before the lights changed and released a flood of traffic coming at me from the other direction.

After three of these fume-drenched sprints, I stopped for lunch. I was not in a rush, or so I thought. Manizales was now quite close, and I wasn't due to meet with Frances until 5.30pm.

My dawdling lunch, however, had a surprise chaser: a climb of 4,000 feet that, embarrassingly, I was not expecting. There it was though – the road rising, undeniably. After a couple of hours, I stopped at what I thought must be the

---

† Orange-fronted yellow finches, I've since learned – well-known in Colombia and Venezuela.

summit for coffee and cake. After which the road rose some more. Each time I was sure this *must* be the top, a bend in the road revealed another climb. Any feelings of guilt about the easy days were well and truly vanquished.

By the time I reached the actual summit, I was trashed. Badly overheated, drenched with sweat, legs shaking with fatigue. And very close to a total sense of humour failure. Really, what a bloody stupid place to build a city. My utter relief at finally reaching the top of the hill – Manizales is built on a high plateau – was immediately followed by the usual running-the-bus-gauntlet, it being rush hour, and lots of chaotic, roadside map-checking in the thick of traffic, most of which was not expecting to see – and hence often didn't see – a laden touring bike crawling through or stopped in their midst. More by chance than effective navigation, I fetched up on the road I was seeking. Disproportionately overjoyed, my milometer ticked over the first 1,000th mile almost exactly as I arrived at the rendezvous.

Amazingly, despite having travelled at a top speed of two miles per hour for the previous five hours, I was only ten minutes late. Francis recognised me among the crowd of people, though, to be fair, there were not many people arriving at the Juan Valdez café dripping sweat and pushing a loaded touring bike. Francis, in her early thirties, long brown hair, kind brown eyes, handed me a bottle of water and then a beer. I downed both and put on a fleece, cooling fast as the sun sank. We talked and talked. And then decamped to a bar. About four beers later, we walked to Francis' flat, a lovely, modern pad in a tall building on the hillside. Her partner, David – curly dark hair, warm, welcoming – hugged us both.

The time with Francis and David was an especially wonderful interlude. Suddenly whisked from my solitary on-the-road life into one of indoor comfort and company, they plied me with food, coffee and beer, indulged my craving for marmite and peas, took me to a bike shop, encouraged me to visit the Manizales sights. Best of all, they talked. What a luxury, to have extended conversations, in my own language, about everything from films to David's avocado company to how cycling in Colombia compared with the UK.

And of course, we talked about Francis' work. This was focused on the south-western department of Putamayo and the behaviour of the oil companies operating there.

'Oil extraction is an environmental disaster in Putamayo,' Francis explained. 'Deforestation, obviously. But also, water pollution, air pollution, gas flares.'

Not surprisingly, the people who lived there wanted the oil companies to leave, Francis told me.

'But the government want them to stay,' she said. 'It's about money, of course. The government has bombed the area, on the pretext that they are bombing the FARC guerrilla. But it's actually to get rid of the locals, who want rid of the oil company.'

FARC emerged in the 1960s as a left-wing, anti-imperialist movement fighting a much-needed fight for the rights of small farmers and the rural poor. It morphed into a brutal organisation funded by kidnap, ransom and the production and distribution of illegal drugs, among other things. Francis told hair-raising stories about her encounters with them in Putamayo, including an occasion when a group of men had come into her room in the night and shone a torch into her face.

'Presumably they were checking I looked as if I could be who the locals had said I was. I just lay still and kept my eyes shut until they left. I was too terrified to move for the rest of the night,' Francis said, grimacing.

But she was still more concerned about the general situation than any personal risk. Colombia, she explained, had the second highest number of internally displaced people globally, behind only Syria. And the figures only included those displaced by war.

'FARC get blamed for much of it,' Francis said. 'You have to say you were displaced by guerrillas to get IDP compensation.'

There were, though, Francis explained, many other things that disconnected people from their homes and livelihoods. Collateral damage when the FARC fought the paramilitary. Army bombing. Landmines deliberately placed close to mining areas to force people to leave. Banana plantations, drug corridors, military bases, palm oil. Climate change disasters. The government spraying coca crops from the air, which also destroyed food crops and poisoned water.

It added up to a shocking 6 million Colombian people displaced since the 1990s. Putamayo was just one – particularly awful – example, made even worse by the Mocoa mud avalanche.

Cycling out of Manizales was an awful lot easier than cycling in, though the long descent left me less than convinced that the new brake pads the bike shop had said I didn't need, wouldn't have been a bad idea after all.

Meanwhile, the terrible situation in Mocoa continued to unfold. The army had been drafted in to help; people were burying their dead. I had been getting emails asking if I was OK. I was, of course, though a friend from the language school had been in Mocoa and would have been there during the disaster had she not left earlier than intended. More than 180 municipalities were now on red alert for flooding, including most of my route between Manizales and Ecuador.

A key part of the journey's concept and timing was to stay in the dry season all the way down. But now I was running a month late and it was beginning to look as if the wet season might catch up with me.

Several days later, an hour or so from the town of Cartago, I was given my first real taste of what this might be like.

Proper drenching number one was a truly torrential rainstorm, complete with thunder and lightning and rain like a sluice-gate being opened in the sky. Within minutes, the road was a river and I had to take my glasses off and so was unable to see much. This included floating branches, potholes, road signs and, indeed, the road.

The storm was spectacular. Despite the inconvenience of such a deluge from a cycling perspective, part of me revelled in the sheer power of it. Yet there was also a sense of something tainted, a sinister undertone introduced by the knowledge that human-caused climate emissions were probably contributing to the storm's ferocity.

I came into the Cartago roundabout in the dark and wet. Asking directions, I was pointed down a dubious side street, allegedly the fastest route to the town centre. Cycling along (and through) what appeared to be small river, I emerged from under 'road closed' police tape with no clue where I was.

Hotel Don Gregorio, when I eventually found it, looked unexpectedly smart. A porter helped me lift the bike up the front steps and in seconds I was transported from the dark, cold, wet, alarming streets to warmth and safety. In my room, Woody on some cardboard in the corner, I discovered that I had huge black smears below both eyes – road dirt mixed with traces of ancient mascara, a legacy from another lifetime.

It rained heavily through the night, but I woke to blue skies and went out in search of a bank. The town had been transformed by sunshine; the main square was full of coffee vendors serving delicious black or latte-coloured brews and people drinking in the warmth.

By the time I left, it was beautifully hot. It was an easy ride out of town, over my friend the Río Cauca and into flat farmland. Occasionally, the largest rigs I'd ever seen went by, trucks pulling five wagons full to bursting with plant cuttings. The road became ever quieter. I stopped at a tiny place advertising *arepas* – small, round, cornmeal-based basics, especially delicious when hot with added cheese – but none were available until late afternoon. I had a glass of *aguapanela* instead – 'Colombian Red Bull' – but the sugar burst rapidly receded, leaving me fantasising about egg and chips.

When I'd ridden in South America in the early 1990s, the main roadside café diet had been beans and rice, often with salad or plantain, so it had been easy to eat vegetarian. Now the beans and rice had been displaced by chicken and chips, and almost all my meals were composed primarily of eggs. In Colombia, portions were generous. On those days when road-side café eggs were the basis of breakfast, lunch and dinner I was probably eating an average of nine a day. Nine eggs a day?! Yet there I was, craving more of them.

It stayed lovely all day. Woody and I climbed steadily upwards, through birdsong and gorgeous landscapes. Complex, angular hills, a mix of forest and farmland, the patched green backdrop studded with the huge star-shaped leaves of small palms. Coffee was growing by the roadside alongside trees with branches festooned with lichens.

I topped out at La Carbonera, which consisted primarily of a military checkpoint, a petrol station and a couple of cafés. The hotel was an utterly eccentric, entirely unexpected place, once rather raunchy, now full of former glory and defunct jacuzzies. Nothing in the hotel actually worked, bar one cold dribble of a shower on the floor below and a single lightbulb in my vast, black and red bathroom. I had a cold shower, hung socks on the balcony to dry and watched a hawk suspended low in the sky. And then went out in search of food.

Returning to the empty hotel later, there was something eerily romantic about being the only guest, coming in and out in the dark with a headtorch. I sat on the balcony under a blanket watching the lightning play silently across the mountains. And then I went to bed and read about Serraniagua, the project I was about to visit.

I'd been connected with Serraniagua, a community-based environmental organisation, and its lead, a man called César, by a European colleague called Daniel, who worked for Climate Alliance in Austria. I was already grateful to him for wonderfully clear directions to El Cairo, off the beaten track, and my map. The more I read, the more interested I became. This project was clearly special.

The soldiers called me over as I left the next morning. They were young lads but heavily armed and searching bus, truck and car passengers at gunpoint. As I pushed Woody towards them, a touch apprehensively, they broke into grins.

'*¿¡Es de bambú?!*'

'*Si.*'

'*¿Podemos tener una foto?*'

I could have anticipated this. Everybody was fascinated by the bike. Everybody. They told me they were checking papers and looking for drugs and guns. One of them tried to lift Woody – he was fully loaded – and the scene descended into banter and photograph-taking. Should you ever wish to travel safely through a still-militarised country solo, I strongly recommend a bamboo bicycle. It's like being on a journey with an instant icebreaker combined with a magic pass.

Eventually, with lots of waving, I pushed off down the other side of the hill I'd climbed up the previous day, where it rapidly became clear that my brakes were no longer very effective at all. The road went down and down through the green mountains for nine or ten miles, often steeply. Then I climbed back up on the other side. It was wonderful to be out there in this lush, peaceful, vibrant place.

Suddenly El Cairo was on the horizon and then, abruptly, I was in the town square. Huge trees in the middle. White buildings with green, orange and red doors and shutters. A dovecot. A mountain backdrop. I cycled slowly around until I found the Serraniagua office, a couple of huge, brightly painted birds – a hummingbird and an exuberant, coiffured 'cock-of-the rock' – adorning the front wall.

The office was shut. Woody and I stood about for a bit. Then two people appeared, Cedric and Sara, an intern and a master's student from France and Austria respectively, both English-speaking. More people began to arrive. Maffi, one of César's sisters, who ran a hostel-cum-pizza restaurant, exactly my age. César's other sister, Claudia, who I was to stay with, super-welcoming, full of stories. A woman called Costanza, who was a Colombian, English-speaking biologist, conducting research on the Serraniagua project. Then someone who worked for tourism, and someone from the local government, and various kids. And then César himself, slim, dark, big brown eyes, unassuming, also about my age.

By now it was around two in the afternoon, and I was ravenous. In Colombia, one of the world's most famous producers of some of the world's most wonderful coffee, it had become fashionable to drink instant. I was often offered a packet of Nescafé, sugar and a mug of lukewarm milk as if I were being given champagne. But this time, the coffee was the real deal. Humboldt called coffee 'concentrated sunshine' and, as I sat drinking delicious *café con leche* and eating cake looking out onto the beautiful square, I thought he had got that just right.

Later, we all had pizza at Maffi's. I chatted with Sara, who was in El Cairo researching the magic point at which shade thrown across coffee plants by nearby trees is optimal for biodiversity, soil, water retention, climate change resilience and cost. She was full of positive energy and enthusiasm, her first time travelling on her own.

'The only thing I don't like here', she said, 'is not being able to walk out of the town alone.'

'Seriously?' I asked, more than a little sceptical.

'Yes,' she said. 'I was sceptical too. But someone was kidnapped recently. Very nearby.'

Back at Claudia's later, a touch tipsy, I learned more about the kidnapping. Another guerrilla organisation had been involved, and Cedric had been motorbiking back from visiting a local coffee producer when he'd come across them.

'The road was blocked with trucks and men in masks,' Claudia said. 'One of them shouted, "Turn around and go back. Now."'

The kidnapping had hit the fringes of the UK news, apparently. I hoped my dad hadn't spotted it.

Next morning, we left early, heading for the Cerro el Inglés nature reserve. César drove, the jeep jolting on a rough track through low cloud. I sat in the front, listening to him telling stories in slow, considerate Spanish. There were various tales about the reserve's name, and where the 'Inglés' bit came from, ranging from treasure hunters thought to be English who were actually German, to a local woman with one leg who fell in love with an Englishman and walked into the reserve never to return.

The reserve lies within a mountainous region called the Serranía de los Paraguas, part of the Cordillera Occidental – the most western of the cordilleras. On its western flank, the Serranía de los Paraguas connects with the Chocó region, a huge lowland area that runs down from the mountains to the Pacific coast. These two regions – the tropical Andes mountains and the Chocó – are among those with the greatest biodiversity on earth. The Serranía de los Paraguas straddles them both, part of an area considered to be a biodiversity hotspot – a global conservation priority.

The idea of biodiversity hotspots developed from a seminal paper published back in 1989 by Norman Myers, a British environmentalist who specialised in biodiversity. Myers of course knew that the biggest direct driver of biodiversity loss on land is habitat damage and degradation. Three-quarters of the earth's ice-free surface has already been significantly altered by human activity, much of it for agriculture. In relation to forest habitats – home to 80 per cent of the world's terrestrial biodiversity and about 300 million people – 40 per cent have already vanished, razed by burning or clear-felling for farming. Logging and mining are also significant. This is ongoing – forests are being destroyed at a rate of at least one football pitch equivalent *per minute*.

Protecting remaining habitats is clearly crucial. Myer's breakthrough idea was to prioritise areas that were both unusually high in endemic species – species found only in one geographical location – and unusually threatened. He called these areas biodiversity hotspots – the richest and most endangered reservoirs of life on earth. The 36 biodiversity hotspots now recognised worldwide represent only a tiny proportion of earth's land surface[‡] yet, according to Conservation International, are home to over half of the world's plant species as endemics, and nearly 43 per cent of mammal, bird, reptile and amphibian species. This in turn makes them crucial to people, thanks to the 'ecosystem services' they provide.

The Tropical Andes Biodiversity Hotspot that I was travelling through in César's jeep is the most diverse hotspot in the world.

'It's an amazing area,' César said, as something of an understatement. Then the conversation shifted to climate change.

‡ 2.5 per cent.

'The seasons have become very unreliable,' César said. 'Plants and animals don't know how to respond. And the rain is much heavier. It damages crops, plants, soil ...'

I thought of the unnerving ferocity of the rainstorm I'd not long cycled through. Climate change is creating huge challenges for plants and animals in other ways, too. It can contribute to habitat loss and disrupt critically important timings and interactions between species. Animals including bears have been emerging from hibernation earlier, for example, only to find that their normal sources of food are not yet available. Physiological stress is another challenge, especially for species that can only live within a certain range of temperatures – like flying foxes, who can't cope with temperatures above 42°C. Thousands die every time there is a heatwave.

As climate change accelerates, it will become an ever more significant driver of biodiversity loss. And the highest rates of climate change-related loss are anticipated in biodiversity hotspots. Developing climate change resilience is therefore central to much of Serraniagua's work. Sara's research into 'shade coffee' was one aspect of this. Another was conservation corridors.

This was an idea I'd first encountered on *The Carbon Cycle* ride.[4] Corridors are important partly because animal populations are affected when habitats are degraded, as well as when they are destroyed: when the quality of the habitat is reduced, for example, by replacing a forest of diverse, native species with a monoculture. Or when habitats are fragmented, for example, by a forest road that cuts off animals and plants on one side from the other. In Banff, Canada, I'd met people working on a project called the Yellowstone to Yukon Conservation Initiative (Y2Y) – a brilliantly ambitious attempt to tackle this by joining up parks, nature reserves, sympathetically farmed ranches and other relatively wildlife-friendly areas to create a 2,000-mile-long habitat corridor all the way from Yellowstone National Park in the USA to the Yukon in Canada.

'Conservation corridors are key here, too,' César told me. 'Climate change means many species must change where they live. All sorts of living things are on the move. Insects, mammals, birds. Even trees are moving! Corridors make this easier. And connecting fragmented pieces of habitat improves that habitat for those who aren't moving, too.'

We arrived at a small farm, still in the clouds. The farm was run by a homeless family who had asked César whether they could live in Cerro el Inglés and look after the farm in exchange. From the money they made serving food to Serraniagua's guests, they had been able to buy a calf. This calf, they said, made them happier than they could ever remember being. She clearly meant more than future milk and meat.

César, a local guide, Sara and I walked down a rough and extraordinarily muddy track and into the forest. There was an immediate sense of immense lushness, aliveness and power, much as I had felt in the Sierra Nevada. Scrambling above a small, fast-running river, we emerged at a viewpoint, overlooking a gigantic bowl of trees. The trees, at first sight a vast mass of unified greenness, gradually clarified into a breathtaking array of different shades and shapes.

The montane forest was part of a conservation corridor project that aimed to connect high-quality habitats right up to the Darién region in north-west Colombia, a region that in turn connects via the notorious Darién Gap with Panama. The Gap was previously much wider. Then, 40 million years ago, the land bridge now known as Panama had risen from the sea and utterly changed the world. For the first time in the history of the planet, species from South America encountered and merged with species from North America, and the Isthmus of Panama became the most important biological bridge in the history of life on earth.

Previously separated creatures crossed into a brave new world; horses and big cats from the north crossing tracks with marsupials, armadillos and mastodons from the south. The giant sloth and the condor-like 'thunderbird' – named for the noise of its pounding, six-metre-long wings – are among those now extinct. But many existing animals and birds on both continents have shared ancestors from that extraordinary mingling of animals,[§] and 'the Great American Biotic Interchange' is considered one of the reasons why Latin America is still so very rich in biodiversity. Panama, like Colombia, is among the most diverse countries on earth – if you choose your Panamanian hectare well, you can find more species of trees, insects and birds there than in the whole of North America.

Later, we all regrouped at the Serraniagua café. There was an espresso waiting for me when I arrived. Then wine appeared. Outside, the sun was setting. Food stalls had been set up in the square, people and dogs wandering between them.

I sat listening to the conversation in Spanish, understanding snippets of it. The café exuded warmth and welcome, the large table in the centre now full of people. César was in full flow, passionate and funny. He held everyone's attention, brilliant, gentle and apparently ego-free.

The Spanish was speeding up, and I found myself disconnecting from the words being spoken and thinking about the project. In these uniquely biodiverse, high-Andean forests, Serraniagua featured a now familiar mix of forest protection and regeneration, and community development and engagement – plus elements I hadn't previously

[§] Some animals crossed earlier than others. The main interchange got underway a mere 3 million years ago.

encountered. These included 'capacity building': strengthening people's ability to think through and debate different conservation proposals, so that local communities were not just 'engaged' with nature conservation, but able to inform and shape its goals and how to achieve them.

Then, of particular interest to me, there was a strong focus on values – and the need for alternative-to-the-mainstream ones. The Serraniagua café, clearly El Cairo's social hub, radiated values to do with hospitality, generosity, good company and sense of community. It needed to tick over, financially, but making money was clearly not the main driver.

It was during a visit to César's farm the next day, though, that I realised quite how different Serraniagua was on the values front. For me, with a background in philosophy, it was a revelation. I've long thought that philosophy[¶] is both supremely important and prone to crippling abstraction. On the farm, I saw a cogent and attractive set of alternative values brought to life in practical ways, perhaps even more than in the café.

The farmhouse was a handsome, elderly building, with white walls, a terracotta-tiled roof and a shaded porch with blue-painted railings. Flowers erupted from hanging planters and pots. Two gleeful dogs welcomed us. We set off across some fields and through the forest to a river for the kids to play in. Later, we walked to the coffee-growing area. Dense clusters of glossy, cherry-sized, oval beans grew straight out of the woody stems, some still the green of an unripe tomato, others a rich collection of reds. Ripening coffee is always beautiful. It was the rest of the plants that were

[¶] Understood as the ability to unearth, and think critically about, the assumptions our societies are built on.

unusual.** Instead of the fields of coffee in rows I'd seen on other plantations, the coffee here was in among a lush riot of shrubs and trees – the so-called 'shade plants'.

'If you plant a coffee tree every square metre or so,' César explained, 'you get more coffee and more money. But what do you lose?'

We looked suitably quizzical.

'You need to use chemicals to control the pests. So you lose your fertile, healthy soil and your biodiversity. You lose your water. And you only have coffee. So you lose a farm that also provides fruits and other food.'

César paused for effect.

'Whereas if you plant the coffee among plantains, avocados and different trees that give shade, the coffee matures more slowly and tastes better. You don't need pesticides. You keep the soil, water and biodiversity. And you have a generous farm with much to offer guests.'

On César's farm, the coffee grew among a mix of plantain, avocados, oranges and cacao. It was indeed a wonderful place to be a guest. And beyond the area that produced food for people, there was a wilder, forested area where a multitude of other species lived alongside the farm and its human community.

Using the term in its widest sense to mean something like 'the things we think are important, that we aim for and that guide us', César talked about the contrast between this and the values embedded in the profit-driven version of coffee farming dominant across Colombia, and in the wider, consumerist culture influencing us all. Success defined almost exclusively in economic terms; competitiveness that pitched farmers against each other; a certain kind of self-centredness;

** Unusual on my journey. Shade coffee is well-known and widely rated as superior bird habitat, for example. It was the first time I'd encountered it, though.

the exaggerated role of money and possessions in the mainstream account of what a good life looks like. And he talked about the importance of finding communities where other values could still be found and nurtured.

For César, this had been a long-term project. He had grown up in and around El Cairo and never wanted to leave. His father had insisted he go away to university and then come back if he still wanted to. He had been living and working in the area ever since, on his own farm and with local farmers, encouraging them to resist 'modernisation' and instead strengthen traditional practices compatible with biodiversity. Serraniagua was now a couple of decades old.

'During this time,' César said, 'people's focus has shifted from their own profit to the health of the community and pride in what they've achieved together. And they should be proud,' he continued, as we walked past a sign reading, 'COFFEE: Friend of the birds.' 'We've shown that traditional farming methods, generosity, delicious organic coffee, a good life in general – plus nature conservation and climate resilience – can all go hand in hand.'

'When I write about my visit here, what's the most important message from Serraniagua to the world?' I asked César later, back on the colourful front porch. There was a long silence.

'Have you seen the men in the mountains wearing ponchos?' he asked, eventually. '*Los Montañeros en los Paraguas. Los Paraguas* is where the biodiversity-rich Andean and Chóco systems meet. *Los Montañeros* are the people who live there. Also, their culture, tradition, history, spirituality. Serraniagua is about bringing this together, the culture and the nature. Nature is not a luxury add-on to culture. It's vital to it. And if you want to preserve nature, you have to change the cultures that are destroying nature.'

'And instead?'

'Seek out traditional cultures built on the values of respect, gratitude, appreciation, friendship and community.

Cultures that know that people, animals and nature can all coexist. Work with these cultures to strengthen and reinvigorate these values. Then they will spread out from there.'

'I love it here,' I found myself thinking, suddenly, with both unexpected force and sadness.

How often would I encounter that bittersweet mix, the tantalising promise of connection and belonging, at odds with the compulsion to keep moving?

Were I not on a journey, I could happily stop in El Cairo, I thought. There was work that I could join in with, feel part of something worthwhile, even feel at home. And I didn't think I was feeling this in a rose-tinted way. El Cairo and everything around it had an edge I was alert to. But I still felt I had found my place – my place in Colombia – were I ever lucky enough to come back.

The next day, I sat trying to catch up with my journal while enjoying a latte with even more appreciation. Definitely shade coffee, I thought, with the smugness of the newly informed. I was trying to write about the way Serraniagua offered a possible answer to the all-important question of how we change values, and what we should change them to. It seemed an approach that could be used widely, beyond traditional farming communities.

'How replicable is Serraniagua, though?' I asked Cedric, who was working at a nearby table. 'I mean, isn't César unique?'

'Well, he is very special,' Cedric replied with a grin. 'But there are other César's out there. And the market is changing. More and more consumers want organic coffee with this sort of backstory.'

'Of course it is replicable,' César said, when I asked him the same question.

'But you are not normal,' I said.

'You neither,' he replied. Touché.

I had a great send-off. The guide from the Cerro el Inglés visit had returned with two friends who were bike mechanics. They worked with me to figure out how to change the brake pads on Woody. I had not used disc brakes before and was daunted by them, much to my exasperation. These lads restored my confidence as well as my brakes, arriving as if summonsed by that random god of travellers who intermittently produces things at exactly the point they are needed. Both the pads and the brake cables were, they pronounced, trashed. They choked with laughter when I said, 'But they've only done 2,000 kilometres.'

In the café, Cedric made me a final latte. Maffi gave me cake for the road. César offered me a poncho with the Montañeros en los Paraguas logo, a present from everyone. Finally, at around ten, I climbed onto Woody and wheeled away. César cycled with me to the outskirts of town. Two other local cyclists accompanied me to Alban where we were joined by a neighbour on a racing bike. A man came over with a jar of honey, saying that he had heard about my journey and wanted to give it to me as a present as he thought environmental issues were important and too often neglected.

And then I rode on alone, mulling about how much the visit had meant. It was to prove a brilliant foil for the next visit, too: one equally impactful and equally about values, though in a different and much darker way.

5. ALL THAT GLITTERS ...

No to mining, yes to water and life.
COSAJUCA CAMPAIGN SLOGAN

If someone offers you papaya, take the papaya.
COLOMBIAN SAYING

After centuries of deliberate isolation from the rest of the world, the Kogi suddenly made contact in the late 1980s. The Kogi understand themselves to be the guardians of the world, and they came out of isolation to deliver a warning, from them, the Elder Brother, to us, the Younger Brother. The warning was stark:

> Younger Brother is killing the world. He – we – must stop our violent over-exploitation of nature and learn to look after the earth, if any of us are to survive.

This uncompromising message was delivered via an extraordinary film made with the Kogi by Alan Ereira.[1] It raises profound questions. How do the Kogi, having spent centuries in a corner of north-west Colombia with no links to the rest of the world, know that the whole world is in trouble? What justifies their claim of impending ecological collapse? Is it like 'Western' evidence? And, most importantly, are they right? An unplanned visit to a gold museum was to lure me further down this trail of enquiry, a trail that began with enticing smoothness, before degenerating into rocky confusion.

After leaving El Cairo, I had decided to zigzag back across country to the town of Armenia, to renew my Colombian visa. The visa was due to expire in ten days – many fewer than I needed to reach the Ecuadorian border. Trying to leave Colombia without a visa was unlikely to go well, either. No, renewing the visa was a must, and Armenia was the closest town with an immigration office where I could do this. As it turned out, the detour, as well as leading me to the gold museum, was also to take me to one of the most impactful encounters of the whole journey; an encounter that was to smash open a whole new set of questions, as if I'd been cycling on ice and had suddenly broken through to a different and darker world beneath.

Visa renewal in Colombia, it transpired, is a process that used to be conducted by humans but that has recently migrated online and is now impossible to achieve unless you happen to have booked into a hostel that employs a highly savvy, computer-gifted and super-helpful teenager. Which by great good fortune, it turned out that I had.

I'd left Don Gregorio's in Cartago with lots of friendly goodbyes and photo-taking, Woody leaning calm and assured against a wall. I headed out of town and off onto the hilly route to Armenia. It was overcast and hot. I played tag for a while with a man riding bareback on a lively chestnut horse. With a fast walk/trot, the pair overtook me on the hills. I overtook them on descents and flats. They stopped at a rubbish pile, and the next time I saw them there was a thin layer of cardboard between the man and the skinny, sweating animal.

I cycled through drizzle into the outskirts of Armenia the next afternoon. Armenia had originally been called Villa Holguin, renamed later – according to some – in memory of the 250,000 Christians who died in sectarian massacres in Turkish Armenia in 1894. The outskirts of modern-day Colombian Armenia are, frankly, a bit grim. As usual, I found city navigation nerve-wracking. A kindly motorbiker stopped to ask where I was going as I struggled with the challenge of proceeding straight at right-hand junctions without being taken out by buses. It was with relief that I arrived at the hostel – and received a friendly welcome.

'The visa office is a short taxi ride away,' said Daniella, on reception. 'I would get there by 8am.'

I was sharing a room with a Colombian mother and daughter who got up at 6am so this advice was unexpectedly easy to follow. I was there before the office opened, in fact, with only two people in the queue in front of me. *This is going to be quick*, I thought, rashly. It was, but only because the official I dealt with told me I had to fill in the visa renewal form online, and that under no circumstances could I do it there. I was out of the building in minutes. I walked to the nearby Bolívar Square and enjoyed a much-needed caffeine hit in the sunshine. Then I caught a taxi back to the hostel, armed with a document setting out each step in the online application, in English. How hard could it be?

Sitting at a table near reception, I was soon thwarted, the correlation between the instructions and what appeared onscreen being less than exact. My whimpers of frustration were detected by Rosa from Mexico, who got me as far as the section that read 'now upload your documents' and handed me over to Daniella, who spent an hour figuring out how to perform that apparently simple task.

Clutching my laptop, I raced back to the office. Finally, just before noon, it was my turn. My uploaded documents were perused. Additional questions were asked and answered, photographs taken, passport and alleged onward travel details checked. Then, in Spanish, 'OK. You need to come back at 2.30pm to get a signature from my boss. Then you are done.'

'2.30pm? Not now?'

'No. 2.30pm. He is not here now.'

Back at the square, I worked on my email backlog in a very nice café until it was time to return.

'I have come for the signature,' I said with a hopeful smile to Mr Official.

'OK. You need to come back at 4.30pm to get a signature from my boss. Then you are done.'

'4.30pm? Not now?'

'No. 4.30pm. He is not here now.'

On the brink of exploding with exasperation, I decided to make the best of the situation and go somewhere I might not otherwise have made time for. Contributing generously to the local taxi economy that day, I caught a cab to the Quimbaya Gold Museum,* on the edge of the city.

The Quimbaya, a pre-Colombian Amerindian culture, lived in the regions around the Río Cauca between the western and central ranges of the Colombian Andes. People have inhabited this region since about 12,500 BCE – for over 14,000 years. Somewhere between 3,000 and 7,000 years ago, nomadic hunter-gatherer tribes transitioned to agrarian societies. Settlements were established, crops grown and pottery and other crafts appeared. About 1,500 years before I cycled through their regions, the Quimbaya were at their zenith. Like the Kogi's ancestors, they were especially talented at working with gold and had figured out how to create *tumbaga*, a fantastically versatile gold/copper alloy. From it they produced some of their most spectacular pieces. And some of those were in the gold museum in Armenia, most likely having arrived there relatively recently, thanks to tomb robbers – such a large-scale activity in '70s Colombia that there was an attempt to establish a tomb-robbers' union.

* Museo del Oro Quimbaya.

Many of the artefacts were small animals; some realistic, some stylised, all beautifully crafted.

'The insects in a state of metamorphosis ... were possibly symbols of change,' read an interpretative signboard, while 'snails represent metaphors of ideas we know nothing about.'

There could be many snails ahead, I thought.

As well as the animals, some long, heavy-looking mouth ornaments of abstract design were splayed out below the glass. And then there was the gold *poporo*.

Poporos are small, handheld items used by indigenous South American cultures to store lime, which they produce from crushed and burned seashells. They have a roundish receptacle for holding the lime, about the size of a small melon, and a lid which merges into a long stick-shaped extension that can be used to reach into the lime, allowing the person holding the *poporo* to dab the lime on the wad of coca in their mouths without having to touch it. The lime activates the coca, and chewing the coca supresses sensations of tiredness and hunger, enabling people to work for immensely long hours, walk incredible distances, tolerate higher altitudes and go without sleep. For many indigenous peoples, chewing coca is a sacred act, and *poporos* tend to be associated with both mystical powers and social status. The gold *poporo Quimbaya* in the museum was significant enough to be the star attraction.

Poporos made from the dried fruit of a gourd plant are still used by the Kogi, or at least by the men. After they have licked the stick, they wipe it, 'in a contemplative manner', on the top of the gourd. The gradual build-up of a thickened, yellowish 'calc' of lime and spittle has been described as a book.

'We write our thoughts in it,' the Kogi say.[2]

In a culture without writing that puts an immense importance on ideas, this build-up of lime and spit has a significance incomprehensible from the perspective of cultures that have written documentation. In what sense can it be compared with words on a page? Could someone other than its author read a 'spit-book'? The more I found out about the Kogi, the more I was confronted with these chasms of understanding. Something I learned about them would seem semi-graspable, then morph into something unfathomable. Like the Kogi's beliefs about gold.

On one level, the Kogi believe that gold is vital to the earth and that digging it up has dire consequences. Comprehensible, even if we're hazy about the causal mechanism. But unpack this belief, and the illusion of understanding dissolves like snow on hot tarmac.

The gold that the Kogi are most concerned about is gold that was skilfully crafted into artefacts by their ancestors, then placed in pots and buried in tombs. They believe that the artists who created them worked, not just in what we would call the real world, but also in *Aluna*, which is, roughly speaking, a spiritual world that parallels the material one.

These artefacts, how they were made and where they were placed are all of supreme significance. They are 'focal points' for a spiritual power known as a 'Mother' that nurtures and protects specific forms of life – such as macaws or yucca or even the Younger Brother.[3]

Gold has a further significance, too. The Kogi believe that the world is a living organism and that gold, formed when the earth first became fertile, is the blood of the living earth, vital to the life of the 'Great Mother'.

Hardly surprising, then, that digging up these gold items, created and placed with such care and to such purpose, is held to have terrible consequences. It is the Kogi's view that when the Mothers associated with the gold are unearthed – whether by grave robbers or archaeologists – it's not only the particular life form they protect that is put at risk, but the whole of nature.

'As the Mothers are dug out of the ground and taken from their pots, the forces which they represent are dissipated and the world slides towards chaos.'[4]

The explanation behind the Kogi's belief that digging up gold is harmful, however fascinating, is completely at odds with a Western worldview. But, for the Kogi, their analysis is constantly verified: shown true by the disappearance of the animals and plants of the Sierra Nevada that were protected by the gold figures in the buried pots.

'[T]he Kogi have no reason to doubt their understanding of the world,' Ereira writes, 'because that disappearance is taking place before their eyes.'

Since the Kogi believe that the Sierra Nevada is both the 'heart of the world' and an ecological mirror of it, they also believe that this ecological decline must be happening everywhere.

Watching Ereira's film is a mind-stretching experience. From the Kogi's perspective, it is our response that is unfathomable. Having put out their message to the world they waited for us to heed the warning and to change. Nothing happened. Eventually, years later, deeply perplexed, they contacted Ereira again and asked him to make another film. Even leaving aside the difficulty of trying to communicate across worldviews, they were up against a by now well-known challenge. Conveying information, however accurate and pressing, isn't necessarily enough, whoever you are. Al Gore, for example, writes about his early days as a climate change activist when, having read and understood the science, he went – horrified – to the US Senate, convinced that all he needed to do was pass on the information and change would follow. That's not quite how it panned out, of course.

What is really unnerving is that the Kogi are right. While writing this book, two further 'Living Planet Reports' were released. They revised the 2016 figures upwards, the 2020 figures showing that the earth's populations of wild animals have now declined by a shocking 68 per cent since 1970.[†] The main driver continues to be habitat loss, followed by over-exploitation and various kinds of pollution. Mining, including mining for gold, is heavily implicated in two out of the three.

The Western-trained scientists who authored the report are working within the context of belief systems profoundly different from those of the Kogi. But their conclusions are terrifyingly similar. Mining, among our other excesses, is killing the planet.

After an utterly fascinating hour or so in the museum, I caught a taxi back to the visa office. Finally, not long after 5pm, a troupe of people in identical uniforms entered the building. One by one, the small, defeated looking group of visa supplicants were called and visa extensions were, at last, duly signed, with a complete absence of flourish. As I was leaving the building, I said quietly to the person ahead of me:

'You know what, they never asked me to pay.'

Visa extensions usually come at a not insignificant cost. She turned and, with a huge grin and without hesitation, no doubt thinking of the lost day we'd all spent, said: 'Run!'

If someone offers you papaya, take the papaya.[‡]

[†] James Borrell argues that 1970 is an arbitrary starting point and that if you go a few hundred years further back, the decline would be more like 90 per cent or even higher, Royal Geographical Society EXPLORE Weekend 2021.

[‡] A friend who knows South America well has suggested that this saying has connotations I probably don't intend. I was innocent of them at the time.

On the way back to the hostel, I bought the best and most expensive chocolates I could find for Rosa and Daniella. Having delivered them, I sat in a quiet corner of the kitchen area trying to figure out what I should do next.

It felt as if Colombia was closing in around me. The news announced that Manizales, not far behind me, had suffered torrential rain and mud avalanches, leaving eleven people dead. I sent messages to Frances and David, worried, yet knowing they almost certainly would be OK. In cities, it is mostly the poor who are hit by mud avalanches, their houses or shacks built in places that no one with options would choose. Ahead of me, Ecuador had just had a serious earthquake. I had a growing sense of dodging disasters. Torrents of mud, the El Cairo kidnapping, news about instability as FARC pulled out of various areas. Yet I also understood the unlikelihood of any of this impacting on me. I was skimming along in a privileged, protected state, and this was feeling increasingly uncomfortable.

My problems had more to do with the number of options that kept arising, in ever-increasing tension with my need to keep moving southwards. I had already been in Colombia weeks longer than planned. If I was ever going to reach Patagonia, I would have to start saying no to amazing opportunities: a seemingly inescapable dilemma created simply by devising a journey with a destination fixed in time and place.

There was one story I really wanted to be able to tell: Cajamarca, the town that had said no to a major gold-mining company. It felt important, and different from anything I'd encountered so far. It also felt complicated. Now an avid user of Michelin online maps – complete with contours – I knew it was one of those 'only 30 miles away' journeys from Armenia, involving a humongous, 6,500-feet ascent in each direction. If I cycled, it would be three days before I was heading south again.

The next morning, I met with a local environmentalist at a café near the university. Adriana talked about the proposed mine, and how long it had taken to organise Cajamarca's resistance to it.

'It's been a decade,' she said. 'The people are farmers. Now they understand what it means, they absolutely don't want mining there.'

Many calls later, it was sorted. I would be picked up early next morning and driven to Cajamarca by a man called Alejandro, who spoke English. At Cajamarca we would be met by a woman called Jennifer, who could tell us about the campaign. Alejandro would translate and drive us back the same evening.

I was at ease with the car decision, having clarified my own rules about cheating. I would do my utmost to cycle every inch of the route south. But diversions to the west, east – like Cajamarca – or, indeed, north could be made by other means. That ease turned into positively gleeful relief when we were underway. In addition to the monstrous climb, the road between Armenia and Cajamarca was utterly truck-infested, had no hard shoulder and, on that day, was river-like with torrential rain. At no point in the journey did I feel the slightest regret that I wasn't crawling up those wet, fume-drenched, tarmac-mountain switchbacks on my bike.

It had been an even earlier than needed start, as my otherwise lovely hostel roommates were up at five, packing to leave with much loud-whispering and bustle. My own alarm went off sometime after six and I met Alejandro at seven in the pouring rain. Alejandro – late twenties, short, dark hair, brown eyes, big smile – opened the door of a small, grey Renault; one of those kind, assured, young Colombian men that immediately put you at ease.

Soon we were driving out through Armenia and onto the road to Cajamarca. This is also the main artery to Bogotá from the port town of Buenaventura on the Pacific coast, and it was packed, nose to tail, with cars and trucks. Some of the trucks carried blue, yellow or red containers from cargo ships, names I now recognised, like CMA CGM and Hapag-Lloyd.

Seeing the containers gave me strange pangs of nostalgia that clashed with my growing sense of the sheer craziness of the quantity of stuff constantly moving around the world, at such environmental cost. How much of the goods in these containers would be things people really needed?

At traffic lights, men hit the truck tyres with wooden batons.

'They do it to test whether the tyres have enough air,' Alejandro explained. 'They get a few coins from the driver.'

Alejandro had spent about eight months in Brisbane, Australia, as part of his degree. We talked about his life and about Cajamarca. And about the need for change.

On a previous journey, he had been as shocked as I had at the state of the favelas on the Ciénaga road; at the total lack of basics, water, sanitation; at the rubbish. He thought it was disgusting that such poverty still existed in Colombia – and that it was partly strategic.

'Politicians are elected by promising food,' he said, in response to my raised eyebrows. 'They never invest in education. Not just to save money – they keep the poor, especially, undereducated, so nothing changes.'

Chile had said not dissimilar things when I visited Colegio Los Manglares. I suspected Che Guevara would have said the same.

I was just wondering how Che might have worded it when Alejandro announced we were nearly there. He made a phone call.

'We should go to the main square and wait for Jennifer there,' Alejandro relayed.

It had stopped raining and we stepped out of the car into warmth. A cream and terracotta-coloured church loomed over a line of small, yellow taxis. In the square's centre, a metal statue of a farmer wielding a hoe stood on a plinth decorated with metallic crops and the heads of cattle.

Jennifer arrived about ten minutes later. She was young, with long dark hair swept back and up.

'Let's go somewhere they do organic coffee,' she suggested.

This turned out to be a stand by the roadside – one edge of the square was also the truck-beleaguered main road to Bogotá that we'd driven in on. We sat with our coffees under a red umbrella and talked in between pauses when the trucks grinding past made it impossible to hear.

Cajamarca hit the news on 27 March 2017, less than a month before I visited and the day after a near unanimous 'No' vote in a *consulta popular* – public referendum – against the development of an immense gold mine. This, Jennifer explained via Alejandro, was thanks to Colombia's Referendum Act, which applied when a whole town had a vocation, such as farming, and someone or something proposed to change it.

'The people here are farmers,' Jennifer said. 'But if the mine came, they would have to become miners.'

The company involved, AngloGold Ashanti, is one of the largest gold-mining corporations in the world. Based in South Africa, it has been accused of human rights violations and was operating in Colombia with its own armed security team and the blessing of the Colombian government. Against them, like some sort of David and Goliath cliché, a handful of activists, average age 21.

AngloGold had initially tried to persuade the people of Cajamarca and the surrounding villages that both they and the gigantic mine – dubbed La Colosa – would be a good thing. They flooded the area with positive messages about the safety of the mine, the jobs it would create, its environmental respectability and the compatibility of mining and farming. They offered funding for schools and recreational areas.

'The young people watched the older people thinking they couldn't do anything against all this,' Jennifer said. 'But we didn't feel like that. We knew the truth: the mine would be a disaster. And somehow, we still believed we could save the town.'

They founded Cosajuca, a collective focused on social justice and the environment. Then they spent years carefully countering the propaganda from the mining company, educating the townspeople and the outlying villagers about the real implications of having La Colosa in their backyard. In the process, they strengthened the community.

'I was born in Bogotá,' Jennifer continued, 'but my mum came to Cajamarca with me when I was eight. She fell in love with the town. Many young people now feel the same. A lot go away, of course, but they often come back. It's quiet, safe and beautiful' – she gestured to the surrounding mountains – 'and a good place for farming. The soil is rich and fertile.'

'And if the mine came here?' I asked. As Alejandro translated the question, Jennifer gave a quiet sigh, as if she'd reviewed the answer many times and still found it distressing.

'There would be two main casualties,' she said. 'First, farming. Because of pollution to the soil, contamination of crops and impacts on biodiversity.'

'How does gold-mining cause this?' I asked.

'Well,' Jennifer paused. 'For every 0.82 grams of gold you have to move about 1 tonne of rocks. La Colosa would involve blasting about 1,257 million tonnes of rock.'

Alejandro interjected, in Spanish, despite himself. '*¿Seriamente? ¿1,257 millones?*' Whether he was checking his translation or the number I wasn't sure – either way I was deeply grateful he was there.

'Yes,' Jennifer confirmed. Alejandro resumed his role, his smile fading. 'Almost all of this rock would have been underground until then. Much of it contains sulphides. When rock is brought into contact with air and water, the sulphides in the rocks become sulphuric acids. These acids then dissolve harmful metals and semi-metals, such as arsenic, into the soil. And this process carries on for *hundreds if not thousands of years* after the mining is finished.'

'It's basically an ecocide?' I suggested. Jennifer nodded.

The other main casualty, she said, would be water. The páramo ecosystem, critical for water supplies, would be badly damaged. On top of that, mercury is used to extract the gold, and the potential for pollution from the leaching pools where this is done was mind-boggling. The pools at La Colosa would be at almost 10,000 feet, right above a major watershed.

Jennifer took a slow mouthful of coffee.

'Then there are the tailings ponds where toxic waste is stored,' she continued. 'A disaster if it seeps into the groundwater or if the dams burst, which does happen. And not only that: water is needed in vast amounts for the gold extraction process. The mining companies always get the water they want. Meaning it's diverted from more vital uses, such as drinking water and farming. Every gram of gold takes just over 1,000 litres of water.'

'One thousand litres *per gram*?' It was my turn to check. Given the scale of the mining operation, the amount of water implied was difficult to comprehend. Jennifer nodded. For a while, nobody spoke.

I thought of the Kogi, and their emphasis on the cycle of water as vital to all life. Of the approach their ancestors had taken to the Tairona gold. Of their belief that to work with gold was to work with the very essence of life, something that had to be done correctly on every level, spiritual as well as technical.

The approach to gold-mining proposed for Cajamarca seemed the exact opposite.

'But you won the referendum against the mining, didn't you?' I asked, eventually.

'Yes!' Jennifer said, breaking into a grin. '26 March 2017. Victory day. It was brilliant. There was a huge march. It changed lots of minds – everyone saw this wasn't just a bunch of young hippies but a group of all ages, very well behaved. People thanked us, the police marched with us for safety, and no one ended up in jail.'

'And now?'

'AngloGold have said they are determined to continue with La Colosa mine, despite that the result of the referendum is legally binding in Colombian law. Everyone is holding their breath, waiting to see what they – and the Colombian government – do next.'

'And what will you do?' I asked.

'We will do everything we possibly can within the law,' Jennifer said. 'But if it comes to it, we will fight.'

The young people of Cajamarca against one of the biggest mining companies in the world. A company that has been accused of contaminating water in Ghana, and of financing paramilitary groups in the Democratic Republic of Congo,[§] now working in a country that, according to Global Witness, was then the third most dangerous in the world to be an environmentalist, with mining the deadliest industry and protest against.[5] This would not end well.

It had already been deadly for at least two young people. Our coffee finished, we headed for the Cosajuco office, a small, green-painted building on a corner. Inside, there were colourful campaign posters on green walls, and a computer on a desk. And two large photographs of young men, Camilo Pinto and Daniel Humberto Sanchéz Avendaño, who had been involved in the campaign. They had both died under highly suspicious circumstances in 2013 and 2014. The photographs were worded unambiguously.

'*Asesinado*', murdered.

And yet there was Jennifer calmly saying she – a young woman with a young daughter – would carry on with the work.

Are you the bravest person I have ever met? I wondered, silently.

[§] AngloGold have denied these allegations.

'What's the main message you want to send from Cajamarca to the world?' I eventually asked. Jennifer thought for a while. 'Two messages. One is the importance of perseverance, hope and being united. The big companies are monsters. They don't care what they damage and destroy in the process of taking what they want. But a united community can stand against them.'

I nodded. 'And the other?'

'That this isn't just a local issue. The Cajamarca area provides food and water to swathes of Colombia, including Bogotá. If water and soil are contaminated here, it has consequences across the country. And what we are fighting for here is relevant worldwide, too. We're fighting for the need for water, life and community to take priority over corporate profit.'

Alejandro and I exchanged glances.

'So a key message is about corporate power?' I checked.

'Yes,' Jennifer said, emphatically. 'The way corporations can persuade supposedly democratic governments to make terrible decisions: that's an issue wherever there are corporations. Corporations take, trash and leave. Everywhere. Their power must be reined in. That's the main message Cosajuca would like to send to the world.'

'What help can we offer from the outside?' I asked.

'Keep on telling the world what is happening here.' Jennifer said, without hesitation. 'That makes it much harder for the corporations, and easier for us. We received so many messages of support during the referendum. It really helped.'

I told Jennifer she was brave and amazing. And to let me know if they had to fight, and I would come back and join them – knowing full well that if this happened I would be utterly terrified.

Alejandro and I drove back to Armenia, the first hour in silence, moved and shocked by what we'd learned. I'd come to South America to ride my bike in big mountains and explore biodiversity. As the little car navigated the slick-wet roads between the hordes of trucks, I thought about what I'd expected on this journey. I supposed I'd mostly imagined myself hanging out in beautiful nature reserves, feasting my eyes on amazing animals and plants and learning about why we need them. And to an extent it had been like that. But I hadn't had the slightest notion that I might end up grappling, albeit from my privileged and safe distance, with the dark, glittering, dangerous world of mining; with the immense costs to life, land and people of the quest for the vast monetary wealth that comes – to a few – from extracting gold from the earth. And while I must have known, at least in theory, that some big corporations were unscrupulous, I hadn't thought that I might actually encounter this, with all the emotional difference that crossing the chasm between theory and reality makes.

Nor had I anticipated the impact of simultaneously meeting individuals for whom 'inspiring' is a wishy-washy, inadequate word while colliding head-on with these darkest of stories. Thinking about it, I supposed the juxtaposition was not really surprising: the one invoked the other. It was still breath-taking.

Mostly, I sat in the car and grappled with a sense of disbelief. How could corporations be allowed to get away with this kind of behaviour? Destroying ecosystems and livelihoods, even murdering people, for a substance whose main use is decorative, most of it not even in circulation? According to the Environmental Justice Atlas: 'Only 7 per cent of the gold extracted is used for electronic devices and tooth implants. Less than a tenth of it has industrial value and could be replaced by other alloys. 45 per cent of the extraction is kept in vaults as treasures ... and the remaining 47 per cent is used to make jewellery,' much of that unworn and locked up.

No wonder the Kogi saw us as 'moral idiots, greedy beyond all understanding'.[6]

Gold, water and life. Three things we value. It's blindingly obvious which should take priority. And yet, Jennifer and her group had been forced to put their lives on the line to try and prevent a government-backed corporation prioritise gold over water and life. Before them, many thousands of indigenous peoples had lost their lives and livelihoods thanks to the conquistadors' rapacious hunger for the stuff.

The way in which long cycling trips reconnect you with the saner version of the gold/water/life priority was, I thought, one of the best things about them. A bike journey swiftly disabuses you of the value of glittering, unnecessary things and makes vivid the value of basics. Food, water, shade, warmth, somewhere safe to sleep: all become inestimable treasures on a ride of any length. In Alejandro's car, I decided that this must be one of the all-time paradoxes of life. Appreciating these basics is such a simple route to happiness *and* sustainability. And yet it is so hard to stay in touch with that appreciation in so-called normal existence, at least for the privileged.

What troubled me most, though, was a question I was beginning to glimpse under the already dark surface of the day's events. *How exceptional was the Cajamarca story?* Was it a story about a rogue corporation, operating in a country still struggling to figure out how best to function in the relatively new context of peace? Or had we somehow created an entire system that incorporated these bizarrely – dangerously – distorted values? A system in which the power of corporations like AngloGold Ashanti means that extracting gold for economic gain can routinely be prioritised over water? Over life? *Routinely*, not just as an occasional, appalling mistake. That would be terrifying indeed.

I woke to grey skies but no rain, and a feeling of 'Yes! It is time to go!' rather than, 'just one more day here …'

It had been such a great basecamp. Had I not landed there, I would have been legging it for the Ecuadorian border days ago. Thanks to Alejandro, leaving Armenia was easier than arriving.

'Right just before the bridge,' he'd advised. 'Then left at the roundabout.'

Whoosh, out. The road went gently downhill for miles. No wonder I'd found it harder work than it looked like it should be coming in. After that the road really did flatten out, as everyone had said it would. A flat road with no headwind! I'd long thought such a thing was mythical. I found myself relishing the feeling of fast forward motion and, reluctant to stop, ate oat bars on the move. After all these years of cycling in mountains, perhaps I was about to be converted to cycling in horizontal places. I was riding down one side of the wide flat plain that is the Cauca Valley, the immense strands of Andes on either side. There were stands of sweetcorn, high and spiky. White cattle egrets glinted in the bright sun. I put on music for the final miles of the day and rode along, singing out loud. Finally, with KT Tunstall belting fabulously in my ears, I turned off for Palmira.

Hotel El Dorado was delightful in a battered sort of way. It was old and full of brown and cream paint, wood and former glory. A series of sepia photos showed Palmira in its former glory too – the huge cathedral under construction, streets full of horses and carriages, and Hotel El Dorado on the corner, looking magnificent.

I ate a slice of pizza – it had survived two days in a hot pannier with only a hint of impending fizziness – in a chair on a balcony overlooking the café across the road, and the motorbikes, taxis and bicycles in the street below. Then I finished writing journal notes from Cajamarca. The shock was fading, but the sense of disbelief was growing stronger. Long, sweltering days on the bike reinforced the utter madness of exchanging water for gold. Surely this proposed exchange, and all the other craziness of La Colosa, was an aberration?

I fell asleep conscious of having had a good day, despite the questions. And of the irony of writing about gold-mining resistance in a hotel called El Dorado.

6. BAMBOO IS BRILLIANT

Travellers often become enchanted with the first country that captures their heart ... For me it was Colombia.
WADE DAVIS

'Bamboo is one of the fastest growing plants on the planet,' Javier said. We were in his lovely workshop in Cali, all kitted out in bamboo with an upstairs living area above the work benches, reached by a bamboo staircase. Two young men called Jonnathan and Milton were fitting a bamboo bike into a stand.

'You can harvest it constantly without affecting its root system, or causing erosion,' he continued, gesturing me towards the bike. 'It's incredibly strong and versatile. It's used for scaffolding all over the world. You can build, well, everything from bikes to buildings with it.'

Having spent an evening at Mamá Lulú's, a spectacular, entirely bamboo-built eco-hostel on the way to Cali, I knew he was right. I climbed onto the saddle of the bamboo bike in the stand and started to pedal. The bike was connected to a juicer. The juicer started to whir and, pedalling hard to much laughter, the orange fruits inside became a thick orange liquid.

'*Tomate de árbol, si?*'

'*Si.*'

I'd approached Cali with a sort of quizzical caution. Various people had warned me about Cali. I should expect to be greeted by teenagers trying to nick stuff off the bike at traffic lights, they'd said.

I arrived at Constanza's house untroubled by light-fingered youth, but hot and exasperated, nonetheless. Constanza, the biologist I'd met in El Cairo, had invited me to stay. She, and her teenage daughter Anna Marie, had not long moved to a rural location south of the city, adjacent to land being reclaimed for conservation purposes. It sounded idyllic. What she hadn't mentioned was the steep, uphill approach, on a road that turned into a rough track which, inevitably, I pushed my way up in the hottest part of the afternoon.

The overheated, ungrateful state of mind I arrived in was soon dispelled by the welcome. It was to be another great basecamp, the location forcing me to figure out the Cali bus system for a couple of project visits. Javier's workshop was the first of these.

Javier was an associate of an organisation called Ecocultura, a fabulous mix of social entrepreneurship and bamboo bicycles. It aimed to create work that had minimal negative impacts and multiple positive social and environmental ones. They were passionate about using sustainable, renewable resources – like bamboo – to build job capacity and entrepreneurial skills. For Ecocultura, developing the potential of young people ran alongside environmental sustainability, strengthening the local economy, promoting healthy living, building community and supporting cycling initiatives.

I had been keen to visit for the obvious, bamboo bike-related reasons, but also because, like El Cairo, Ecocultura was a real-world example of positive values being brought to life. Jonnathan and Milton had both been unemployed when they came to work there. Now they were the lead bike builders of a company called BambooCo, selling their beautifully built bikes at low prices to encourage more people to cycle.

Later, surrounded by the arresting bamboo architecture of the nearby 'School for Life', they showed me some of the different varieties of bamboo, including a richly green species which provoked a brief, disloyal fantasy about Woody II.

The Ecocultura website describes bamboo as 'vegetable steel', a phrase I loved. 'Properly treated and finished, bamboo has a higher tensile strength than carbon fibre, aluminium, and steel,' it explains.

Like a steel bike, a bamboo bike can absorb vibration and bumps, giving a smooth, but not too flexy ride, as I could attest to. Bamboo grows fast, and is excellent at sequestering carbon: better than your average hardwood tree, according to Ecocultura. Tough, attractive and adaptable, with fabulous environmental properties. What's not to like? It would be so good to see bamboo used more widely in the UK and Europe, I thought. Colombia was miles ahead of us.

I headed back to the north of the town the next day. The Cali bus system was unquestionably impressive, and Cali buses seemed less prone than Medellín ones to filling the air with black, stinking smoke – plus they stayed in their own lanes. They also had pre-decided stops and excellent maps that even I could understand.

Arriving early, I wandered the streets for a while, thinking about the meeting ahead. Asociación Calidris works to conserve Colombian birds and their ecosystems. Their focus used to be aquatic birds – Calidris is a genus of Arctic-breeding, migratory wading birds; the sandpipers and others that form huge mixed flocks on coasts and estuaries in the winter, running along the water's edge and probing for prey in the sand. Calidris now works on behalf of all Colombian birds, with a mix of habitat protection, research and monitoring and community engagement.*

'The community engagement piece hasn't always been there,' Kendra, my host, told me, soon after I'd arrived at the Calidris office. 'Calidris was started by a group of biology students, about 25 years ago. They were mainly doing baseline data work and then they got involved in conservation. It was a very top down, "we'll-go-in-and-tell-the-locals-how-to-do-it", model. They gradually realised it didn't work. Involving local communities is now central to what we do. Our new slogan is: "For the birds, with the people."'

It was a familiar story, told by nature conservationists across the world.

The office was a large room with books and files stacked on metal shelves. In the centre was a huge wooden table, covered with bird posters and a map of Colombia. We stood by it, looking at the table-width map, the three enormous mountain chains running down through it like fantastically elongated, knobbled fingers – or talons.

Another aspect of Calidris' work, Kendra explained, was ecotourism. Originally from the US, she was now working on plans for 'bird trails' in five different areas of Colombia.

'One of the trails is here,' she said, gesturing. 'A small town in the mountains further north. El Cairo.'

'El Cairo!' I exclaimed. 'Yes! I can absolutely see how that would work.'

The bird trail tours were pitched at an international market.

'Why?' I asked. 'Doesn't that just create problems with expectations – and high-carbon flights?'

Kendra said she thought the positives outweighed the negatives. The upside of international guests was that with

* Calidris is itself one of several partners working with the international organisation BirdLife.

fewer, higher paying people, you could bring a greater amount of money into the local economy. Calidris always used local guides, offering good reliable incomes.

She told me about a couple who had both been miners and who were now bird guides.

'People don't necessarily want to be miners,' she said. 'But if you grow up in certain regions here, what else do you do? You mine, or you move to the city. This kind of work can provide an alternative. And the people who actually live in these areas are always more knowledgeable about where to find particular birds than we are.'

In Costa Rica, a birder had told me that, if you saw a medium-sized bird, described its colour and then added the word 'tanager', you had probably correctly identified it. Green and gold tanager. Blue-grey tanager. Orange-headed tanager. Scarlet tanager. I was recounting this to Kendra when we were joined by Eliana, who had studied biology and was Cali born and raised.

Eliana was working on community conservation plans for endangered species, including the gold-ringed tanager and the Cauca guan, a bird she described as 'like a big turkey'.

'The tanager is suffering from lost habitat,' Eliana said. 'With the guan, it's that and hunting. There's probably fewer than a thousand left. It's true of other birds too.' She sighed. 'Take the black and chestnut eagle. They need tall forests to nest in, and these are now rare. Plus, farmers don't like them, because they take small farm animals as well as squirrels and monkeys. It's the same with the jaguar. We destroy their habitat and then don't like it when they are forced into farmland to eat our animals instead.'

Kendra nodded. 'It's obvious why we need a participatory approach to conservation. Farmers, university students, community leaders. They all need to be involved. But it makes for a lot of politics.'

'Yes,' said Eliana with feeling. 'And that's before we even start talking about the park system.'

National and local governments were both involved with running parks, Eliana explained. Los Farallones National Nature Park, close to Cali, was a good example. Like most parks, it had been dreamed up in an office. It ranged over 150,000 hectares, with many people living inside its boundaries. It had all of six rangers. Obviously, the rangers could not police it properly. Illegal mining was commonplace.

'The only way to look after such a park,' Eliana said, 'is to work with the community of farmers. It's huge. But only a few are engaged with conservation. Most know little about it, or don't feel they have time to care. Plus, there's little access for visitors – and it's not really encouraged. It's very different from the North American model.'

It was time to go.

'Helping people feel connected to nature is perhaps the most important thing we do,' Eliana said, in closing. 'Birds are great for that. Even in the backyard. When you find out that the small bird in the bush behind your house has migrated there from Africa, you feel good and it increases respect and wonder. And that will help protect the most valuable thing we have – the earth.'

It was a lovely discussion. I was glad that I'd stayed long enough to have it and hadn't given in to the urge to race off, to keep moving south. Though that was how I felt about every visit. Patagonia remained a long way away.

I left Constanza and Anna Maria's at the end of April, nearly three months after I'd started cycling. Three months: the same time as the entire *Carbon Cycle* journey, and less than half the distance. *I've done a lot more project visits, though*, I thought, consoling myself after the round of goodbyes. Costanza, Anna Maria – with whom I'd had numerous, brilliant conversations – the neighbours, a woman known as *Abuelita* – Granny – who refused to believe that Woody had no engine, and several dogs. Was this always, inevitably, part of a journey? The gift and the thorn of travelling; the friendships that flare up in brief timespans, made vivid and precious by the inevitable goodbyes.

A couple of days later, I woke late in a hostel in Popayán with a tired, slightly vulnerable feeling. I'd ridden a relatively low distance the previous day. It had been hilly, but not *that* hilly. Heat? A bit of altitude gain? Not enough food? I had been moving very slowly. It was a long old day to achieve all of 40 miles and still be knackered.

I walked from the hostel into the beautiful town centre. Popayán, the capital of the Colombian department of Cauca, was rich with churches, many rebuilt after earthquakes, and white colonial-style buildings, gracious despite what they represented. Spanish conquistadors had marched into the indigenous village of Popayán in January 1537. Not much seemed to be known about what happened to the people living there at that time, though we know that the indigenous inhabitants were murdered and persecuted by the European colonisers across South America and that Popayán became increasingly important during the colonial era as a transfer point for stolen gold en route to Spain.

I had been put in touch with a conservationist called Adriana, and we'd arranged to meet in a café. She was young, with long dark hair. Visibly hot, it turned out she had arrived by bike. We drank coffee and talked. Of all the people I met in South America, Adriana was the best communicator. She spoke no English, but simply worked with my poor Spanish, talking slowly and constantly searching for different ways of saying things until I appeared to have understood. Failing that, one of us would reach for Google Translate.

Adriana was working for an Andean bear project. It had been awarded funding from the Rufford Foundation, a UK-based charity that gives money to nature conservation projects around the world. Stage one was to register the presence of these bears in the forests and páramo ecosystems in an area to the north-west of Popayan.

'We use camera traps, bear tracks and direct observation,' Adriana explained. 'We are studying the bears and their behaviour, learning how they use their habitats, and how many there are.'

The Andean bear, relatively small as bears go, is also known as the spectacled bear. White or cream markings circle the bears' eyes and sometimes their noses, extending down the shaggy brown, black or reddish fur of the bears' throat and chest. Every bear has unique markings, like a fur fingerprint. They are found in Venezuela, Colombia, Ecuador, Peru, Bolivia and northern Argentina. Paddington Bear is almost certainly the best-known representative of the species.

It's also endangered, in the vulnerable category on the IUCN Red List. The only bear native to South America, it would be desperately sad if it were lost and, Adriana told me, disastrous for the rest of the ecosystem. As Calu had explained at the aquarium, we used to think that apex predators were basically supported by the many animals and plants below them in the food chain. But everything we've learned over the past couple of decades tells us that top-down influence in ecosystems – by bears as much as by turtles – is as important as bottom-up. The vital roles that Andean bears play include seed dispersal and helping with vegetation succession by creating clearings that new seedlings can grow in.

And like the cotton-top monkeys, the Andean bears are an umbrella species. If you protect what they need, you will help many other species as well – including humans.

The bears are endangered because of the loss and fragmentation of their habitat, often caused by the expansion of agriculture.† And this in turn has brought the bears into conflict with people, the reduction in their habitat forcing them to eat maize and other crops. There had even been a case of a bear attacking a cow.

The cow attack was disturbing – as was the response. The farmer had cut the bear's paw off and sent it to the national park office with a note: 'If you don't control the bears, we will kill them.'

In Adriana's view, part of the of the solution lay in better farming, especially using less land per cow. 'Plus, we need to move away from monocultures, to diversified farming with crop rotation,' she added.

The other part was environmental education. Adriana's work included trying to figure out the best ways to share information about the bears with the local community – explaining why these bears are so important and how they live, inspiring the community to help protect them.

'This isn't a protected area as such,' Adriana said. 'We need local people onside if we are to look after these animals, and the ecosystems they are part of. Then there's all the regions we weren't able to go into previously because of the conflict. There's likely to be more species than we know. It's a weirdly precarious situation, now there is peace,' she said, echoing Tinka, 'with so many multinationals eager to exploit new regions for coltan, gold, oil ...'

Lunch finished, we headed back into town. Adriana offered an impromptu tour of the city's most beautiful buildings and an ancient cork tree in the main square, its trunk thick with flaky grey bark and wider than my arms could circle.

By this time, I was whacked. But Adriana, a fount of generosity and energy, kept suggesting new things to do. We

† The IUCN claims there are between 2,500 and 10,000 bears left in the wild, and the number is decreasing.

were heading towards a viewpoint to watch the sunset when we walked past a theatre where, it transpired, the bike company Specialized were about to run an event. Adriana asked if I could make an appearance, and the next thing I knew we were legging it back to the hostel. I changed into a cycling top, emptied my panniers and pushed Woody back to the theatre, wheeling him onto the stage in the huge auditorium, where I talked briefly about my journey – with help from a translator plucked from the audience. This was followed by a filming session for a local TV station, including a shot of me riding in erratic loops I couldn't control, having forgotten what happens when you first ride without heavy panniers you've become accustomed to.

The next day, we headed off to the area that Adriana was working in. It was a longish drive in the small Cotrans minibus – part of the local transport system – in and out of rain, winding up a steep road into the mountains.

There were few trees. Much of this ecosystem was degraded, Adriana explained. Extensive cattle farming and potatoes were the main problem, alongside an increase in strawberry production.

We came over the brow of the hill and saw a plain.

'There used to be a lake here,' Adriana said. 'Now it's often dry. No trees, no water.'

We set off up a small track. The track was very wet: wet enough to be classified as a small river, in fact. We climbed slowly up through the rain, away from the degraded landscapes towards the páramo.

Going into the páramo was magical. It was like visiting a habitat on a different planet. The sweeping landscapes were dotted with tall, large-headed plants called frailejóns – 'big monks' – scattered around us like herds of very upright animals. Crouching close-up revealed a morass of hidden colours, tiny red leaves, delicate yellow and blue flower petals, many kinds of orchid, a hidden world of patterns and intricacy. It felt wild, and wonderful, and new. I loved it.

Among all that diversity of plant life, the frailejóns were my favourites. They grow at roughly a centimetre a year and many were taller than me. They had thick trunks, coated with dead leaves and topped with a mop of live leaves, an exuberant, pale olive rosette of tongue-shaped, upward-curling foliage. Mice, apparently, lived in the warm dry spaces under the mop and humans sometimes used the insides of the flowers as ear warmers.

Vital to the páramo ecosystem, frailejóns take more water from the atmosphere than they need, then emit it from their roots, feeding rivers and providing fresh water for human communities below. In theory, all páramos in Colombia are protected because of this. In practice, this doesn't always happen.

'This one was heavily grazed until about ten years ago,' Adriana said. 'It was badly damaged. It's regenerating now though.'

We had a brief, cold lunch on a windy slope overlooking a beautiful scatter of small lakes. Then we headed back down, drenched and scrambling over wet rocks.

I left the next day. A morning of climbing in rain was followed by a long, wonderful but bone-chilling descent into a wide, hot canyon of a valley, two long flat hilly ranges on either side and the usual, angular green mountains beyond. Road workers in hi-vis waterproofs rode on motorbikes with broomsticks sticking out sideways, like modern-day witches.

When the sun came out, I stopped to eat lunch. A man in a truck stopped, reversed back and offered a lift to Pasto.

'Thanks. I'm happy cycling, though.'

And I was happy, eating my picnic there in the warmth, feeling a bit on the edge of feral, hungry, with wet gear spread around me from an earlier cloud burst, the air lively with butterflies and birdsong.

I rode for days, the landscape gradually changing. More and more cacti appeared until they were everywhere: tall cowboy-movie ones and round-leaved fat ones, some in flower, brilliant red bulges at the end of their pear-shaped leaves. Then the cacti retreated and the big, green mountains moved back in, some crazily spiky, like a kid's drawing of pointy summits.

I reached Pasto, where the mountains and rivers converge, in another downpour, the rain washing salt and grime into my eyes, stinging so much I couldn't see, and settled into the Koala Inn hostel for a final couple of days of catch-up before the border.

I wore my Movistar cycling top in honour of my last day in Colombia. I felt some excitement at the thought of Ecuador ahead. But mostly, despite having spent a month more than intended in Colombia, I felt reluctant to leave. I was enamoured with the country, for all its still-dark underbelly, and my three months didn't seem anywhere near enough. I loved the feeling, however surprising, that I knew people; of having new friends and networks scattered across the mountains and valleys. I loved having figured out how things worked – basic, boring things, like phones, banks, shops. Above all, I loved the exuberance, the vitality, the music, the colour, the heat, the diversity of this constantly astonishing land. I loved the feeling, however strange that I should feel it, that in Colombia I was somehow in the right place.

As I pedalled towards the border, much of this was to change.

PART 2

Ecuador and Peru

7. INTO ECUADOR

ECUADOR

Happy Planet Index score: 58.8
Rank: 5th out of 152

To understand the struggle you need to know what is at stake.
CARLOS ZORRILLA

I crossed into Ecuador feeling low. The border was uncomplicated. A friendly official in booth no. 9 asked whether it was my first time in Ecuador, whistled when I said I'd cycled here 25 years ago and gave me a three-month visa with no further questions. Woody and I pushed off into a whole new country. It was grey and overcast. In the town of Tulcán, I grappled with a familiar set of mildly irksome practical challenges. Exchanging money, adapting to a different currency – in Ecuador, US dollars. Figuring out how to acquire a local sim card, etc. After Tulcán, we climbed steadily for hours. Around us, acres of greens and occasionally ploughed brown-greys were sliced and divided into the classic, crazy-paving patchwork geometry of a heavily farmed landscape, slewn across hills that were still complex but more rounded than their angular Colombian neighbours. If there were mountains, they were out of sight.

As famed for its biodiversity as Colombia, Ecuador has strong indigenous cultures* and a reputation as an environmental champion; a country known as a nature paradise, that had formally adopted the Rights of Nature as part of its constitution. The road was regularly embellished with signs reading *La Naturaleza es Vida, Cuídala!* – 'Nature is life, look after her!' Why wasn't I joyful? I tried to analyse my deflated mood. Perhaps it was partly because I'd left Colombia abruptly, without ceremony. Colombia, a country I loved, was now full of friends and I knew how to function there. Ecuador felt different and a bit perplexing. There were no friends waiting for me here. Then there was the initial bleakness of the highly cultivated landscapes and the invisible mountains. Occasionally, I caught a hint of distant, patchily forested slopes, but for the most part, any mountains and volcanos remained stubbornly hidden in the clouds.

Perhaps my lowness had nothing to do with any of this. Reading my journal notes later, with their frequent references to a feeling of precariousness, a sense that the journey could end abruptly at any minute, I was shaken by the thought that this mood had been a premonition.

The email arrived while I was in Otavalo.

'I want you to know that everything is fine,' the message from a friend read, ominously. 'Everything is fine, but ...' The kind of email that every traveller dreads.

Everything was fine, but Chris, my partner, temporarily forgetting it was me that was supposed to be having the adventures, had been involved in an accident, sea kayaking off Walney Island with friends. The friends were unharmed, if shaken, but Chris was in hospital.

---

* About 6–7 per cent of Ecuador's 17 million citizens are of indigenous heritage, hailing from at least a dozen distinct groups with as many languages.

'Nothing to worry about, they are just erring on the safe side,' the email continued.

Two days later, he was still in hospital with hypothermia. *Two days in hospital? That was quite some case of hypothermia.* He must have been in the water for a considerable length of time. The friends in question were not experienced at deep-water rescues, so it sort of made sense, assuming Chris' roll and self-rescue had both repeatedly failed. But what could have led to such a scenario? Chris was a highly experienced sea kayaker, and he wouldn't have taken this group to this location in anything other than a mild sea state. In which case he would have been able to roll or get back into his boat. Unable to contact Chris in person, I had to guess at what might have happened, and how Chris was now, from opaque, carefully worded messages sent by well-meaning friends. And then to try and put it aside. *Chris is being looked after. He will be fine. Get on with it.*

Carlos Zorrilla was a man I really wanted to meet. He was a cloud forest champion and anti-copper-mining activist who had had various run-ins with the Ecuadorian government. He lived on a small farm, off into the mountains, and had sent me instructions explaining how to reach him by bus. The journey began with a long wait at the station. This was no hardship: I had just discovered Kate Raworth's *Doughnut Economics*. It set out two aims to shape any economic model. First, stay within Rockström's planetary boundaries, the outer ring of the doughnut. Second, provide quality human lives, the inner ring. In sum, economics should operate within the doughnut. So simple, yet so powerful. And so very different from the dominant model, with its fixation on never-ending growth. I was hooked.

When the bus finally pulled away from the town, the route was spectacular. A switchback road dove into the forested mountains with big, plunging drops. A couple of hours later, the conductor indicated that I should get off. There was a small school with a track running alongside it, just as Carlos had said. There was a river, fields and trees. As I walked it became increasingly forested. It was so very lovely to be off the road.

Carlos' house lay, as promised, at the end of the track, a slightly battered but welcoming looking construction of white-painted wood. Two golden retrievers came to greet me and then Carlos himself. In his sixties, I guessed, grey beard, glasses, blue sweatshirt and jeans, yellow wellies.

I liked Carlos immediately. He was gentle and easy to be with, and he asked lots of questions – though I was more interested in hearing his stories.

Carlos had been leading campaigns to resist copper-mining in the community he lived in – not far from Junín, in the Intag region – which had been at the centre of the mining struggle since copper was first found there in the 1990s. A few years before my visit, he had been denounced on TV by the then president of Ecuador, Rafael Correa. Correa had accused Carlos of trying to destabilise governments like his and declared Carlos' work to be anti-development – and hence anti-Ecuador. He had asked fellow Ecuadorian patriots to 'react' to the threat that Carlos was said to pose. In response, Amnesty International, concerned for his life, issued an international action alert and Carlos was compelled to go into hiding.

This was not for the first time. Back in 2006, an arrest warrant had been issued for a crime he had not committed.

'My neighbour had seen a group of armed police heading down the track towards my house,' Carlos told me. 'He had been trying to ring me, to warn me. It was in the days of dial-up when you got on to the internet via your phoneline. I was doing some research, and he kept getting the-line-is-busy tone and couldn't get through.'

'Then what?' I asked, already gripped.

'By sheer luck, I must have disconnected at the last minute,' Carlos said, with a half-grin, half-grimace. 'The phone rang. My neighbour just had time to shout, "They are coming, get out!" I grabbed my laptop, threw it into the hen house and ran into the forest. The police never found me. Instead, they harassed my son and my neighbour, ransacked my home and stole some money – the only time I've ever been burgled in all the decades I've lived here. I stayed in hiding for weeks.'

Carlos said that he was later told that the plan had been to have him arrested and then killed in jail.[1]

'Two years later, the lawsuit that kick-started the incident was ruled to be malicious,' Carlos said. 'It had been filed by a woman from the USA, who had been paid by the Canadian mining company that was pursuing the copper-mining concession.'

It transpired that this had been just one part of a wider plan to deal with opposition to copper-mining in the Intag region.

'The plan was drawn up by an international security firm,' Carlos explained. 'Two weeks after they tried to have me arrested, they sent in paramilitaries with attack dogs, machetes and tear gas.'

'Seriously?' I asked, still attached to the idea of Ecuador as the champion of the rights of nature.

'The confrontation with the communities was filmed,' Carlos said, gently. 'You can see the footage in a documentary we made, *Under Rich Earth*.'

The second time the paramilitaries turned up, the community managed an entirely peaceful citizen's arrest of 50-plus armed men, despite some of them having used pepper spray and guns against them previously.

The Canadian company eventually pulled out of Ecuador. But the copper, of course, remained. Since 2012, the companies seeking to extract it were primarily Ecuadorian and Chilean. The change of nationalities had made no difference. It was, Carlos told me, the identical strategy used by mining multinationals in the face of community opposition all over the world. First, the company would offer jobs and improved infrastructure – roads, schools and so on. If that failed, they would move to more brutal tactics.

In the Junín community, there had been other arrests.

'Then,' Carlos said, 'the Ecuadorian government sent in a 400-strong elite police force to intimidate the communities just as the Canadians sent in their armed security firm in 2006.'[2]

'And you are still here,' I commented, wondering, as I had done with Jennifer, at the bravery of this kind and gentle-seeming man.

Carlos smiled. He was Cuban, originally, and had moved to Ecuador decades ago, from the USA – though now when people asked he would say he was from the cloud forest. He had been looking for a community – a human community – that he could be part of and contribute to. The community he had found in Intag was one of his reasons for staying, despite the ongoing threats.

The other was the ecological community.

'The cloud forest here is near pristine,' he said. 'And it's incredible.'

The heavy rain that had been providing the background soundscape to our conversation had eased. Carlos suggested a walk.

'Cloud forests are the lesser-known cousins of rainforests,' he said, as we headed down a narrow, mildly muddy track between the trees. 'Rainforests get the attention. But cloud forests are even more diverse, botanically speaking. They should get more press.'

Only a tiny proportion of the earth's tropical forests are cloud forests, Carlos explained as we walked, yet they play vital roles in the water cycle, including regulating the flow of thousands of rivers and streams. And the extra-humid nature of these forests are key to their exceptionally rich biodiversity, and high rates of endemism.

As a general rule, the closer you get to the equator, the more biodiversity there will be. This is partly thanks to the stability of the equatorial climate. In Ecuador, Carlos said, it is also to do with the complexity of the mountain ranges – the many changes in altitude creating numerous microclimates. The humidity comes into play because of the relatively high altitude at which cloud forests in the Andes are found – between 1,500–3,000 metres. They have a relatively cool, damp climate all year round, and sunlight doesn't penetrate very far.

'This,' Carlos explained, 'means that, in a cloud forest, you need to think about biodiversity in a vertical way.'

To cope with these conditions, many species of plants live in and on the trees. Mosses. Orchids. Bromeliads. They are known as 'epiphytes' or 'air plants'. They don't have their roots in the ground but take water from the air.

'Look up and you'll see what I'm talking about,' he said.

I looked up. There were plants growing on plants, all the way to the canopy. Spiky arrays of leaves, like pineapple blades, sprouted from branches and trunks in thick clusters. Generous clumps of moss hung among trailing lianas and giant ferns. In the dense, green confusion of leaves and branches it was often hard to tell which was a tree rooted in the ground and which an air plant.

Back at the house, Carlos showed me a book he had written about the forest. It was full of photographs; exquisite details of plants and animals gleaned from 30 years of patient, curious observation and coexistence.[3]

Then we talked about how the stunningly diverse forest I'd just wandered through would be affected if copper-mining came to the region.

'It would be devastating,' Carlos said.

A study had been done by a mining company, he told me, for a much smaller, open-pit copper mine. It had predicted massive deforestation, as well as contamination of rivers and streams with heavy metals. Endangered species such as jaguars, spectacled bears, mountain tapirs, mountain toucans and brown-headed spider monkeys would be put further at risk. The Cotacachi Cayapas National Park, which runs adjacent to the Intag cloud forest, would be badly impacted.

'That reserve is one of the most important in the world, biologically speaking,' Carlos said. 'It comes ahead of the Yasuní National Park in the Amazon in terms of irreplaceability.'

Grim social impacts had been predicted, too. A rise in crime. Four communities, including Junín, would have to be relocated. And this was a study for a small mine. The much bigger mines threatened the entire region with profound environmental and social upheaval. It would, Carlos said, be one of the most environmentally damaging mining projects in the world.

'Copper is at least useful, though, right?' I suggested, thinking about the largely decorative uses of gold despite the devastation left in the wake of extracting it. Copper is an excellent conductor, and we use about 22 million tonnes of it a year, and rising, in phones, cars, houses across the world.

Carlos' view was that the usefulness of copper didn't justify wrecking some of the world's most ecologically important, biodiversity-rich forests and violently displacing human communities, to extract it. Not least because a large proportion of the copper we genuinely need could be sourced by recycling and upcycling.

One of the things that worried him most was, perhaps ironically, the predicted transition to electric vehicles. They are lower on carbon and other harmful emissions, at least at the tailpipe – but require roughly three or four times as much copper.

'If the whole world shifts to electric cars, it will be disastrous for the communities fighting copper-mining,' he said. 'And do we really need so many millions of private vehicles?

We should be thinking about better public transport infrastructure, affordable for all.'

'Tell me more about DECOIN,' I requested, as the late afternoon slid into evening.

DECOIN stands for Organización para la Defensa y Conservación Ecológica de Intag.[4] Researching for this visit, I'd read that Ecuador, one of the world's seventeen 'megadiverse' countries, has the highest rate of deforestation anywhere in South America – the continent that has the highest rates of biodiversity loss anywhere in the world.[†] In Ecuador specifically, more species of animals and plants face extinction than in any other country on the planet.

'Well, DECOIN is basically a grassroots effort to reverse this trend,' Carlos said, as I recounted this.

Ecuador's extinction rates are driven by the same range of human activities as they are across the world. Land use change, especially deforestation and the conversion of grasslands, savannas, forests and wetlands to agricultural land. Logging, both legal and illegal. Climate change. And industrial activity, including the so-called extractivist industries that take large quantities of natural resources from the earth – oil, gold, copper, lead, silver, trees, fish, industrialised agriculture and so on. After that, Carlos said, you should throw in hunting and the introduction of invasive species. Plus, a lethal fungus that was impacting amphibians everywhere.

He paused.

[†] It's only the highest now – according to WWF, the destruction of northern hemisphere biodiversity largely took place more than 30 years ago.

'Here in Intag, we face the most environmentally destructive economic activity of all, though. Large-scale, open-pit mining.'

Part of DECOIN's work, supported by Rainforest Concern and others, is to help local communities purchase forested land for conservation purposes and then manage these areas. They support community-managed watershed projects too, which, in some ways, Carlos suggested, were easier to advocate for – unlike biodiversity, everyone understands the importance of water.

Then there was work focused directly on resistance to mining. This included publishing a handbook designed to help any community that was facing a damaging extractivist industry in their area.[5]

'No wonder you were not President Correa's favourite citizen,' I said, with a grin.

'Indeed,' Carlos agreed. 'But actually, I think he was more threatened by the positive alternatives to extractivism we created than by the critiques and resistance. Good, small-scale, community-based livelihoods – that really worried him.'

With extractivist industries, Carlos explained, most of the resources are exported as raw materials, processed elsewhere and sold on. 'Extractivism', a word I hadn't even heard of before, was used as a term when countries or companies depended heavily on these industries to create wealth.

It had, he said, taken off 500 years previously, when countries like Spain removed resources in huge quantities from the lands they had colonised and shipped them back to Spain. Now extractivism was often driven by transnational corporations (TNCs).

'Conquistadors or TNCs,' he said. 'In either case, the countries where extraction takes place are dominated by outside forces, who benefit from the extraction much more than their hosts do.'

'What about when the companies have been nationalised, or at least are South American?' I asked.

Extractivism's advocates, Carlos said, argued that the revenue was needed to alleviate poverty and improve living conditions – for development, in other words. Some of it had indeed been used for this, including, famously, by President Correa. But even nationalised extractivism brought social problems. Working conditions were often unsafe. The wealth was unevenly distributed. Communities who lived where the extraction was taking place rarely benefitted and often experienced worsening living conditions, conflict or even forced relocation. And the environmental impacts were the same whoever owned and ran the mines.

'But still, the huge, short-term wealth can be used to improve quality of life for some, and that's always how it is justified,' Carlos said.

'So, really, by challenging the mining, you threatened Correa's entire view of what development *is*,' I said.

'Yes. The extractivist, top-down, material-wealth-based notion of development,' Carlos agreed, 'and the immense income streams and power that comes with it.'

The thought that we might need alternative models of development to achieve nature conservation objectives – to protect habitats and bring biodiversity loss to a halt – seemed suddenly both obvious and startling.

After a meal at the nearby ecolodge DECOIN used as a base for educational visits and ecotourism, we went back to Carlos' house. He opened the bottle of wine I'd brought as a gift. We talked on.

Part of the solution, Carlos maintained, involved international campaigns to expose what the mining companies were doing. And then to threaten their finance streams.

'If these projects were not financed, they couldn't happen. We need to lobby governments and banks so they cut off the funding to destructive, unethical industries. Reach out to shareholders and stakeholders – national and international.

In a way, it's about exposing the dark side of Ecuador. Part of Ecuador's mythology is that it is an ecological paradise. Ecuadorians themselves have no idea how much of their country has already been given over to mining concessions, nor how badly the mining companies treat the people and the land. Tourism, an increasingly important source of revenue, is also based on this myth. Exposing it could have big consequences.'

We were in new territory here, I thought. No one else I'd met on this journey had talked about recrafting financial structures, or development models, as a necessary part of protecting nature.

Carlos reached over to refill my glass. The wine glinted red in the quiet room. There was a long pause.

'I think the point about development and what it means is probably the most important,' Carlos said, eventually. 'Across South America, you see the dominance of extractivism – governments depending on it to generate wealth, to fulfil election promises of good lives for their citizens. We need different models. Other ways a country can create the income it needs, without the environmental and social costs. And just as crucial, different versions of what counts as a "good life", of what living well as a human really entails.'

He paused again. 'What do you think of when you hear the word "wealth"? Most people think of money and material wealth. But there are many other kinds of wealth ...'

The indigenous notion of *sumak kawsay* – translated as *buen vivir*, or good living – was relevant here, he said. It understood 'living well' as being primarily about living in harmony. With yourself, with your human community and with nature. Real wealth was harmony, not things.

'I think these "alternative" models are relevant all over the world,' Carlos said, 'in so-called developed countries like the UK as much as in so-called developing ones, like Ecuador. We *all* need different, saner, fairer, more sustainable ideas about what quality of life means.'

It was getting late. 'What's the main message you'd like to send from Intag and DECOIN to the world?' I asked.

'Wake up to cloud forests,' said Carlos. Most people don't realise how valuable they are – unbelievably important from a biodiversity perspective – nor how threatened, nor how much our lives are interconnected with them. We need a PR campaign for cloud forests, and a worldwide network of communities working to protect them.'

Carlos thought for a while. Then he continued.

'We all need nature, including forests. And we are all part of the system that's driving its destruction. Even without meaning to, we're taking ever more from it, faster than it can recover. Or some of us are. The ecological footprint of rich countries is much higher than poorer ones, of course.'

'Did that "three-planet" analysis get discussed in Ecuador?' I asked, thinking of some powerful early work by WWF. They'd shown that, if everyone on earth enjoyed the lifestyle of an average Western European, by 2050, we'd need three planet Earths.

'Yes, it did,' Carlos said. 'And that's it. Over-consumption is at the heart of the problem. So we need to ask, how do lifestyles in the West and elsewhere impact on places like Ecuador? We need to make changes so we can reduce those impacts, while still living well. To prioritise a healthy, biodiverse environment and peace in everything we do.'

'There's only about twenty nights a year you can see the stars,' said Carlos, as he pointed me in the direction of the cabin I was to sleep in. 'I love the forest, but I do miss the stars.'

A stick insect swayed benignly on the toilet wall as I cleaned my teeth. I'd agreed to meet Carlos early the next morning to walk into the forest where we might see the spectacular Andean cock-of-the-rock birds. These birds, a vivid scarlet with a high head piece not dissimilar to an Elvis coif, gather in groups known as leks – the avian equivalent of a male singles' bar – to try and attract females with a loud and extravagant display of bobbing and leaping. First, we sat and drank coffee on Carlos' front bench, watching soft clouds move across a delicate sky. A woodpecker was audible from somewhere in the closest group of trees. Strolling into the forest, we could soon hear the cock-of-the-rocks, too, though we never saw them.

'These birds are just one of many biodiversity superstar species,' Carlos said. 'For example, the longnose stubfoot toad. It was thought to be extinct. Then a biologist we hired to do a biodiversity survey found four of them in a community forest. It's a tiny creature, but it could become very powerful if the Ecuadorian Rights of Nature ever came to mean something in the courts.'

I would have loved to stay longer. To talk more, and to help out on the farm. I was feeling something powerful for this man. Not in a romantic sense, but some other kind of love; love for who he was and the work he was doing. And love for the place. Like most tourists, I'd come to Ecuador eagerly anticipating volcanos. I found I was falling for forests. But I'd also picked up a message that Chris was out of hospital, and I wanted to get back to town so I could talk to him. And so, I left, with a copy of Carlos' book, a heart-warming hug and an agreement that we'd stay in touch.

I was mulling on the visit while I searched out a café with good Wi-Fi for a Skype call with Chris. It was a long and distressing call. Holy crap. It had been a very near miss. Chris had nearly died. He had been in the sea – which, off the UK coast in June is almost at its coldest – for about an hour and 45 minutes, all the time working hard, physically, to get himself and a capsized friend to shore, while the current swept them parallel to the coast and steep, chaotic waves made it impossible to get back into their kayaks. By the time the lifeboat picked him up off the beach, Chris' core body temperature was below the bottom of the thermometer. He had lung and kidney damage and incipient pneumonia. He owed his life to the friend who had called the coastguard and the lifeboat that responded when the coastguard upgraded the call to a mayday.

The scenario was an out-of-the-blue weather change and freak tidal behaviour. Even the lifeboat folk agreed they had never seen the likes before. It couldn't have been foreseen and Chris had done everything he could to deal with the situation and keep everyone safe. The hospital, the coastguard, the lifeboat and the friends involved had all been brilliant in different ways. But Chris was, of course, unnerved by what had happened, and by the closeness of the call. I was thoroughly shaken too, not least by the suddenness of it all; by the way life can change out of the blue, so fast, in such shockingly unpredictable directions.

'I can come back,' I said. 'I've already figured out the fastest way to do that.'

This was true. I had been eyeing up buses to the nearest international airport for days now.

'Please don't!' Chris replied. 'You are the least sympathetic person I know when it comes to illness and injury.'

This was fair.

'I'm recovering just fine, anyway.'

This was reassuring.

'I'd much rather you stay where you are and keep going.'

I believed that Chris meant it. Nevertheless, it was the first time that it had felt properly hard to be away.

I spent the afternoon in the café, doing mindless email work and trying to get my act together. I did some research on the route ahead, and places to stay in Quito. My mind was all over the place. There was no one to talk to. I could have used a hug and a stiff whisky.

Back at the hotel, I read some of Carlos' cloud forest book in bed, taking consolation from the gorgeous images of moths and birds. Drifting asleep, I awoke in complete darkness, thinking about Chris in the water, struggling to sort out the situation, getting ever more exhausted and cold, yet refusing to give up. It was not a good night.

By ten the next morning, I was packed and back on the road. The hotel, recommended by Carlos, had been beautiful but freezing. I gradually took off layers as I climbed out of Otavalo, on the massive, new, doubtless extractivism-funded road that had started close to the border. Alongside it was frequent, visible poverty. Villages full of half-built houses and huge rubbish tips pushed their chaos up against the road's edge. In between, the heavily used landscapes, the dark and damaged hills.

> Colombia Reports *says that great chunks of Colombia are either under red flood warnings, or already suffering floods and mud slides. It feels as if Colombia is closing in behind me, and that there is no way back. The incident with Chris has left me somehow uncertain. Part of me just wants to quit and go home. I am intermittently on the edge of tears, and feel strangely vulnerable, even though it wasn't me that nearly drowned. The other part says don't be ridiculous. Chris is fine now. He has friends around him. Don't let this wreck the amazing opportunities ahead. You are in Ecuador, for goodness sake! Ecuador!*

Ecuador. *República del Ecuador* in Spanish – Republic of the Equator. In Quecha, it's *Ikwadur Ripuwlika*; in Shuar, *Ekuatur Nunka*. I reached the equator not long after two in the afternoon on 17 May and leaned Woody against the wooden signpost. Patagonia, Argentina, Chile, Lima Peru,

Colombia, Mexico. The sign declared their distances as well as the direction they lay in. Patagonia was a daunting 6,157 km, presumably as the crow flies or, at least, as the Pan-American Highway unfurls its tarmacked self.

The next day, I cycled through an enormous toll spanning multiple lanes on the shiny new road. Close to Quito now, large, expensive-looking cars glistened among the battered trucks and roadside adverts promoted the installation of garden water features. There was clearly substantial wealth in Ecuador, for some, at least.

Coming into Quito was predictably horrible. My route fed me through an immense flyover and onto a main road packed with traffic, with no hard shoulder. It was also uphill, and the sun had come out, so I cooked slowly in the heat and traffic fumes. I was cut close, constantly, by every kind of vehicle. There was nothing to do but keep going, all fingers crossed. I was tired, harassed and mildly heat-stricken by the time I arrived at the hostel.

El Cafecito was another lucky find. I hunkered down for a few days of refuelling, catching up and meeting folk. Among these was Ivonne, who worked for Acción Ecológica, one of the main environmental campaign organisations in Ecuador. Carlos had told me that President Correa was trying to shut it down.

'It's true,' Ivonne confirmed.

I liked Ivonne as immediately as I had Carlos. An English-speaking friend of a friend, she had a powerful presence and strong views. We went out for coffee and sandwiches. Acción Ecológica was a left-leaning organisation that worked on extractivist issues, alongside questions of wealth distribution and social justice. I was hungry for her views, but we were joined by a fellow campaigner who spoke no English, and, out of politeness, the conversation switched mostly to Spanish. I lost a good half of the insights that seemed to swirl around me, just out of reach.

What I did gather was that Correa had widely been considered the 'best ever' president when he first came to power.

'He invested heavily in education, health and infrastructure,' Ivonne said. 'The poverty rate fell dramatically. But the money came from extractivist industries, primarily oil. And that brings a huge set of issues with it, not least environmental ones. We've been slow on the uptake here, on the left. For most of our history, we've been focused on wealth redistribution. We've only just begun to realise that it matters how wealth is created in the first place – and that sometimes the environmental and social costs of this are simply too high.'

'Interesting. I'd say it's the same in the UK,' I said.

'Yes,' Ivonne agreed. 'We talk about the old days of the "brown left" now – as opposed to the "green left", which we like to think we've become. All over the world, the brown left embraced extractivism and forgot about nature.'

Another problem, Ivonne said, was that Correa had developed increasingly oppressive tendencies, shutting down one of Ecuador's oldest environmental groups, the Pachamama Alliance, and trying to shut down Acción Ecológica.

'He attacks anyone who challenges his vision of "progress" and "development",' Ivonne said, echoing Carlos, 'despite that his extractivist vision is flawed in relation to social justice as well as environmentally. Across the world, we've seen that for the people who live where the resources are found, poverty goes up, not down. It's known as the "resource curse".'[6]

Despite this, Ivonne explained, Correa continued to pitch his pro-mining, pro-extractivist arguments in terms of tackling poverty. And this meant he could position environmentalists and anyone else who opposed extractivism as 'anti-poor', thus demonising them and splitting his opposition.

Ivonne paused. 'Of course, it's greatly complicated by the relationship between Ecuador and China,' she continued. And then the Spanish sped up beyond my grasp.

I was to find out more about the role of China when I went to the Amazon. Another friend of a friend, Javier, had recommended a travel company, and I'd booked a trip to the Napo Cultural Centre, run by an indigenous community, in the Yasuní National Park. It sounded wonderful. I was to leave, by bus, late that same night.

8. RAINFOREST

We can't be beggars, sitting on a sack of gold.
PRESIDENT RAPHAEL CORREA

Rainforests are home to at least half of all the remaining plant and animal species on the planet.* They run in a thick, vital, green belt around the equator, stretching to ten degrees north and ten degrees south of it. To look out across the Amazon rainforest is to look out across a green ocean of treetops, a few giants breaking through, shadows playing across the canopy formed from the curved heads of thousands and thousands of trees. The inhabitants of the rainforest present themselves slowly – it's not like being in a David Attenborough documentary, suddenly surrounded by big cats and see-through frogs – but the sheer scale is immediately astonishing. There are 2.1 million square miles of Amazon rainforest. If it were a country, it would be the seventh largest in the world.

I was in the Yasuní National Park, in the Ecuadorian Amazon. I'd travelled there from Quito on an overnight bus that plunged downhill from the city into tortuous, writhing roads. Curled up on a double seat, I had eventually managed

* An incredible figure, given that they only cover 6 per cent of the Earth's surface, according to *National Geographic*.

to sleep despite a heavily snoring man in the seat behind me. Waking in darkness, I swayed down the bus to find that the toilets were locked. Then, getting out at what I took to be a toilet stop, I found we'd arrived. It was 4am. Coca bus station. Finding an unoccupied corner, I joined the piles of people dozing with their heads on their bags on the floor.

Later, I sat looking out over the sides of a long motorboat at the wide, flat, coffee-coloured Río Napo. We were heading for the Napo Cultural Centre. Touches of blue-grey were reflected from the sky, and everything was watery and washed out, as if we were travelling through a faded water-colour painting. Except for the horizon, where a line of green became more and more vivid as it moved towards us. The green line became vegetation, then a chaos of diverse, distinct trees. A clearing appeared. The boat crunched onto the shallow beach. The small group of passengers not heading further upriver were handed ashore. Handed across the wet, wooden boat and the tiny waves into another world.

The Napo Cultural Centre is an ecolodge run by the Kichwa Añangu indigenous community.

'They set up the lodge about fifteen years ago, as an alternative to working for the oil industry,' said Andreas, the guide who had escorted us there. 'These are people who retain many of their customs. The forest is fundamental to who they are and how they live. But they've also decided to engage with the "modern" world, to some extent.'

'There are other indigenous groups in the region?' I asked.

'Yes,' Andreas replied. 'Including at least two uncontacted tribes. The Tagaeri – they may or may not still be here, no one really knows – and the Taromenane. If you come across them, they won't say hello or ask any questions. They will simply kill you. You could see them as the true defenders of nature.'

The centre was luxurious. Walkways lead to a row of thatched cabins, the walls painted red and salmon pink.

Inside were huge beds, mosquito nets, hot water and functioning lights. The thatched restaurant area looked out over a flat clearing to the forest edge. It served fabulous food in huge quantities, and it had a bar. We were greeted with a glass of iced tea, met our local guide, Gabrielle, a member of the Kichwa Añangu community, and walked a short distance into the hot and humid forest to a viewing tower.

'The whole focus is sustainability,' Gabrielle explained. 'Logging is banned across the 21,400 hectares of forest that we manage. So is hunting and fishing. As a result, it's relatively easy to see wildlife here, and that helps bring the tourists in.'

Any profit was, he explained, reinvested into the community: into education, health, community centres, conservation work and so on. It was so different from mainstream models of development, I thought, with the relentless pursuit of economic growth and theories about how financial wealth would 'trickle down'. Here the community used its income to invest directly in what it most valued; in the things that would improve quality of life for everyone. They were completely uninterested in economic growth for its own sake.

We stood on the tower in an outbreak of sunshine, looking out over the ocean of trees, the light changing on the billion shades of olive and green, and the Río Napo – which flows east to the river Amazon, in turn flowing east to the Atlantic – glinting in the distance. Through a telescope we could see two black and white toucans sitting together on a branch, their huge beaks just touching. We could hear the clear 'plink plink' calls of an oropendola bird, like drops of water on stone, and the occasional harsh cry of parakeets. I felt happiness flooding through me, and a sort of relief. Relief to be somewhere wilder, away from the roads and the traffic and the city. The sense that the green ocean was teeming with diverse life was overwhelming. As was the vastness: trees upon trees stretching out for miles and miles.

Suddenly, there was noise in the forest below us, close by. Crashing and grunting.

'Wild pigs!' said Andreas. 'Shall we go find them?'

We clattered back down the tower steps, out of the sunshine, dropping into the many layers of shady foliage. We could hear the pigs, lots of them. Then, abruptly, they burst out of the thick morass of leaves and branches, a confusion of fast moving, small, grey, bristly, grunting creatures, crashing through the undergrowth. It was impossible not to smile.

The next morning, the motorboat took us to a jetty further down the river, where we would switch to a long, engine-free, grey-painted wooden canoe. As a kayaker, I was excited at the thought of paddling in such a place, but it became clear that we were to sit on the green and blue wooden seats in the centre of the boat and that the joy of propelling the canoe along the river, using lacquered wooden paddles with elegant, diamond-shaped blades, was reserved for the guides.

Pushing off into a narrow tributary, we slid deeper into the forest, rain bouncing off the brown river. On either side of us were complex root systems and multi-shaped leaves seemingly growing straight out of the water. The river was hemmed in by a morass of trees, bryophytes growing on trees, lianas hanging from trees. Masses of a brighter green, small-leaved plant on the water surface were sometimes vivid against the edges of the river, reflecting a duller green from the tangle of foliage above. An occasional flash of azure from the wings of a blue morph butterfly was the only relief from the green/brown/khaki palette; the noise of tree frogs, quacking like small ducks, the only sound bar the rain. It was not pretty. It looked messy, tangled, uninviting. Yet it felt so very *alive*.

Gradually, perhaps as we tuned in or perhaps as the boat glided further upriver, details began to emerge from the morass. Animal shapes in the myriad leaf, root and branch patterns became animals. Three toucans flew above us, their bills in silhouette against the low grey sky. Andreas saw a boa wrapped around the branches which, after a time, we thought we saw too. A fast movement in the branches became a chestnut-coloured squirrel monkey, almost flying through the trees.

'That's a hoatzin,' said Andreas, pointing to a fox-red and grey bird with a wild, spiky crest sitting on a branch like a small buzzard. 'Also known as a stinky turkey. They are ancient creatures, related to dinosaurs. They still have a claw on their shoulder, a remnant from when they were reptiles. If the young birds fall out of their nests, they use the claw to climb back up.'

Our guides paddled on. It was silent bar the sound of the blades dipping into the river, and of rain on leaves and water. Then, without speaking, Gabrielle held up his hand. The canoe stopped. He pointed to the edge of the river. We saw lianas, leaves, roots, rain. And then eyes. Unblinking, river-coloured eyes.

'Caiman,' Gabrielle said, quietly.

The eyes, their pupils a motionless, dark, thin, vertical line, sat at the top of a long, mostly submerged snout. Behind them, a rugged grey and brown back stretched out like a winter tyre tread just above the water.

'He's basically a small crocodile,' Gabrielle said. 'Though not always so small – black caiman can grow to be more than four metres long.'

Our character was considerably shorter, but still exuded an ominous air, despite the absolute lack of motion.

'He's in a sort of suspended state. Neither awake nor asleep. Waiting. They eat mostly fish. I wouldn't put your hands in the water, though,' said Andreas with a grin.

Later, as we paddled slowly back, we began to hear a strange, low-key roaring noise. To us visitors, it was an utterly otherworldly sound, slowly rising to a crescendo. Two troupes of howler monkeys were coming close to one other, Gabrielle explained. We saw nothing, the by now deafening roaring in the trees as if made by a bizarrely localised wind or hefty-lunged ghosts. The hoatzin, the caiman, the howler monkeys, the blue morph butterflies; such different ways of living in the world, all sharing some needs, behaviours and emotions with us, yet all so fabulously, powerfully, intoxicatingly *other*.

> *It is weird being a tourist, on a boat. Not allowed to paddle, not responsible for anything, taken and shown. It feels small part luxury and relief – no decisions to make, everything done for me – big part frustration. I so want to make a wilder journey, to be on this huge, brown, forest-edged river in a boat I am propelling, to go deeper into the forest, to be immersed in the vast and vibrant life that is here. But I wouldn't know how. Where to land? What are the currents? What can I eat? What might eat me? Could I learn these skills, acquire this knowledge? How long would that take, to be able to move easily and safely here, and know what I was looking at? Or would I simply disappear into the jungle and never return?*

Back at the centre, we sat on the covered porchway listening to water plummeting from the roof and watching rain sweep across the clearing in thick broken stripes of silvery grey. Even for a rainforest, it was a spectacular downpour.

'Rainforests don't just create rain,' Andreas said. He'd begun to explain this extraordinarily complex, beautifully calibrated process to us earlier. 'They also move it around.'

We sat talking about the aerial flows of water vapour and thinking about the invisible 'sky rivers' above our heads, taking water to ecosystems hundreds of miles away. A single large rainforest tree on a sunny day can pump over 1,000 litres of water up from the soil to the air. All the rainforest trees together add up to a living system vital to the weather patterns over the past 12,000 years of stable climate known as the Holocene era.

'One thing people don't always understand', Andreas said, 'is that all our living systems are robust, but only up to a point'.

The rainforest was, he explained, no exception. We could degrade it and cut more and more of it down, and it would still function. But there would be a tipping point. And once we'd crossed it, changes to the system would not only accelerate but be *irreversible*. If that happened, the rainforest would become a savannah.

Walking to the motorboat the next morning, we stepped carefully over a line of leafcutter ants. A constant stream of them scurried across the path and straight up a tree trunk, as if it were horizontal. Each ant in the returning stream carried a portion of bright green leaf, clipped out with their jaws, many times larger than the ant itself.

'They take the leaves to chambers in their nests, mix them up with ant saliva and ant shit and cultivate fungus in what you might call underground "gardens",' Gabrielle said. 'Then they eat the fungus. The ants are farming, basically.'

He told us about the complexity of the leafcutter ant society and how different ants perform different tasks within it, including the cleaner ants who check the leaves for parasites before they get admitted to the nest and the guardian ants who protect the marching column.

'Then there is the ants' role in the wider ecosystem,' Andreas said. 'There are birds whose main food source is the other insects that the marching ants displace, as well as birds and other animals – like, well, anteaters – who eat the leafcutter ants themselves.'

The whole thing was, they explained, fantastically well-balanced: a wildly intricate, dynamic system in a constantly changing equilibrium, but an equilibrium, nonetheless.

'Take the anteaters, for example,' Andreas said. 'If they find an ants' nest, they eat loads of them, but they always leave enough for the colony to recover so they can come back and eat them again another day. Sustainability in action.'

'There's a lesson in there for humans,' Gabrielle added, with a smile. 'Never eat all the ants ...'

We boarded the motorboat for a trip to a clay lick, a steep bank where the clay is exposed, said to be frequented by hundreds of parrots. Rounding a corner in the river, we saw two crested owls, perched one above the other on a steeply angled branch, their crests like enormous eyebrows slanting out above their big yellowy orange eyes, beaks neatly tucked into pale breast feathers. Something about them was painfully moving. They were so very beautiful, so together and so still against a many-layered backdrop of gently swaying palm fronds. So in place. And yet the owls also seemed to symbolise the astonishing, fragile, complex beauty of it all – playing their part in this immense, interconnected system, the millions upon millions of individual plants and animals that, some argue, deliver up to a third of all the oxygen released on earth through photosynthesis; that move a fifth of the entire world's fresh water through its rivers and vastly more than that through the transpiration of its trees; that locks up immense quantities of the carbon dioxide we are still pumping, recklessly, into the atmosphere. A global life support system. Our life support system.

As we got closer to the clay lick, we heard a new sound, something like the white noise of high-powered radio static. As we got closer, we began to discern abrupt, individual bird chatters within the wall of noise. And then we saw the source of the cacophony – hundreds and hundreds of green parrots, perched on foliage around the clay lick, flitting from leaf to leaf, chattering and squabbling, leaves swaying and bending as they landed and took off. The parrots' feathers were the exact same green as the leaves. In flight, they flashed a blue turquoise from their underwings – the mirror opposite of the morpho butterflies. The parrots were gradually descending. Leaf by leaf, branch by branch, they dropped down through the wet leaves towards the clay lick.[†] They racketed around, closer and closer, but didn't land.

'They are worried about something.' Andreas said. 'Perhaps us.' You have to be so cautious to be a long-lived bird in a rainforest.

We sat in a hide, being quietly mauled by mosquitos. Even without seeing the parrots land it was mesmerising – the noise and the movement of parrots in their hundreds, working with and around each other, all focused on doing their thing, whatever exactly that was.

Witnessing this wild, visibly organised yet profoundly foreign activity was intensely moving, despite our lack of comprehension. The early 20th-century writer and naturalist Henry Beston captured this when, talking about what a mistake it is to judge other animals in relation to ourselves, he referred to them as 'other nations'.

> In a world older and more complete than ours, they move finished and complete, gifted with the extension of the senses we have lost or never attained, living by voices we shall never hear. They are not brethren, they are not underlings: they are other nations, caught with ourselves in the net of life and time, fellow prisoners of the splendour and travail of the earth.[1]

The parrots were not just a critical part of the natural system that supports all life, but members of another nation. One that seemed, here at least, to be flourishing; one we lived alongside and were completely interdependent with. Watching these colourful, alert, intensely alive creatures I was swept with a sudden, furious clarity. Extractivism

---

† Thought to aid digestion and to neutralise toxins in seeds and fruit.

wasn't just a group of industries, it was a mindset; part of a wider worldview that reduced all this vitality, these vivid, sentient creatures with their complex and mysterious lives, to a set of resources. A set of resources we'd mistakenly assumed to be infinite, alongside the unquestioned belief that humans – or some humans – were entitled to exploit them without limit.

Listening to the intricate cacophony of sound that is many hundreds of parrots communicating, I knew with certainty that their value as the providers of resources, however critical, or even the wider aspects of 'what they do for us', could not possibly be the whole story about why other living beings and living systems matter.

Back in the boat, buoyed up by the parrots, we chugged further down the wide, brown river. Swallows dipped to the water and, looking back, we could see a vulture on a promontory, wings outstretched, drying out in the sun. And then, rounding a long, slow bend, we saw flames. Strange, candle-shaped flames, shooting into the sky from somewhere higher than the tree canopy.

'Flares,' said Gabrielle, quietly. 'Gas flares.'

Even at a distance, we could see that they were huge.

'The flares attract insects, and the insects attract birds,' Gabrielle said. 'They all die in the flames. The dead are so many you have to weigh them, not count them.'

There was a silence. 'It's cheaper for the oil companies to burn off the gas than to use it.'

A little further on, a huge, flat-bedded barge went by, carrying trucks.

'Also associated with the oil exploration,' said Andreas. 'They simply shouldn't be here.'

'Here' was within the Yasuní National Park, a UNESCO Biosphere Reserve, considered to be one of the most biodiverse protected areas in the entire world. When oil was discovered underneath it, President Correa made a deal – the oil would be left safely under the ground, in exchange for a multimillion-dollar payment from the international community. Agreed in 2007, it was an unprecedented arrangement, the first time a national government had committed to forsake oil wealth to protect biodiversity, the climate and indigenous communities. The countries involved in financing it were rich, Western ones, who had already destroyed their own forests and burned the most carbon. It was like a gigantic carbon and biodiversity off-setting scheme. Known as the Yasuní-ITT (Ishpingo, Tambococha and Tiputini are the names of the oil fields), the deal won him plaudits around the world.

'But in 2013 the deal was rescinded,' said Gabrielle. 'Correa said it was because the international community hadn't come up with the money. He needed to tackle the poverty in his country, and the oil – so-called 'black gold' – was the obvious way to do that. He said he didn't have a choice, that we couldn't just sit on that wealth while so many people lacked the basics.'

Having seen something of the poverty that Ecuador still endured, I could understand the dilemma.

'Some say this was predictable and could have been avoided,' Gabrielle continued. 'They say that the problem was the minister he appointed to oversee the deal. Others say it was always his intention to let the oil exploration go ahead.'

Gabrielle fell silent. After a while, Andreas picked up the story.

'Infrastructure for exploration and drilling was already being put in place *before the deal collapsed*,' he said. 'The concessions were given to a Chinese company. And Ecuador is massively in debt to the Chinese.'

There was silence except for the gently chugging engine. Gabrielle was looking at the water.

'We used to be able to pull fish from the river with our hands,' he said. 'Even if you caught something now, you wouldn't want to eat it.'

It wasn't a new story. Alongside *Doughnut Economics*, I had been reading Joe Kane's brilliant book *Savages*. Kane spent time in the Ecuadorian rainforest with the Huaorani people – the title is of course ironic – trying to make sure the Huaorani's views were heard as oil industries competed to drill on their lands in the early 1990s. Oil was discovered in the Ecuadorian rainforest by Texaco in the late 1960s, with a 312-mile pipeline over the Andes to the Pacific coast completed by 1972. But with no environmental regulations for oil production, there was virtually no attempt to assess its environmental impact until a North American woman called Judith Kimerling came to the rainforest in 1989, travelling by foot, canoe and truck, and sleeping in the homes of indigenous people. It was Kimerling who reported that the pipeline had ruptured at least 27 times, spilling nearly 17 million gallons of raw crude into the complex and vital network of rivers – more than the oil tanker *Exxon Valdez* released into the ocean when it sank off the coast of Alaska in that same year. Her other findings included that the industry was 'dumping 4.3 million gallons of untreated toxic waste directly into the watershed every day. Malnutrition rates near oil-producing areas were as high as 98 per cent. Health workers reported exceptionally high rates of spontaneous abortion, neurological disorders, birth defects, and other problems linked to contaminants. They predicted an epidemic of cancer.'[2]

Not surprisingly, the Huaorani and other indigenous peoples had been keen to avoid similar scenarios. It had turned into an extended – and ongoing – battle. Chevron acquired Texaco in 2001, including all its assets and civil liabilities. One of the liabilities was 'a 1,700 square-mile disaster in Ecuador', otherwise known as the 'Amazon Chernobyl' because of the cancers and other human health impacts. Texaco eventually admitted that it had knowingly discharged 72 billion litres of toxic water into the forest and that this toxic water had ended up in people's water supplies. It also admitted constructing hundreds of unlined waste pits in the forest floor. Thousands of excess cancer deaths were, as predicted, among the foreseeable consequences.

A coalition of indigenous peoples and local communities took Chevron to court and, after *eighteen years* of legal battles, won $9.5 billion in damages. But Chevron – a company with a stock market value twice the GDP of Ecuador – refused to pay up.

The boat engine sputtered and coughed into the silence. A clash of worldviews was at the heart of *all* of this, I thought. For many indigenous and other communities, human quality of life is not to be found in pursuit of economic wealth but by living in balance with the earth and other species. From this perspective, real wealth, as Carlos had also suggested, is found in flourishing, peaceful communities, human and other-than-human. Humans are part of nature, not separate and superior beings, entitled to extract relentlessly from the rest. Nor is nature viewed primarily as a commodity but, as Aldo Leopold, the North American hunter turned conservationist, put it back in the 1950s, a community we belong to on much the same terms as any other species.[3]

No, it was not a new story, though it is a heartbreaking one. Nor is it a story unique to Chevron or the Amazon rainforest. People are and have been fighting oil companies in relation to oil spills, other varieties of environmental devastation and human rights abuses pretty much wherever there is oil or other highly valued natural resources. Carlos had argued such conflict was inevitable. And for what?

Moi, a Huaorani leader quoted by Joe Kane, sums it up beautifully: 'Must the jaguar die so you can have more contamination and television?'[4]

Back at the lodge, we were escorted across to a tall, palm-thatched wooden building – a reproduction of a typical Kichwa Añangu dwelling – and invited to watch a display of dance and drumming, and to learn something of their traditional ways of life. The Kichwa Añangu are named after the leafcutter ants – 'because they work hard and ceaselessly' – and some of the details were fascinating. The three small stones in the building's central fire represent the family; the smoke from the fire coats the palm leaves and makes the roof waterproof; as in many cultures, most of the food preparation is done by women. But these bits of information seemed like disconnected fragments, stray threads plucked from thousands of years of history and an entire worldview. And the dance, a women-only affair, seemed stiff, the dancers slightly embarrassed. There was something dispiriting about both the display and the thought that it would shortly be repeated for another group. After it had finished, we were ushered into another building to look at bracelets and other crafts for sale.

That evening, after we'd all attempted to hit a not very distant papaya on a post with a shortened, tourist-friendly blowpipe – an activity that brought home the extraordinary skill involved in using a full-length blowpipe to hunt a distant and fast-moving monkey – I blew an extravagant quantity of money buying a round in the bar. I was craving conversation and an attempt at understanding. What we'd been shown had mostly left me feeling even more distanced from the Kichwa Añangu than before I'd arrived. Not that this was surprising. How could a complete outsider hope to gain any kind of insight into such a different way of life from a staged display of dance and sellable crafts, at such a linguistic and cultural remove? I wanted to know what the Kichwa Añangu thought about the horrific impacts on their forest wrought by the oil industry. I wanted to know how they understood the relationship with people and nature; what sort of value they attributed to other species; who their gods were.

Gabrielle provided glimpses. 'I think of nature sort of like a father,' he said, in answer to one of my many questions. 'A father who shows me how to hunt, among other things.'

The Kichwa Añangu's original nature-based gods had, he told us, like those of many indigenous people, been either displaced or supplemented by a Christian God after decades of Catholic missionaries.

Heading back to the bar for a top-up while I processed this, I returned to find that the conversation had moved on to sustainability. 'There's no perfect tribe of eco-people out here,' Gabrielle was saying. 'We're all different. Some of us are better forest guardians than others. A few of us are in favour of oil and mining and the money it brings. Many oppose it. Some violently oppose it. Communities and even families have been split over this. Either way, it's a mistake to romanticise us.'

'But we depend on the forest,' Gabrielle continued, 'in a way that's direct and evident. You also depend on the forest, of course, but you typically have no idea that this is so. We know the truth of this on a daily basis. That's one reason we've developed sustainable practices over the thousands of years we've lived here. And why there is a huge correlation between intact, protected forests and areas where indigenous peoples still live. We're all different, but overall, we're still the best forest guardians, by far. Well, we are if our rights to the land are recognised, that is.'[5]

Later, I was to wish I'd pursued this further, but at the time I was still troubled by questions from the visit earlier. Was it really of value for us to be here, watching a staged show? Could ecotourism ever provide an income that could rival oil? Andreas, passionate about the ecolodge concept, leapt in at this point.

'Yes and yes, definitely,' he said. 'The oil industry used to provide at least 85 per cent of Ecuador's income. It's now about 65 per cent. It's all about diversification. If we are to tackle the oil industry, we have to find other sources of national income, to diversify further. Coffee, cacao, bananas, mangos. Ecotourism is probably our best hope, though.'

He paused. 'Of course, it's also a dilemma – there's currently no non-oil-based plane transportation. And the vast majority of our visitors arrive here by plane, from Quito, often after a long-haul flight from somewhere else.'

I was clearly an aberration, having arrived by cargo ship, bike and bus.

'Nevertheless, we need this kind of tourism to increase,' Andreas insisted. 'And not just because of the income. People, everywhere, need to experience the forest if they are to understand how precious it is.'

I got this. In an average hectare of the Yasuní National Park, for example, there are more species of animals and plants than in the USA and Canada together, and approximately 100,000 species of insect. In the whole of the UK, we have somewhere around 50 species of native trees, compared with about 2,200 species of tree in Yasuní. But none of this knowledge had prepared me for how powerful the feeling of actually *being* in the rainforest would be, for the vivid sense of aliveness, of being somehow within the heart of the planet, surrounded by and immersed in such a wild and astonishing diversity and abundance of life. To be there was to feel the utterly irreplaceable, supreme value and importance of the forest.

Gabrielle nodded. 'And this kind of experience-based ecotourism is good for our community, too. Both sides see a different perspective. And of course, it's crucial to create alternatives to working for the oil industry.'

On our last night, I'd sat on the steps of my cabin, listening to the vast hum of the forest. A million cicadas were singing. An occasional bird, wingbeats sounding loud in the darkness, flew invisibly by. Above me, the sky was unusually clear and bursting with the immense, chaotic scramble that is the southern stars. I recognised scarcely any of them, save the plough, which was upside down. What I did recognise was the now-familiar feeling of bittersweet happiness; the beauty of the moment shot through with the knowledge that I was about to drag myself away, just as I was beginning to feel I'd started to arrive.

On the Quito-bound bus the next day, I felt a sadness verging on downright grief to be leaving the forest for the city. What the grief was for, exactly, I struggled to name, but it was related to a sense of being wrenched away from somewhere I wanted to remain, a sense of loss of a kind I'd often experienced descending from a long day in the mountains, or returning to the world of towns, tarmac and traffic after an extended sea-kayaking trip.

It struck me that the grief wasn't just about leaving a particular environment, or people or places I'd enjoyed spending time with on a journey, though that was real enough. It was also grief in the face of a fundamental disconnection; grief at what you see when the mirror that is coming out of the forest or down from the mountain is held up to so-called normal life. We are never actually disconnected from nature, so long as we breathe and eat. But I'd constructed even this journey, despite its biodiversity focus, so that most of my time was on roads. It was an apt metaphor, perhaps, for the way the rest of life so often keeps us on a very human-focused track, distanced from nature, at arm's length from other living beings. How very much, on some level, we miss their company.

There was another grief, too, swelling in the face of a conclusion I'd been trying to not to reach. Jennifer's story of the almost unbelievable environmental and social impacts of large-scale gold-mining in Colombia was not an aberration. It was not a tale of corruption and damage that was, however dark, nevertheless a one-off. I had now heard the same story – environments devastated and people persecuted – multiple times. I'd heard it in relation to Carlos and copper, oil and the Amazon. García Márquez had been writing about it 50 years previously in relation to fruit companies; Che Guevara in relation to poverty and mining. Francis had talked of exactly this when I met her in Manizales. And suddenly, on that long bus journey back over the mountains, this appalling realisation punched me in the guts.

We *had* created a system in which unconscionable and ultimately self-destructive amounts of environmental damage were routinely inflicted in the process of extracting these all-too lucrative substances. Routinely, not as an aberration. A system in which companies and governments who stood to gain were willing and able to inflict violence, to displace communities, to maim and kill. A system at once insanely destructive of life and terrifyingly powerful, with a seemingly inexorable momentum. A system in which, I was beginning to understand, extractivism was itself embedded in wider stories of colonialism and control, values and power, capitalism and indefinite growth.

The forces that Jennifer and Carlos and the Huarani and the Napo and so many others were up against – that we are all up against – were colossal. And the response had to include something that sounded like a cliché: system change.

9. VOLCANOS OF THE MIND

When nature is perceived as a web, its vulnerability also becomes obvious.
ANDREA WULF, *THE INVENTION OF NATURE*

Environmental destruction is inseparable from relationships of racial and colonial domination. It stems from the way we inhabit Earth, from our entitlement in appropriating the planet.
MALCOM FERDINAND, RESEARCHER AND AUTHOR

Those who get to tell the stories rule the world.
MIKE CARTER, *ALL TOGETHER NOW?*

The taxi ride to the station where I'd caught the bus to the Amazon had taken in a chunk of the main road out of Quito. It was narrow, rammed with traffic, very hilly and infested with spaghetti junctions. It would be fair to say I was not looking forward to cycling it.

In fact, leaving Quito became one of my best days in Ecuador. I was in El Cafecito deleting emails when Javier arrived. Javier, attired for the occasion in a smart blue cycling top and shorts, turned out to know the owner, who insisted we all had a second breakfast. Quantities of omelette, fried potato, fruit salad, yoghurt, pineapple juice and coffee later, we set off, fabulously full, into sunshine and 'cycling Sunday'. Swathes of roads across Quito were, Javier explained, closed to traffic and taken over by cyclists of all ages on a regular basis. It was super-well organised. And there was a celebratory air, as riders ranging from Lycra-clad racers to tots on tricycles filled the normally fast, loud and dangerous combustion-engine-dominated streets.

'It's been going on for at least ten years,' Javier said.

He was the perfect guide, knowing as much about Quito history, volcanos and all things Ecuador as he did about the best cycling routes. Not only were we able to pedal out of town on traffic-free roads, but we took in the sights en route while bypassing the steepest climbs, which was just as well. In the few days I'd spent at sea level, I'd lost almost all the altitude acclimatisation I hadn't realised I'd gained and was struggling with breathlessness at the slightest incline. Quito, at 9,350 feet, is the second highest capital city in the world.*

A couple of long climbs took us steadily higher, past the entrance to an agricultural research station and back into the patchwork of green, divided fields that ran along both sides of the main road, stretching out towards the volcanos we still couldn't see, but that Javier insisted were there.

'This road is known as "Volcano Avenue",' he said, as if that proved it.

Several hours later, we swept down a glorious, curving descent, relishing the effortless, grin-making speed. The descent bottomed out at a petrol station that sold sandwiches. Then Javier, kind, helpful, knowledgeable Javier, waved his goodbyes and headed back up the hill. I watched him disappear round the curve, considerably faster than he would have done had he been pacing me. Woody and I carried on south. South! It was good to be on the move again.

---

* After Bolivia's La Paz.

That evening, in the town of Machachi, I lucked out in a small hotel and was given a cheap room with a panoramic view across the town towards where the volcanos should have been. I wandered out and soon came upon a market, half-inside an immense barn-like building, half-outside. There were tables stacked with fruit: bananas, apples, melons, granadilla, papaya, pineapples, oranges, grapes, strawberries, lulo, tree tomatoes and a spiky red fruit that looked like a kind of lychee. Women dressed in vibrant colours with dark hair in plaits below their bowler hats vied for the many customers. Beyond the tables of fruit were vegetables and fabulously pungent herbs. Then stalls selling cooked food, stacks of rice, chicken feed and various kinds of dried beans. It was busy with people and dogs and cargo bikes, packed with colour and life. I meandered through, relishing the vibrancy of the place, the diversity of food on offer and the apparent success of local, small-scale farming with wonderful produce sold direct to local customers.

The sun had set by the time I walked back with my spoils. I sat in the window and caught up with the journal while I grazed on fruit, nuts and cheese patties. Intermittent snatches of music floated up towards me, leavened with traffic sounds. I could see yellow tables in a food venue in a covered market below; a small, tilted, waning crescent moon hung above it. It was a wonderful spot to have landed, after such a good day. I'd left Quito in much better humour than I'd arrived. Perhaps this was the beginning of a better relationship between me and Ecuador.

The next day, I found myself thinking about the market again. On the ride out of Quito, I'd asked Javier what other big Ecuadorian biodiversity stories I should be telling.

'Agriculture,' he said. Ecuador, he explained, had a growing population in a small country. More and more land was being taken over for farming.

'You can see the fields going high up to the volcanos,' he said. 'Farming is supposed to stop at just under 12,000 feet to protect the ecosystems above. But no one is monitoring it and it's not policed. It means the páramo is being damaged, despite that it's critical for freshwater.'

A few miles later, when the road had flattened out and I had a bit more breath, I'd asked him what he thought the solutions were.

'More fertile lower areas need to be available to small farmers,' he said. 'They get pushed uphill because the big agribusinesses dominate the land below, producing crops such as bananas and coffee for export rather than to feed people here. And we should diversify our crops. That research station we passed – INIAP, the National Institute for Agricultural Research – has an agricultural seed bank. It includes 600 kinds of potatoes. How many do you think we actually use? About twenty, tops.'

Ecuador was not alone in any of this. Across the world, 95 per cent of the energy we get from food comes from a mere fifteen types of crop, and over 60 per cent comes from just three – rice, wheat, and maize. The more we lose species variety within this limited number, the more the lack of diversity in the plants we eat leaves us vulnerable to pests and disease, and to a rapidly changing climate. Diversity = resilience. A generalisation, but more often true than not, in domestic as well as wilder contexts.

And of course, agriculture has an immense impact on wild species: on ecosystems and on biodiversity. In fact, agriculture, and especially meat and dairy farming, is one of the biggest drivers of habitat loss and degradation across the world, as well as of climate change, and it's one of the biggest users and polluters of fresh water on the planet. About half of all the habitable land on earth is used for agriculture.[†] Of that half, a whopping 77 per cent is used either for farm animals or to grow crops to feed to farm animals. In fact, the vast majority of mammals on earth – 96 per cent – are now either humans or their livestock. Only 4 per cent of the mammals on this planet are wild.

'And that's the most important change we need to make,' Javier concluded. 'More efficient use of land, by changing our diets. Ecuadorians eat more chicken than beef. But we need to alter our eating patterns to further reduce our meat and dairy consumption.'

No, Ecuador was not alone in needing saner, genuinely sustainable approaches to farming and food. The impacts of agriculture seemed especially visible here though, and Javier's words were often with me as, following the high road that tracks the Ecuadorian Andes north to south, I cycled for days, usually upwards, through khaki-green landscapes. Field boundaries criss-crossed the land up to the edge of the clouds and (to my novice eye) the páramo did indeed seem damaged, with scarcely a frailejón in sight.

Perhaps this was partly why, for the first time in my life, I was beginning to find a mountainous terrain tedious. For months, I'd been riding happily at an altitude of between 8,000 and 12,000 feet. Here the road seemed constantly to fall and then climb again, fall then climb. It felt like some sort of betrayal of my mountain-loving self, a weird and disorienting shift in identity, but I was no longer relishing the climbing. In fact, I was cold, tired and fed up. I was becoming careless on descents, too, sitting at 30 mph paying no attention whatsoever.

On one such descent, one of the emotions I was grappling with had suddenly crystalised out of the murk of confusion. What I was feeling was *trapped*. Trapped in this ride,

[†] See Chapter 4. A total of three-quarters of this ice-free land has been significantly altered by some form of human activity.

in these unnecessarily, ridiculously, overwhelmingly immense mountains. Now, back on my bike again, I realised I had had enough. But there was still so very far to go. Most of the journey, in fact, still lay in front of me, stretching endlessly south down the long spine of the continent. I'd had enough; but giving up would be awful, rendering the whole journey pointless.

Over a bowl of avocado, cheese and potato soup – an Ecuadorian highlight – I tried to take a step back and look at this emotional morass from the outside. I did not like what I saw one bit. I was in danger of becoming a prime representative of an approach to 'adventure' I detested; the one where a privileged white person puts themselves, entirely voluntarily, into a situation where they endure self-imposed hardships and then whinge about them. By the second bowl of soup, I'd come up with a change of plan. I couldn't quit. I had to keep heading south, and trust that this dark, sulky, joy-sapping mood would change. But I didn't have to stay at altitude. Instead of continuing on the high road, I would drop down from the mountains and ride towards Guayaquil, the city where I had ended my first South American ride back in 1992. Then I would cross the border and go back up into the mountains in Peru. But first I would visit Chimborazo.

Chimborazo is the volcano that the 19th-century, fabulously eccentric German explorer and naturalist Alexander von Humboldt climbed in 1802, when it was thought to be the highest mountain on earth. It is not the highest mountain on earth, of course, though it is, at over 20,000 feet, the highest mountain in Ecuador. And because it is so close to the equator, the summit is the furthest from the centre of the earth. So, it sort of *is* the highest mountain in the world: a thought that makes more sense to some people than others.

I knew from Wulf's *The Invention of Nature* that I'd often been on the same route as the one Humboldt took from Colombia to Peru. The thought of his expedition – especially the mules, the crates of live parrots and monkeys and the scientific equipment – frequently left me appreciating my own journey as wonderfully simple. When Humboldt got to Chimborazo, he climbed it with an infected foot in inadequate footwear. The scientific equipment he heaved up the volcano with him included a barometer, a thermometer, a sextant and a cyanometer – for measuring the blueness of the sky.

Somehow, despite the bad foot, bad shoes and the fact that he and his three companions were carrying all the gear because their porters had refused to go on, Humboldt got to within 1,000 feet of the summit. There, at 19,413 feet, in freezing winds, he had a revelation about nature that still influences how we think about it today.

'He saw the earth as one great living organism where everything was connected,' Wulf wrote.

You could say that he came up with the Gaia theory about 170 years before James Lovelock did. His lightbulb moment of realisation on Chimborazo was that vegetation zones 'were stacked one on top of the other', and that what grew in them was influenced by altitude and climate *in the same way all over the world*.

I was super keen to reach Chimborazo and to see Humboldt's vegetation zones. Cycling into Riobamba, the closest town, the volcano was still invisible.

I set up basecamp in the Estación Hotel, a tall, narrow, colourful, friendly establishment, with multiple floors around an open staircase adorned with train artefacts, paintings of Chimborazo and sofas. From the hotel roof, I was assured, there was a view of Chimborazo itself.

Walking across the railway tracks into town the next day, I saw two heavily laden touring bikes leaning against some railings. Going up to say hello, I recognised the cyclists – a couple from Bulgaria who I'd previously encountered in Quito. They'd started cycling in Alaska and, living on $3 a day, were on the hunt for free camping. I'd recently discovered the iOverlander app and found a few suggestions. In return they pointed me to a tourist office where I could get free maps.

It was one of those lucky breaks. Among the stands of brochures, I got chatting to a young man, with dark hair and fluent English.

'Do you want to go to the summit?' he asked.

'Well, yes. But probably no.' I said. 'Really, what I want to do is walk around the mountain and see the different ecosystems and some of the wildlife.'

The man, whose name was Alex, broke into a grin. 'Well,' he said, 'as it happens, I can guide below-summit walks. We could go through the different vegetation zones.'

It couldn't be a coincidence. 'Vegetation zones? As in Humboldt?'

'Yep. As in Humboldt.' Alex grinned some more.

The clincher was a vegetarian picnic as part of the deal. We arranged to meet at the bus station at 7.20am the next day. I just hoped my legs would cope. It had been a while since I'd walked for eight hours and, judging by my breathlessness when inspecting the roof terrace – more building site than manicured area with sun loungers – my altitude acclimatisation wasn't fully restored either.

Back at the hotel, I hauled myself up to the roof again, on the off chance. And there, to my astonishment, was the white head of the volcano, the massive dome soaring above the turquoise, orange and white buildings of the town. One half was still in cloud, but the other was so clear you could see the gullies sweeping down from the summit. It had to be a good omen.

The bus took us to the gates of Chimborazo National Park, where we emerged into thick cloud and cold air. A man in a battered car asked if we wanted to drive up to the first refuge. This would save us 'a boring, fairly horrible walk', according to Alex. We agreed. He charged us $10, and the car stank of petrol, but it saved us hours of slog on a dirt track with no view. We were almost there when the clouds dissolved away and Chimborazo was suddenly visible, a long, graceful ridge line leading up to the immense, rounded, cone-shaped summit, scattered knuckles of dark showing through the snow.

Alex's plan was to contour around the mountain, rather than head up towards the next refuge on the main tourist track. Our path – the Templo Machay route – took off over loose, grey lava rock. A couple of vicuñas – wild relatives of llama and alpaca – were silhouetted on a distant ridge.

Chimborazo, brilliantly white against the blue sky, was a constant, shining, wildly alluring mass of snow and rock that arched above the grey zone.

'She is my true girlfriend,' said Alex. 'I have another girlfriend in Riobamba, but Chimborazo is the one ...'

He had good taste. The mountain was utterly compelling. I had to stop, again and again, just to look. Wrenching my eyes back to the path, occasional flowers poked through the grey shingle; tiny, delicate, purple ones that Alex said were a kind of violet and short, upright branches of green, spiky leaves topped with gorgeous flame-orange tufts of flower, called chuquiraguas. The vicuñas' coats were much the same colour, and both looked fabulously right in this place. The Inca had valued and protected vicuñas but, after the Inca Empire ended, they were hunted to extinction. Now, Alex said, reintroduced in the 1980s from Peru and Chile, they were protected again and thriving.

We contoured round the slopes of the volcano at about 15,000 feet. Uphill sections were hard work, but on the flat I felt fine. We scrambled up a steep slope to reach our trail's namesake, Templo Machay, a cave set into the southern flank of the volcano. It had been the site of offerings for hundreds of years, though to which gods, Alex was not sure. The modern offerings on the dark, dank floor were large plastic containers of drinking water. A classic cave-rock profile framed the view of our descent as we looked out.

Below us, chuquiraguas provided dashes of orange against the grey lava.

Eventually, we headed down. My descent was nothing compared to his but, in my long-suffering trainers and holey bamboo socks, slithering on the tricky, loose sections, I couldn't help but think I was there in a Humboldt sort of spirit.

Looking back a while later, we could see the banded layers on the mountain slope: the snow-capped summit, the lava field, the section dominated by chuquiraguas, then a belt of grass, then sand with an astonishing mix of flowers in it and finally a mass of yellow and red flowers speckled white with big daisy-like plants. Humboldt's 'vegetation zones' were stacked right there in front of our eyes. We could almost envisage him among them, limping down from the snow slopes with his infected foot and heavy load, looking at the different ecosystems or habitats – as we would now call them – laid out below him.

'Nothing, not even the tiniest organism, was looked at on its own,' writes Wulf. '"In this great chain of causes and effects," Humboldt said, "no single fact can be considered in isolation." With this insight,' Wulf continues, 'he invented the web of life, the concept of nature as we know it today.'[1]

It was the perfect Humboldt day.

We paused on a grassy slope. I sat down. I wanted nothing more than to stay there, on the flank of the mountain, with the birds and flowers and the unbelievable view of snowy Chimborazo. Alex said that many people, himself included, came to have a sort of obsession with Chimborazo – the 'monstrous colossus' Humboldt had called it – a love that had a gnawing, compelling quality. I suspected I was already falling under the spell.

We did come down, of course. Past a solitary tree and into a wild meadow, the short grass studded with daisies and dandelions. A bird of prey hung, silent, in the blue and cloudy sky.

'Now for something completely different,' said Alex, as we reached a blue-painted wood and stone building. A huge triangular porchway echoed the shape of the volcano towering up behind us, set off with an impressive set of antlers. 'Welcome to Chimborazo Lodge.'

Inside, the blue and terracotta walls were festooned with more antlers, guns, snowshoes, crampons, ice axes, skis, goggles and other climbing paraphernalia, much of it of some considerable vintage. Photographs showcased mountaineers who had climbed Chimborazo – all men – including the UK's Chris Bonington, who also lives in Cumbria and is practically a neighbour.

At the hotel after we'd hitched back to town, we had that best of pints – the post-mountain-day pint – and agreed to stay in touch. Then I ground to a halt. I tried to settle into reading, the idea of climbing Chimborazo, which Alex had suggested was within reach with a bit of training and kit hire, had somehow taken hold. It kept resurfacing, interrupting, niggling away.

I thought I'd come to terms with the notion, dismissing it as large part ego and making peace with the rest. But I woke up with it still alive and definitely kicking. I walked up to the roof terrace. My legs were, emphatically, tired, and I was breathless on the stairs. An intermittent toothache that had been bothering me since Quito had returned in the night. Chimborazo was back in the cloud. The whole thing was crazy. *Let it go.*

The climb up out of town was gentler than I was expecting. There was a mild sun and even a slight tailwind. But my flat, deflated mood had returned. Perhaps it was being back on a road, in traffic. Perhaps it was the toothache. Or perhaps it was Chimborazo. The thought of going back to that mountain, of going higher, of being up there in the ice and snow, maybe even standing in the wild wind close to the frozen summit with the immensity of views that Humboldt himself had seen, had a power that went way beyond sensible analysis. Alex was right. Obsession had taken hold, after only one day on the volcano's slopes.

The next day, I woke coughing with a sore throat and a swollen face. The toothache had properly kicked off, and pain ricocheted around my mouth, lighting up different teeth like stations in a gambling machine game, before settling on a back molar, with a *ker-ching!* If I accidentally bit down on that tooth, the pain made me yell out loud. I swallowed a handful of aspirin and ate my breakfast on one side of my mouth in between visits to the toilet to cough up globules of green, infected gunk. *High altitude mountaineer, my arse.* If there was any lingering doubt about the Chimborazo decision, the state of my body dispatched it.

Back underway after detouring to a chemist for painkillers, the road leaned uphill into the low clouds. It was a long, relentless climb, with several tormenting, false summits, which toothache did not make any easier. It seemed that the tooth had been enlarged by a cushion of pain – if the teeth above touched it or even came close, it hurt like hell, as if a malign life force were hitching a ride in my mouth. *If it is infected, I guess I have millions of lives in there, in fact.* Around me, the khaki stripes of farmland stretched up steep slopes on one side; more apparently trashed, frailejóns-free páramo on the other. Finally, I reached the top. Another false summit. I climbed some more. At long last, after a near miss with a total sense of humour failure, the road tilted down. A brief looping rise, and then I was plummeting down the mountain like a stone.

Given my infected face, dropping down from the mountains now felt more like a luckily timed smart move than a cop-out. Wanting to make the most of the new route, I had confirmed a visit to a nature reserve on the edge of Guayaquil – and this part of plan B proved to be an accidental ace. After mentioning I had toothache in an email exchange with Eric Horsman, the head of the reserve, he had booked me an appointment with his dentist and arranged for a truck to pick Woody and I up on the edge of the city and deliver us there, with a translator. I couldn't have been more grateful.

But first I had to get there. The relief of going down rather than up was soon eroded. Long descents can be hard work. This one was cold and required concentration. The road surface was a mess, and there were lots of warnings – I had seen at least ten dead dogs on the side of the road that day. I couldn't recall having seen any in Colombia. Perhaps Ecuadorian dogs were not as traffic smart. *I need to be more Colombian dog than Ecuadorian dog.*

It was the next day before I reached sea level, descending through increasingly lush landscapes, the air thick with oxygen and water vapour. Feeling almost drunk on oxygen and warmth, I bottomed out on a wide road that ran through small, rough-around-the-edges towns, a church often prominent among half-finished houses. There was a strong sense of poverty. And a series of maddening road signs, with slogans like: 'Water is life, take care of it, don't pollute it.' *Maybe stop drilling for oil in the Amazon, then?* And then: 'The earth is timeless, take care of her,' right in front of a banana plantation.

After that, the banana plantations dominated. For miles and miles, on both sides of the road, the monoculture of plants stretched out like stubby palm trees across the flat land. It was as if I had been presented with a caricatured case of 'biodiversity bad guy'. I had read about the impacts of these plantations; now they were literally in front of my eyes. The plantations were biodiversity deserts. Nothing grew beneath those plants. Each bunch of bananas was wrapped in single-use plastic and, to top it off, they were sprayed with something hideous from a yellow plane that appeared from nowhere and roared over me, leaving a multiple-stranded tail of noxious substance low in the air behind it. I vowed never to eat a non-organic banana again. When I finally emerged from the planation to ride alongside a wetland, it was like switching a screen from monochrome to full colour. The air was suddenly full of dragonflies; the roadkill comprised mostly of frogs. A startlingly white egret paused graciously mid-step and two large green-and-cream-coloured snakes slewed across the road to one side of me.

Are banana plantations a form of extractivism? I wondered as I pedalled through the delicious if muggy heat towards the next one. Historically, they'd certainly set a precedent for extracting value from the earth, no matter the environmental and social impacts.

'[T]he banana importer United Fruit created the blueprint for how global corporations wield influence and power at nearly any cost' is how it was summarised on the cover of Peter Chapman's book, *Bananas: How the United Fruit Company Shaped the World*.

The combination of heat, tiredness and toothache were not conducive to clear thinking, let alone in relation to something so dark and complex. Yet, cycling into another battalion of pesticide-drenched, plastic-wrapped banana plants in the heart of once-colonised Latin America, I couldn't help thinking that, despite this, I was in exactly the right place to be grappling with it.

The academic/activist Malcom Ferdinand takes a view not unlike the one I'd reached in the rainforest – that we should see colonialism as a mindset or worldview or, more accurately, a way of being in the world. Colonialism, he writes, is not just a historical era, but 'a certain way of inhabiting the earth, from some believing themselves entitled to appropriate the earth for the benefit of a few ... a violent way of inhabiting the earth, subjugating lands, humans and non-humans to the desires of the coloniser'.[2]

Ferdinand calls it 'colonial habitation'. It comes with a sense of entitlement; the assumption that the coloniser is superior to, and more important than, the colonised. Violent exploitation of people and land was therefore (in the minds of the colonisers) justified.

For Ferdinand, colonial habitation also drove environmental degradation. And he argues, the legacy of colonialism in the form of colonial habitation as a way of being in the world – 'the mentality of appropriation and hierarchy' – is not a thing of history, but endures to this day.

On the banana road, this was all beginning to make sense. The modern multinational 'extractivist' corporations whose impacts and behaviour I had been witnessing and learning about were responsible for a startlingly similar set of consequences to those that the colonisers had been. Exploitation of nature and people. Immense environmental and social impacts. Wildly differential benefits, the corporations, like the colonisers, gaining more than the communities whose resources they were exploiting.

And they were informed and legitimised by pretty much the exact same philosophical assumptions.

Thinking about the treatment of Jennifer's group and the ecologically devastating impacts of gold-mining; of Carlos' ecological and human community at Intag; of the Huarani in the Amazon rainforest and so many others by the gold, copper oil and other extractivist industries, Ferdinand's analysis seemed horribly telling.

Extractivism, I realised, was the ultimate example of colonial habitation as a way of being in the world. And the 'yellow gold' of the banana plantations were as powerful an example of all this as the 'black gold' of the oil fields.

I lay back on the dreaded blue chair. I have always been unreasonably, disproportionately scared of dentists. Dentists speaking at speed in a language you only partially understand can, I discovered, propel this fear to another level. Tania from the reserve, in white T-shirt and grey hill-walking trousers, her dark hair held back in a ponytail, was struggling with the translation task she'd been assigned. This much was clear, though: whatever was going on in my mouth, the dentist, who despite my terror I noticed was wearing turquoise shoes, was not impressed. Eventually, Tania phoned Eric, and the dentist spoke with him. Then I spoke with Eric. The problem was an infected root canal. Several injections later and, just as I had resigned myself to a complex set of doubtless still painful procedures, the dentist discovered an abscess on the surface of my gum. Not a root canal issue at all. He squeezed puss from the swelling, scraped a chunk of calcite off the offending tooth, wrote a prescription for antibiotics and charged me £40. Then we left, me with a swollen and senseless mouth and a very big grin.

We had driven to the dentist in a reserve truck, Woody in the back, me and the panniers in the middle, Tania in the front with Suzanna and Lorenzo, who was driving. The bridge over Río Guayas was much wider than I remembered, and Guayaquil was huge. Modern, evidently wealthy sections, complete with shopping malls and a Sheraton, rubbed up against areas of obvious poverty, the squalor heightened by the juxtaposition.

If the banana plantations were the biodiversity bad guys, the Cerro Blanco reserve, run by the award-winning Fundación Pro-Bosque and supported by the World Land Trust, was up there with the best of the good guys. It had been tasked with protecting one of the last remaining tropical dry forests in Ecuador – 6,000 hectares of forest that butted right up against the western edge of the biggest city in Ecuador, a city of over 2.5 million people. Like cloud forests, tropical dry forests are among the most threatened ecosystems in the world, and a worldwide conservation priority. Unlike cloud forests, the species that live in them are adapted to cope with many months without rain, followed by deluge. They are also highly biodiverse, and this one, despite its city-side location, was no exception. An astonishing array of plants and animals lived within the reserve – 54 mammal species had been recorded there, including five species of big cat, plus 21 species of bat, 221 species of bird and uncountable numbers of amphibians, reptiles, insects and plants.

'One edge of the reserve is mostly modern development, malls and so on. Things are pretty controlled there,' said Eric, when I met him in the reserve office later. Eric sat at a paper-strewn desk in blue jeans, a beige, short-sleeved shirt and well-used trail shoes. I had gone to the office in the early afternoon to thank him for arranging the dentist visit and to ask if we could arrange an interview. We ended up chatting well into the evening.

'But the other edge', Eric continued, 'backs on to an area that is much poorer. People are constantly building temporary buildings and then moving in.'

Eric had been studying for a degree in journalism when he'd discovered that 'international studies' was a degree topic in northern California and switched. After a visit to the Galápagos, he had done everything he could do get into the park service. When that didn't work, he'd come to Ecuador with the Peace Corps instead, and had then returned to run this reserve. Twenty-five years later, he had an Ecuadorian wife and a son. Guayaquil was, in his view, a special place.

'I was welcomed here. They never judged me as an outsider,' he said.

It had been an eye-opening experience, nonetheless. Arriving, in his words, 'young and idealistic, with black and white ideas about how nature conservation should be carried out', he'd been rapidly confronted by reality, or a particular version of it.

'What would you do, if you caught someone hunting in the reserve but you knew he was hunting for food for his family?' Eric asked, as he sketched out for me some of the challenges and dilemmas of managing a reserve on the edge of an ever-growing city, in a county with so much poverty.

'Or if you found someone building a shack within the reserve boundary, to claim squatting rights and evade homelessness a little longer? My Peace Corps training was in forestry. They didn't train us in social mediation.'

The rangers were central to finding solutions, he told me, constantly negotiating good relations with the communities that crowded up to the edge of the reserve. And that was just one aspect of their work.

The next day, one of those rangers, Armando, picked me up in a truck. With us was Ruben, a young Colombian who was doing a PhD on fragmented forests. Armando steered us out of the reserve gates, along the main road and then back into the reserve on a road that became ever rougher as we climbed.

First stop was a semi-derelict building that had previously been used to house birds waiting for release into the wild. We clambered out of the truck. The former aviary was clearly long empty. Ruben and I looked at the sagging wire mesh, not sure why we had been brought here. Then, 'Look, behind you!' Armando whispered, urgently.

Across a clearing, high in a tree, sat two of the largest, rarest parrots in the world: great green macaws. The birds sat close together, constantly interacting, touching each other's beaks with their own. It was hugely moving to see them, something to do with their relationships as well as their rarity – to each other and to the trees that surrounded them. They were as green as the leaves, and almost exactly the same shade, like another piece of the forest. They had come from the forest, evolved in the forest and were now a feathery, conscious part of the forest.

Back in the truck, we lurched further along the now deeply rutted track to a viewpoint. A huge vista of wildly lush primary forest stretched away below us, so close to the city, and yet so very alive. This forest was on a completely different scale from the Amazon, but it was every bit as clear that you didn't need numbers – of species, of hectares, of biomass – to know it was precious. As the three of us stood looking out over the jade and olive and emerald hues in the mass of rounded treetops, the value of the forest was visceral; a feeling of vibrancy and vitality that was almost as powerful as it had been in the Amazon.

We ended the tour back at the reserve nursery. A variety of native trees were nurtured there, from delicate green shoots curling up out of the growth to strong-looking saplings.

'The reserve plants thousands of trees every year, to help with forest regeneration,' Armando said, with evident pride.

Later, on a self-guided trail, I walked to a viewpoint that confirmed the dual importance of reserves like Cerro Blanco. Across the forest, a cement factory, all grey angles and smoke against the green curves of the canopy, skulked on the near horizon. The reserve was holding the city at bay, protecting a chunk of immensely valuable habitat, teeming with diverse life, against encroachment and diminution.

But it wasn't just about these traditional nature reserve functions. On the walk, I'd passed colourful, carefully crafted sign boards offering information about the forest and why it mattered, asking questions, provoking engagement. There was a jaguar-shaped totem pole, and huge bamboo flutes standing where any slight wind would conjure up a tune. There was a campsite and a meditation area. The reserve's location meant it could offer, alongside protection for its other-than-human residents, the chance for local city-dwelling humans to spend time in the forest too; opportunities for some sort of reconnection with the nature that so many of us are now cut off from, at least in terms of our experience of forests, mountains and wilder places.

'Yes,' said Eric, the next time I saw him. 'On one level, we've never become disconnected from nature. That would be impossible, so long as we breathe and eat. But on another, many of us live in a way that keeps us at a distance from other species, prevents us from experiencing the ecosystems we are part of. South America, like everywhere else, has undergone a massive urbanisation of its populations. Globally, over 50 per cent of the human population now lives in cities.'

I nodded. 'And you think some sort of reconnection has to be at the heart of inspiring care?'

'Yes,' Eric said. 'And here we can offer local people diverse ways of enjoying and connecting with the forest. One of the biggest questions we've been grappling with is how we make these experiences deeper. Richer. More meaningful.'

It became another long and sparky conversation. We covered a lot of ground. The Rights of Nature, potentially one of the most powerful legal frameworks imaginable, a fantastic complement to individual reconnection and a much-needed challenge to extractivist notions of both development and our relationship with nature.

'Ecuador recognised the Rights of Nature in its constitution in 2008,' Eric said, 'the first country in the world to do so. It grants legal rights to rivers, forests, mountains, animals, oceans, air ... It's really quite profound. It means nature is no longer understood as *property*.'

Instead, he explained, the constitution now states that 'nature in all its life forms has the right to exist, persist, maintain and regenerate its vital cycles.'

'It's a radically different way of conceptualising nature from the mainstream, highly instrumental norm,' Eric continued. 'And it has real legal force. Both people and nations are acknowledged as having the legal authority to enforce those rights on behalf of ecosystems, with the ecosystem named as the defendant.'

It was not a new idea – recognising the rights of nature is completely consistent with the traditional views of many indigenous cultures and their practice of living in harmony with nature. But its legal implementation was new to the modern, Western world.

Unfortunately, almost everyone I had spoken to about it agreed it was a fabulously exciting idea, but in practice, an almost meaningless one. Eric included.

'How can anyone ever prosecute against the Rights of Nature when most of the violations are linked to corporations that bring vast wealth to the government, and when the government controls the judiciary? I seriously doubt it will ever be enforced.'

He paused. 'Though it must still have some positive effect on how we think about nature, the kind of stories we tell.'

From there we moved on to the reforestation work in the reserve, and the problem of shifting baseline syndrome, where each generation grows up thinking the amount of biodiversity it enjoys is normal, when in fact it's already dangerously depleted.

'The collective memory of mangroves on the Ecuadorian coastline is a good example,' Eric said. 'People think these denuded coasts have always been like this, without the mangroves as a fish nursery. With no protection from storms. But of course, this is relatively recent.'

The conversation that most fired me up, though, was about the need for new narratives of nature.

'That's on many levels,' Eric said, offering me a top-up of water from the office fridge. 'Take this reserve and how it's described. 'Dry forest' has negative connotations. It conjures an image of brush and undergrowth, something, well, dry and boring. In fact, the dry forest is one of the most vibrant ecosystems on earth. We need to find more compelling language.'

On another level, Eric argued, we needed take a critical look at the whole rhetoric around 'ecosystem services' and 'natural capital'.

'What language do you think we should we use instead?' I asked.

Eric said he thought that, aside from acknowledging nature rights, talking about nature as our life support system was better than as a set of ecosystem services; and that the metaphor of the web of life was also useful.

'It's obvious that if you damage one part of a web, you risk damaging the whole thing,' he said. 'Ecosystems are exactly like that. Those leafcutter ants you saw in the rainforest. If their numbers fell drastically – because of pollution from the oil industry, say, or because the trees whose leaves they use were logged – then that would affect the birds and anteaters that eat them and also, because of their role as decomposers, the health of the soil, and many other forms of life, above and below ground.'

Ultimately, he explained, the loss of leafcutter ants could affect the ability of the entire ecosystem to function. They modified their environment so much they were known as 'ecosystem engineers': losing that engineering would have ripple effects in multiple directions.

'But I'm not sure the web metaphor really helps reveal where humans fit in the story,' he admitted. 'I think the best metaphor is that of nature as a community. A community we are part of, rather than a web. Definitely rather than as a set of resources for us to itemise and exploit.'

'Community in the sense that Aldo Leopold meant, when he wrote those famous lines back in the '50s?' I asked.

'Exactly,' said Eric. '"We abuse land because we see it as a commodity belonging to us. When we see land as a community to which we belong, we may begin to use it with love and respect."'

I could have talked with Eric for a lot longer, relishing the chance to explore ideas with this brilliant, innovative, committed man; and to revisit some of the philosophical terrain that had once been my work. I wanted to ask him what he thought about the idea of 'colonial habitation' as a root source of our environmental and social ills, and about whether the idea of living as citizens in an ecological community could be a viable alternative to it. But it was getting late.

'What's the main message the world needs to hear in relation to biodiversity?' I asked, in conclusion.

'Protect what is still here.' Eric said, without hesitation. 'Despite all the damage we've done – and I'm not downplaying the severity of that – much nature remains intact or, at least, complex, vital and healthy. We must prioritise protecting what we still have. And then focus on restoring the rest, on re-establishing natural processes and allowing those processes to get on with the job.'

I left the next day, thinking hard. The work that Eric and his team were doing on the reserve was, without doubt, crucially important. Protecting and restoring biodiversity and habitats. Reconnecting people with its value. Telling different, more generous, more equitable stories about what nature is, and how we fit within it.

But as I cycled away from the reserve through the sprawling outskirts of Guayaquil and past plenty of both PetroEcuador and 'Nature is Life' signs, I found myself thinking again about all the anti-extractivism activists I'd met, risking their lives to protect wildlife, ecosystems and human communities from the devastation wrought by these industries. And about Ferdinand's analysis, and the way these extractivist industries and the people behind them typified it. It all pointed to the same conclusion. To protect biodiversity – to protect nature, and people – reserves, community engagement, reconnection and all the rest would always be vital, but would never be enough.

Profound transformations were needed: and they had to extend far beyond nature reserves. To out of control corporate power. To the kind of growth economics that demands ever more extraction of resources from the earth. To the model of development based on extractivism, a narrowly understood notion of wealth in solely economic terms and a vision of what it is to lead a good life in which possessions and money play too high a role. The globalisation of commodities, so that a demand for palm oil or beef in the West could drive deforestation in the South. The invisibility of much of this to citizens positioned as mere consumers. Nature understood in purely instrumental terms as a set of services and resources for humans – or some of them; humans understood as not really in nature but separate from it and superior to it. We needed different stories about all of this. And not just stories, actions.

We needed different ways of inhabiting the earth.

10. CATS, MICE AND THE MINISTRY OF SARDINES

With every drop of water you drink, every breath you take, you're connected to the sea.
SYLVIA EARLE, MARINE BIOLOGIST AND AUTHOR

PERU

Happy Planet Index score: 55.9
Rank: 15th out of 152

Eric's words came back to me a couple of days later, at a small nature reserve just off the road. It was late afternoon and the rangers had all gone home. A helpful young woman who was just leaving the office said I was welcome to spend the night and that there was a walk I could take myself on. The walk was solidly uphill, on a trail that got more and more Indiana Jones-like, with creepers over the path, overgrown steps and huge spider webs – occupied by huge spiders – at face height that were strangely hard to see until you had stridden straight into them. There was a lot of rustling in the undergrowth from invisible animals I never saw. In the canopy, numerous howler monkeys howled and, above it, numerous large birds circled. Butterflies scattered the foliage with colour. The trail, or what was left it, was still below the treeline at the top, but between the trees I could see the road a long way down and watery strips of marsh beyond it. The place felt wonderfully wild and, like the Cerro Blanco reserve, it made me think how important conservation areas close to roads and cities are. Holding ground. Providing habitat, even if fragmented. Making space for humans to reconnect. When I felt harassed by traffic on the road, I found myself thinking that this was what so many other animals had to deal with, pushed to the edge, disturbed by our machines, constantly under threat. Here was another small sanctuary. *Protect the nature we still have.*

I spent the night in my tent on a wooden veranda outside a room full of posters explaining the history of the reserve and old photographs, including one of a man on a horse with a monkey in his lap. The tent was free-standing, bought for this trip, erectable on tent-peg incompatible surfaces like soft sand or hard rock. Or verandas. Snapping the tent poles into place, it was immediately, robustly, tent shaped. I fed in my mat and sleeping bag, then climbed in. It felt good to be in there; my very own portable burrow I could erect and retreat into anywhere. It was wonderful to still be outside, able to hear the night critters – birds, frogs, cicadas – and feel the slight coolness in the dark.

For the next few days, I rode through flat, hot, beautiful, damaged lushness. In between the banana plantations and the evident poverty were watery meadows scattered with flowers, visible and bright among the waterlogged greenery. Tall egrets stood still and alert; a handful of giant kingfishers poised on posts and wires. Star-shaped yellow flowers often massed up to the roadside, and something about them, so exuberantly alive, yet so delicate, provoked a sudden image of the astonishing beauty of the world: forests, rivers and oceans full of diverse creatures, the archetypal blue-green jewel of life in black space. And we humans trashing it. The image faded, leaving an immense pang of grief and love.

Around the next curve, as if in confirmation, a dead anteater lay on the tarmac against the flowery backdrop, its long brown and black body stretched out on the hard shoulder, reaching for the woods. It had nearly made it. Then an owl, only just recognisable, smeared across the road, and a tiny black and white finch, newly hit. I stopped to photograph the dead finch, imagining a shot of the small body in my hand, standing for our impacts on all of life. But when I turned to walk back to where it lay, the onslaught of oncoming traffic deterred me.

As the miles fell away behind me, my mood got steadily, if intermittently, better, swinging between a sort of bleak depression and moments of elation to settle on something akin to my normal, generally positive state of being. By the time I was nearly at the border, bowling along through increasing birdsong, I began to feel as if I were coming out from under a cloud. I was a little unnerved by how much I was enjoying not being in the mountains, relishing the heat and the flatness; though my legs were strangely stiff, grappling with some unaccustomed stressor generated by cycling on the level. And while 'escaping' might have been too strong, I had a definite sense of relief at leaving something behind.

I'm sorry, Ecuador, that we haven't got on better. It's me not you. Well, apart from the road signs saying 'water is life, don't contaminate it' while you're opening up the Amazon for oil exploration.

I spent my last night in a cheap hotel in the border town, the only cyclist on the planet to have failed to love cycling in Ecuador.

The hotel had Wi-Fi. I checked in with the world, a surreal experience, as if semi-random fragments of a completely different life were reaching me through the ether, uncategorised. Chris, recovered from his accident, was visiting his grandchildren in Japan. Two Scottish friends, Lee and Ricky, were in North America, competing in the Tour Divide mountain bike race, the length of the Rockies. The USA had confirmed it was withdrawing from the Paris climate change agreement.

It felt good to be riding to the border the next day, though borders always made me vaguely apprehensive. *There will be some piece of paperwork missing, or something else wrong, and they won't let me in.* It was all fine, though very different from the previous border. Absolutely no one was selling anything. Everything was very official and controlled. Within an hour of arriving, we rode off, into Peru. I stopped to take a photo at the entry checkpoint, Woody looking small against the blue and white road toll-like hut, a green *Bienvenidos al Perú* sign suspended from the high glass and steel canopy above. Beyond, my first ever sight of a Peruvian landscape: dusty, scrubby, a hair-breath from flat.

I was excited. I had never been this far south, and I was entering a whole new country. A country that immediately felt huge, almost limitless. The road ran smooth between the shrubby lowlands. There were, if anything, even more birds. A rich chestnut-orange dove in increasing numbers. More white egrets. Numerous shrike-shaped, stripy, brown and cream creatures, long legs below neatly tucked-in wings.

The scrubland morphed into farmland and then the outskirts of the town of Tumbes. A woman driving a tuk-tuk stopped to tell me to take care. The man at the customs' exit had offered a similar warning. I was used to this by now – people telling me to beware the country/town/village/person ahead. Would the warnings prove justified this time? Everyone I'd met so far had instructed me to be wary while in Peru. One particularly eyebrow-raising proclamation had been that, in northern Peru, the *rateros* would kill me for my phone. I was also feeling uncharacteristically nervous about dogs. Many people had said they were unusually vicious in Peru. And I'd read on Instagram that Tara who, with Aidan, was some distance ahead of me, had been badly bitten.

The outskirts went on for miles. The traffic volume ramped up. Tuk-tuks ferried scrap metal and wood. There was a car packed with bulging sacks of red fruit, pressing against the back window and pushing out through the roof like giant raspberries. Small hordes of boy-racers screamed past in beat-up, American-style saloon cars, their wheels raised to a precarious height. Arriving at what I took to be Tumbes itself, I wove off the main road and into the town centre. It felt different and strange in a way that was hard to pin down. I pushed Woody up numerous one-way streets, always against oncoming traffic. Yellow and black tuk-tuks came from all angles. Tangled powerlines swooped low over the streets from the high, angular buildings. It was noisy, manic, hard to read. *Crikey, how do things work here?* I felt simultaneously alert, alive and a bit flummoxed.

At the top of a particularly crazy street there was, suddenly, temptingly, an expensive-looking hotel. Feeling a bit sheepish, I checked in. Throw money at a situation and everything changes. Did my freedom to do that occasionally make the journey in some sense less genuine, or just more viable? I paid using emergency dollars stashed in a pannier. It was not an emergency, and it was way over budget (though considerably less than a night in a cheap hotel in London). I couldn't quite shake the feeling I was cheating. But Woody was safely in my room, and I was soon romping through my to-do list, contacting people about the next visit; drafting a short article for a conservation organisation; summarising the projects I'd visited in Ecuador.

Then I drank my first ever pisco sour on the hotel roof terrace. It was not to be my last. Friends had said they didn't think pisco sours compared well with mojitos, but I was impressed. A sort of margherita taste, without the salt, but with a limey kick. The city stretched out from the roof around me, chunky, rectangular buildings, a hundred shades of terracotta reaching out to the flatlands beyond. Swallows dove through the rapidly darkening sky. It all looked so different, so new. *Soak it up while you can still see it like this. How fast this novelty will fade.*

Cindy Hurtado was tall and slim, with glasses and long, dark hair in a ponytail. In sand-coloured hill-walking trousers and shirt, she looked every bit a conservation biologist. I met her at the rangers' station in the village of Rica Playa in the late afternoon, about 25 miles inland from Tumbes. I'd cycled there after spending the morning grappling with the bizarrely complex set of procedures required to make a phone work in Peru and was relieved to be back on the road.

Cindy had been a girl scout as a child. This, she told me in excellent English, had fostered a love of the outdoors, and from there to the creatures that lived in it. She had studied biology in Peru and the United States and wanted to do a project involving mammals for her dissertation.

'In Peru, everyone goes to the Amazon to do this,' she said.

Cindy decided to look for somewhere more accessible, and less well known. It had only taken one visit to fall in love with the dry forest.

'On my first trip I saw animals it normally takes months to see,' she told me. 'Armadillo, anteater, tropical otter, crab-eating racoon ... Yet dry forests are little appreciated, compared to rainforests. This one is easy to get to. The park entrance is just twenty miles from the Pan-American highway. There's huge potential for developing ecotourism.'

The park was the 151,000-hectare Cerros de Amatape National Park. With help from the Rufford Foundation, the plan was to set up a visitor centre. With that in place, they could support community development, create a base for biodiversity researchers and provide information about the forest for the local community and visitors.

What most motivated Cindy and her colleagues in the organisation she had helped set up, Centro de Investigación Biodiversidad Sostenible (BioS), was research. Dry forests are understudied and there are huge gaps in our knowledge about the wildlife that inhabit them.

'We have some baseline biodiversity data here,' Cindy told me, 'but not much, and not for all groups. There are numerous birds found only in this area. There are likely to be small mammals that are endemic too. We just don't know.'

Every conservationist I had ever met had, at some point, mentioned the importance of establishing a 'biodiversity baseline data' – an assessment of the variety of species and the size of populations in a given area. As Cindy talked about the work that she and BioS were undertaking, I realised I'd never thought about the millions of hours of work that lay behind that casual and frequently deployed phrase.

Cindy's own research was currently focused on a particularly poorly understood group of animals: rodents. Finding something she could use to trap mice that wouldn't first be devoured by ants was only one of many challenges.

We slept inside the station, in our tents on the floor, to protect ourselves from the voracious mosquitos massing outside. My tent, having spent the entire journey bar the veranda night in a dry bag, was suddenly coming into its own.

The next day, we walked into the forest. The aim was to find a possible route for an interpreted tourist trail and places where Cindy could site rodent traps. We set off on an obvious path that took us through attractive, regenerating forest (it had been partially logged and grazed until ten years previously), past Inca archaeological remains, through the constant buzz of insects and up to views of the dry forest proper. We ate lunch by a small river, the vegetation hazy blue with a fabulously abundant flowering creeper, racoon prints – five long toes – on the muddy bank, the tourist-trail potential evident. And then the path disintegrated and we were forced to hack our way through ankle-grabbing, spider-full, increasingly dense undergrowth in the heat. I grinned as I thought of my constantly repeated desire to get off the road. *Be careful what you wish for.*

Finally emerging hot, scratched and bemused from the undergrowth several hours later, we detoured via a lovely, wide, easy-to-follow track to a small beach on the Río Tumbes, home to tropical freshwater otters among other creatures.

'Can we swim?' I asked. In that heat, even I thought the brown water looked cool and enticing.

'Well, you could,' said Cindy. 'But you'd need to keep an eye out for crocodiles. It's pretty polluted, too,' she added, in case that weren't deterrent enough. 'Mercury from gold-mining, among other things. Whenever it's reported, the government says the pollution is below permitted levels. I plan to start taking readings myself so I have figures I can trust.'

That evening, we sat with beers outside the station. Cindy would soon change from studying mice to cats. Big cats. Mountain lions, or puma, also known as cougar.

'I'm a little scared of them to be honest,' she said. 'They're pretty pissed-off when they've just been trapped. You have to think, well, they're just big tabbies.'

The cats would be collared and radio tracked. She hoped to figure out their regular routes, and where they needed protecting or enhancing. Places where the animals had no option but to travel through villages, for example. This would then inform the attempt to develop conservation corridors, which would benefit not just the cats, but many species.

Cindy had, of course, heard of the Y2Y conservation corridor, though not Karsten Heuer's wonderful book about walking the length of it.[1] But I hadn't heard of the Latin American equivalent to the Y2Y, the Jaguar Corridor Initiative, driven by an organisation called Panthera.

It was stunningly ambitious. The initiative's aim, Cindy said, was to link populations of jaguar all the way from Mexico to Argentina. A symbol of power and strength in pre-Colombian Central and South America, they are largely solitary cats that love to swim, famed for their unusually powerful bite. A jaguar can pierce the shells of armoured reptiles or the skulls of mammals. As apex predators and keystone species, they are ecologically important. But across their range, they are threatened by habitat loss and fragmentation, and by conflicts with ranchers and farmers. Panthera have tried to alleviate the latter by figuring out how to make enclosures predator-proof, and by explaining the link between overhunting the jaguars' prey species, and livestock predation.

'Country by country,' Panthera's website explains, 'Panthera's scientists begin by mapping the jaguar's presence and the corridors through which they live and move. A corridor might include a cattle ranch, a canal development, a citrus plantation, or someone's backyard.'

Then they use this information to manage land in ways that work for people, are ecologically sustainable and allow 'safe passage for jaguars and other wildlife'.

I said goodbye to Cindy the next morning. I had loved spending time with her, full of ideas for the future of conservation in the dry forest, fired up by her belief in the value of her research.

A small family of goats watched as I made to leave. Eric in Ecuador had also talked of the importance of conservation corridors – and the need for good cross-border relations to help establish them. I had promised to connect Cindy to him. As Woody and I pushed off back onto the road, I had a sudden image of myself as if from above, a tiny ant, crawling across these vast landscapes, leaving a miniscule trail and occasionally creating connections, joining people up.

It was a joy of a ride back towards the coast, with sweeping views of the forest and little traffic. The villages were full of colour, greens, creams and pinks, more muted shades than Colombia, but still vibrant. Yellow and maroon tuk-tuks were parked on streets of small, rectangular, single-windowed houses. A terracotta and cream church stood alongside a huge, slightly dishevelled palm tree.

Back on the Pan-American Highway it was overcast, though still blissfully hot. For several days, the road tracked the coastline, sometimes looking down to the Pacific from low cliffs, then dropping to run alongside. There was a wonderful, scruffy wildness to the long, nondescript beaches and shallow breaking surf. There were hundreds of orange crabs on the beach, scuttling off down into their perfectly round holes when they saw me. Above us, frigate birds hung in the air, able to soar for weeks on the wind currents.

In a hostel one evening, the owner told me that the stretch of sand out front was used by turtles. The previous year, when a turtle had laid her eggs on the beach metres from his building, he had erected a fence around them and slept there in a tent for 53 nights. When they hatched, he and a circle of local turtle supporters walked with them, shepherding the tiny creatures safely to the waves.

'There were 160 eggs,' he said, smiling, '112 hatchlings made it to the sea.'

I wondered how typical his attitude was and wanted to hear more about these turtle champions. But the conversation moved, with a mysterious logic, to a serious falling out between Peru and Chile on the matter of pisco sours. The pisco sour was, without question, a Peruvian drink, he explained. But it had, scurrilously, been claimed by Chile as a Chilean drink. This outrage had culminated in Peru not being represented at the world pisco championships* because Chile had decreed that Peru couldn't call their piscos, piscos. By this time, two other hostel residents had joined us, and a round of pisco sours was deemed in order, if for no other reason than to underline their Peruvian-ness.

The next day, I raced in the heat to a meeting at a school in the tiny town of El Ñuro. There was an unanticipated hill – not 4,000 feet but steep – and I was late. A man started filming me as I arrived, hot and dripping, on the street the school was on. The street was long and straight, with an avenue of high, angular streetlights, small desert-coloured houses, dusty sand on the roadside edge and the by-now familiar air of a town that that had run out of energy before its construction had quite finished. The school lay behind gigantic gates, and its yard was packed with kids. Many were holding up big sheets of paper, on which they had explained why nature reserves were needed in their area, and how important nature was. One read, '*Bienvenida Kate*' – 'Welcome Kate'. I looked out across the sea of grinning faces, immaculate white shirts and shiny dark hair, and felt an explosion of delight; one of those moments where you are overjoyed to be exactly where you are.

Surging with a sort of energised, focused happiness, I was about to launch into a half-prepared talk when the children were marshalled into a line and one young lad stepped forward to give a speech – translated by Christopher Giordano, a Peace Corps volunteer who had been brought in from a nearby research station especially. It was all very formal and orchestrated. There was a moment when it relaxed and the children crowded around with their signs, eager to chat. Then I was whisked away. Woody was loaded into the back of a truck, and we swept out of the yard and were gone.

* An event I made a mental note to find out about.

The visit had been arranged by the organisation Nature and Culture International (NCI). Their aim was to conserve large areas of biologically diverse landscapes across Latin America, and they had been working with the school for some time. The next visit had been arranged by them too – on the village pier.

The sea around the pier was turquoise, and alive with green sea turtles. Pelicans hustled with the turtles for a position closest to fishing boats where discarded fish scraps were being thrown overboard. The pier was busy with people: hauling the fish from the boats in plastic boxes, hosing and cleaning them in water tanks, sorting them for size. And the sky was alive with frigate birds, like huge black swallows with their forked tails and white heads, hovering overhead. The whole scene was one of vitality; of multiple species, including humans, jostling along together in a rare coexistence.

'This is an important area,' explained Cynthia Ramos from NCI, translated by the ever-helpful Christopher. 'Two ocean currents, the warm Ecuadorian and the cold Humboldt, meet here in northern Peru. So the sea is unusually rich with nutrients.'

Working in countries across central and south America, NCI was calling for the creation of a marine reserve in this part of the tropical Pacific. The rationale for this was as clear as the turquoise water: 'unusually rich in nutrients' was something of an understatement. The meeting of the currents created the most productive cold water upwelling system on the planet, and what that meant in terms of biodiversity was astonishing. About 70 per cent of Peru's marine biodiversity was to be found in this northern sea, home to sea turtles, humpback whales, the endangered Humboldt penguin and many endemic species. And the area supported one of the world's largest fisheries, accounting for nearly a fifth of the world's entire fish stocks.

'Two out of three of all fish eaten in Peru are caught in this area,' Cynthia added.

They were caught, at least by the local fishermen, sustainably. El Ñuro boasted an ancient fishing culture, with a way of fishing that had not changed much since before the Incas. The contemporary fishing fleet no longer used rafts, but the boats were small and many did not have engines, powered only by sail. They used lines or hooks, with no indiscriminate nets. They went out fishing in the morning and came back for lunch. Sustainable, almost carbon neutral and turtle friendly.

I looked at the flotilla of small, colourful boats and their turtle and pelican retinues, bracing myself for what I guessed was coming next.

Both the area and the local fishing tradition were seriously under threat, Cynthia told me. Most immediately, from industrial fishing boats and especially the huge bottom trawlers that dragged up immense quantities of fish, catching juvenile and adult fish of all species, severely damaging the seabed, destroying the habitat for juvenile fish. These gigantic vessels were not supposed to come within five miles of the shore, but they did, and regularly. As a result, fish stocks had fallen drastically. The local fishermen were bringing in smaller catches, and smaller fish. There were devastating impacts on many other forms of marine life too. And no one could do anything about it.

If the area become a reserve, which was NCI's goal, that would change. The locals would be allowed to continue fishing sustainably, but the factory ships would have to stay away.

'We were close to success last year,' Cynthia said. 'Then there was an election and a new government, and it was back to the drawing board.'

Cynthia was optimistic they would eventually succeed. But another big challenge, she told me, was that the oil companies fiercely resisted the idea of a reserve, fearing it would limit the areas where they could explore and drill. And of course, the industrial fishing companies resisted it too.

The local fisherman did not see the reserve as a threat at all. On the contrary.

'We need to have this area protected. It is about job security locally, and food security nationally,' Edilberto Ruiz, one of these fishermen, told me.

Nature conservation for its own sake, he said, was a hard sell in Peru. But the turtles brought in tourists, as did the whales who travelled this coastline with their young between July and November every year. Ecotourism could readily be developed here.

'But at present,' he added, 'the government bodies responsible for marine protection and fishing are only interested in the big players, and only a few species. We call them the Ministry of Sardines.'

According to Edilberto and Cynthia, the sardine and anchovy catches mostly went for chicken feed, dog food and aquaculture, rather than feeding people. And the big fish catches went to Russia and Ecuador, among other overseas destinations. Japanese and Chinese boats came in at night and took huge quantities of fish, too. According to the local fishermen, there was even less control over industrialised fishing than over mining or oil. A disturbing claim, given what I'd been learning about mining and oil.

A key hope, they explained, was that the international community could put pressure on the Peruvian government.

'The people of El Ñuro ask that people across the world tell the Peruvian government why creating this reserve is so important,' Edilburto said.

In this regard, my visit was a warm-up. The North American marine biologist and ocean advocate extraordinaire Sylvia Earle was to visit in August, in less than two months' time.

We piled back into the truck and headed back the way we'd come to the town of Organos, a little beyond where I'd stayed the previous night.

After lunch, I was dropped off at a hotel. They all helped with Woody, and there were lots of friendly good-byes – though I would meet Cynthia again further down the road. Lugging my panniers up to my room, the stairs were crunchy with crickets, mostly dead. On the increase for the past few days, now, suddenly, they were everywhere.

Marshalling the live crickets behind the bathroom door, I worked on a post about the El Ñuro fishing community. And then I went to a café that Christopher had recommended, for veggie tacos and beer.

By now, night had well and truly descended. If any lights were turned on, we were instantly besieged by crickets. They appeared to lack any spatial awareness and collided constantly with faces, hair, laptops. I sat in the dark with a small paraffin flame, thinking and making notes. This wasn't just about El Ñuro, of course: the oceans are the earth's largest habitat, fantastically important from a biodiversity perspective. Most of life exists there, and all life depends on it, including ours.

Yet only about 7 per cent of the ocean is protected and almost all of it is subject to a barrage of impacts. It was no longer a surprise to learn that constraining big business, in this case, extractivism in the form of industrial-scale fishing, would be as crucial in protecting biodiversity at sea as it is on land.

I looked up from my notes to see that May Lin, the café owner, was standing beside me.

'Would you like a final drink before I close?' she asked. 'You look like you could use one, to be honest.'

She brought two shots of something sharp and potent, and we drank them together in the darkness.

11. DARKEST, FLAT PERU

Over-exploitation and ever-expanding agriculture are driven by spiralling human consumption. Over the past 50 years our Ecological Footprint – one measure of our consumption of natural resources – has increased by about 190 per cent.

World Wildlife Fund

I checked out of the Cricket Hotel and cycled off into the hot morning. It was two hours before I passed the turn to El Ñuro, galling given how little time it had taken in the truck. Then a long climb up into a landscape like nothing I'd ever seen before. It was as if I were riding through an immense, abandoned quarry. Miles of chiselled flats and slopes were packed with short, sandy, angular hills, shaped like quarry deposits and partly overgrown with sand-coloured grass. Topping the hill, this strange vista – which I would never have experienced as vividly if I had been shut behind speeding glass and metal – stretched away for miles, scattered with dots of green scrub. I couldn't tell which bits were natural and which had actually been quarried: a landscape I couldn't read. Then there was a run of mule drills, lifting their heads endlessly up and down behind protective squares of high, barbed-wire-topped fencing, just off the roadside.

I stopped at a petrol station with a promising sign reading '*Empanadas*'. But the shop area stank, probably of dead crickets, whose dark, crisp, leggy bodies lay on the floor in their hundreds. When I went to use the toilet – holes in the ground in a line of doorless, outdoor concrete cubicles, the floors smeared with dubious substances and the entrances strung with thick spider webs – it was so disgusting even I, almost totally lacking in squeamishness, had to walk away.

After that, the road became very straight, and the breeze became a headwind. The straightness was as challenging as the headwind, the kind of road where a sign announcing a bend is a major event. I slogged on, the slog made more interesting by the novelty of the land around me. *Why is newness – of any kind – exciting?* I wondered whether I was in desert cat county, another of BioS' interests. I saw no cats, but somewhere on that stretch I was flagged down by a car with Ecuadorian plates and asked directions to the beach.

I am an obvious foreigner cycling in a desert and you are asking me directions to the beach?!

The towns had sonorous names like Talara and Sullana. In between, it became increasingly grim, with run-down hamlets, occasional lone shacks, barking dogs. On the edge of Talara, there was a rubbish dump with people picking over it and plastic blown up against the roadside-edge for miles.

Coming into the city of Piura a few days later, the road surface degenerated into a mess of crater-sized potholes, random upwellings of tarmac and patches of thick gravel. Frustrated traffic trying to dodge the obstacles cut across at all angles, and the air was thick with churned up dust and heat. Flood damage. Areas across South America had been badly affected by El Niño-related heavy rain, falling for months on ecosystems that were drought-dry, and often causing *huaycos* – mudslides – like the one in Putamayo. In north-west Peru, this had been exacerbated by further downpours associated with a 'coastal Niño' – a localised warming of the Pacific. Hundreds had died. More than

100,000 homes had been destroyed and millions of people put at risk of waterborne disease.

Conscious of my luck – the rains had stopped only a month earlier – and my privilege, I checked into a cheap hotel, which turned out to be close to an excellent café. I would spend a few days there, catching up on writing commitments and playing phone-call tag with a journalist called Margarita Vega, who I was keen to meet. But first, I had a date to go bird-ringing.

Up at horrible o'clock to be ready for 5.30am, I was collected by Sol, a research scientist working with the NGO Centro de Ornitología y Biodiversidad, or Corbidi. We drove through the empty streets picking up various volunteers, including Cynthia who I'd met at El Ñuro. Then we headed to the campus of a private university, which had a chunk of dry forest in its grounds, and whose stated values included celebrating our coexistence with other species. Two small Sechuran foxes were playing pouncing games on a lawn in front of the main building as we arrived. Peacocks were said to wander in and out of the lecture theatres.

I walked with Sol as she and the others opened the fine, black nets that had been strung, tightly furled, between trees or bamboo poles. Corbidi was based in Lima, she told me, but had stations across Peru, including in Piura. Their aim was to establish a database of information about the birds they caught.

'There are many questions the data can help us answer,' Sol explained. 'The territories and preferences of different ages of birds. How bird populations vary as the climate changes.'

Every bird they caught would be ringed with a small, metal leg band with a unique number. If the bird already had one, the number would enable researchers to compare what they found on that occasion with information already held about it – its weight, for example.

After all the nets had been dropped, Sol and her colleagues set up a table in a clearing between the trees and unpacked their equipment. There were tools for ringing, weighing and measuring the birds. Data sheets and pencils. Pegs. Then we walked back into the woods to begin checking the nets.

For a while, they remained mostly empty. Our first catch was a tiny Peruvian owl, huge yellow eyes glaring out from the mess of black netting. Sol carefully untangled the small creature, blowing gently into its coffee and cream feathers to reveal where the netting had snagged. When it was finally freed, she put it into a white cloth bag and hung it from her shirt with a peg. Soon, more birds were flying into the nets, and Sol walked back to the table like a bizarre, ornithological bag lady, small white sacks strung across her body. She pegged the sacks onto a line between two trees, where they hung, wriggling and chirping.

Then, one by one, the birds were lifted out from the bags. Measurements were taken and information noted. Finally, the bird would be ringed, if it wasn't already. Then, the person holding it would unfold their hands and the bird would shoot out, chattering furiously as it flew rapidly away.

I had a dizzying sense of how much information could be collected, how many questions asked – and how that stacked up against the small number of scientists and volunteers doing the work. As the morning wore on, ever more birds were caught, their bodies suspended at strange angles, feathers, claws, eyes and beaks poking through the mesh. Among the birds I recognised were tree creepers, a nightingale, another owl, a soft grey-coloured dove and many small, bright-green parakeets who fought furiously with whoever was trying to untangle them. The parakeets had a fierce bite – and who could blame them – which didn't help the process. Sol and her colleagues were professional and organised, doing their best to release the birds and collect the information they wanted as quickly and carefully as they could. But it often took a long time to unravel each bird from the fine, tangled squares of netting, and at least one was injured in the few hours I spent there. Sol placed it gently in a tree fork and, instead of scolding noisily as it flew off, this one crouched, unmoving and silent. For the birds still in the nets, trapped, panicking and visibly terrified, especially when someone was handling them, I couldn't help but feel an acute, distressed empathy.

Was it reasonable to cause that much stress and fear, and risk causing injury, to gain that information? Perhaps the answer depended on the kind of information and what it was for. If it contributed to the conservation of the species the bird belonged to, would the cost to the individual be easier to justify? Knowledge can be valued for its own sake, of course. But, surely, not at *any* cost. And even if the information was useful for conservation, gaining it in this way seemed to go against the principles of compassionate conservation that were increasingly advocated.

Just as we were about to start packing up to leave, a whole bunch of parakeets became badly entangled. It took ages to free them. We were there for several more hours than planned, and by the time the final green bird had flown furiously away, even Sol and the volunteers were subdued. The injured bird was still hunched motionless in the tree as we drove off.

Back in Piura, I retreated to 'my' café. I wanted to process the morning, and I needed to eat. Margarita rang. Margarita worked for a paper called *El Tiempo* and had a well-informed interest in Peruvian environmental and social issues. She was as keen to meet as I was, despite her alarm at the multiple risks my journey, in her mind, entailed. Arranging the meeting, though, was complicated. Margarita spoke in fast Spanish I couldn't follow, or incomprehensible Spinglish, and always rang when I was somewhere noisy. There was the usual confused exchange. We agreed to meet that evening. To aid communication, she would bring her English-speaking husband, which seemed like a very good idea.

It turned into the first of two evenings. Guido and Margarita, dark-blonde hair tied back, glasses, a yellow sweater and brown slacks, met me in a café. While I finished a pile of salad and green veg, she and Guido told me about Tambogrande: a town an hour north-west of Piura, the first community in Peru to have held a *consulta popular* like the one in Cajamarca, Colombia. There was deep concern about the environmental impact of the copper mine proposed by an international mining company, particularly the risk it would pose to lemon and mango farming, the main source of employment. Tambogrande had also voted 'No'. It was, Margarita told me, one of the more organised centres of copper-mining resistance in Peru. Frustratingly, I wouldn't have time to visit.

'There's a town on your route, though,' she said, 'after Huaraz. Cerro de Pasco. You must go there. There is a lead mine right in the middle of it.'

I spent the following evening with them too, which ended, after another lovely mix of laughter and serious exchange, with Guido driving us in his aged VW Beetle to the newspaper office to pick up a copy of *Semana – El Tiempo*'s Sunday supplement – complete with a story about *The Life Cycle*; and then down Avenida Grau, revealing an area of posh, smart Piura that I hadn't known existed.

Margarita and Guido dropped me at my hotel after hugs and mostly cheerful goodbyes. Margarita, while she had quit warning me about thieves, men, cars and disease, was still not convinced I would survive the journey ahead. It was invariably local people, not other foreign cyclists, who were worried for my safety.

Back in my room, an inspection of Guido's notes in my journal revealed, among various maps and the names of towns and rivers polluted by mining, Guido's favourite saying: '*¿Por qué tanto brinco si el piso está plano?*' 'Why so much jumping if the floor is flat?'

Closely followed by '*¡a su macho!*' which, according to the notes in capitals, could be translated as 'OH MAN!'

Stuffing a few things into panniers, in a tipsy token effort at packing before morning, I thought how my short time in Piura had been completely transformed by the humour, warmth and care – for me and for the world – of these two people who only three days before had been total strangers.

From Piura, I headed almost due south out of town and into the flood-prone Sechura Desert. Thin horses. Poor settlements. In places the road was barely ridable and in others the water had eaten away at it, leaving the hard shoulder more like a crevasse than anything you might want to cycle on.

I knew from Cindy that this was desert cat territory: they can be found across South America, from northern Ecuador to Southern Argentina. But little was known about these small cats in in Peru. BioS was trying to remedy this, using radio collars to find out how many there were and where they ranged. I wondered whether the cats were as distressed as the birds when they were trapped, and whether the aim of the research – to figure out which areas were key to the cats' survival, so they could create effective conservation corridors – made the trapping easier to justify.

Pedalling slowly through the heat, I realised that what troubled me most was any assumption it was *obviously* justified because it was science. At its worst, this was accompanied by the view that compassion for individual animals was unscientific, sentimental and somehow illegitimate. The rise of compassionate conservation was challenging this outlook. But perhaps, I thought, the regular pedal strokes for once aiding thought, a bigger question lay underneath all of it.

In the context of biodiversity loss, how much more research did we really need? Conservationists were constantly being told they needed further information to justify habitat preservation. Where information was needed to do this effectively – like the corridor research – that seemed legitimate. But where it was said to be needed to justify taking action in the first place, there was surely something in there about shifting the burden of proof. In the age of the sixth great extinction, habitat conservation should be prioritised pretty much above everything less. Shouldn't we start there, and require proof from anyone who thought otherwise?[1] Didn't we need, above all, not more knowledge, but more action?

The questions seemed both important and baffling. Between attempts to find a way through them, my mind flitted to immediate challenges. Chris would join me in Huaraz, and that reunion was now only two weeks ahead. It was 500 miles to Huaraz – definitely doable in two weeks, but they included 10,000 feet of ascent. A slow climber, much as I was now looking forward to heading back up into the mountains, I needed to keep the daily miles as high as possible while I was still on the flat.

Even more immediately, the stretch of desert I was riding into had absolutely nothing on it for well over any distance I could cycle in a day, or possibly two. By 'nothing', I meant no hostel, café, shop, petrol station or house. It was a case of either hitching a lift to the next town from wherever I had reached at sundown or biting the bullet and camping wild.

I had been worrying away at this in my mind when a touch of South American magic flickered onto the scene. Another cyclist appeared, with panniers, heading in the opposite direction. A woman! We stopped. Two solo *ciclistas*,

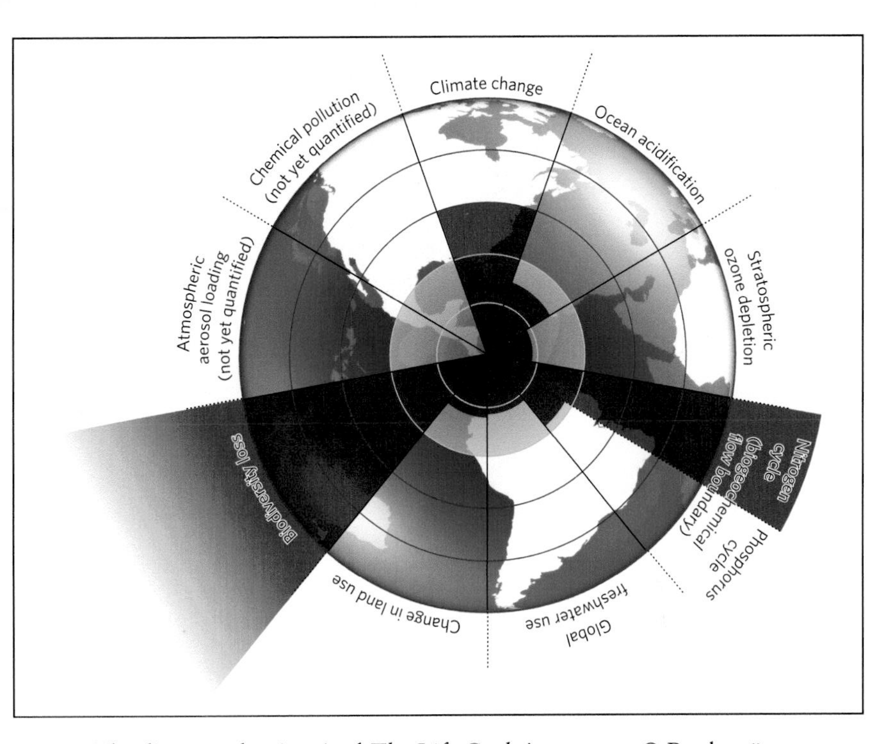

The diagram that inspired *The Life Cycle* journey … © Rockström et al., 'A Safe Operating Space for Humanity', *Nature* (2009)

Building Woody at the Bamboo Bicycle Club. © Lizzie Gilson

Woody's joints look like fibreglass but are actually made from strips of hemp soaked in a vegetable-based glue. © Lizzie Gilson

Above left: The cargo ship *Fort St Pierre*, my transport to Colombia.

Above right: Woody being lifted aboard. I was amazed the panniers didn't fall off.

Below: Early days. Leaving Casa del Ritmo hostel, northern Colombia.

Above: With Carlos Zorrilla, cloud forest champion extraordinaire, Ecuador.

Right: Punta Olimpica, the Peruvian Cordillera Blanca in the background. I loved this section! © Chris Loynes

A banana plantation in Ecuador. Each bunch was wrapped in single-use plastic and sprayed from the air.

Above: Zigzags!
I was relieved to be descending rather than climbing them. Cordillera Blanca, Peru.

Left: *Puya de Raymondi*, the biggest bromeliads in the world.
© Chris Loynes

Above: The lead mine in the middle of the city of Cerro de Pasco, Peru.

Below: Wonderful colours in just some of the incredible variety of sweetcorn and potato species at the Potato Park in Peru.

Above left: Atahualpa, the Inca warrior king murdered by the Spanish.

Above right: Woody at the end of the Eduardo Avaroa National Reserve section.

Below: Eduardo Avaroa National Reserve, Bolivia – fabulous but tough.

Above: Salt flats at sundown, Bolivia.

Below: I became expert at finding gravel pits to camp behind in the Atacama Desert, Chile.

Above: The desert in flower ... Chile.

Below: Patagonia. Welcome to Cochrane, 'The Final Frontier'.

delighted to meet each other. There was a fast exchange of hugs and information. Frederika from Sweden had just been where I was going – the desert.

'It's wonderful, camping in the desert,' she said. 'There's no one there. So, what is there to fear, from no one?' I smiled. 'You will love it,' Frederika continued, and then, as if she knew what would most convince me, 'I saw a desert fox. And the birdsong in the morning is incredible ...'

Afterwards, savouring the gift of unexpected reassurance, and its astonishing timing, I couldn't quite believe she was real.

As sundown approached, I'd been looking out for possible places to discretely pitch my tent for about an hour. A track appeared, leading away from the road. I waited for a gap in the traffic and then dived down it. Off to the left there was a patch of dead sweetcorn that someone had tried to grow in this arid place. Perfect! The corn was on flat ground and easy to pull away. I cleared a bit of space. Soon the tent was up, completely hidden from the road among the tall plants. There was a small pool of water nearby, and a short, shrubby tree full of white egrets that looked mildly ridiculous, perched on the undersized twigs. I sat on a pannier, grinning with pleasure at where I was, anxiety completely dissolved, Woody behind the tent and padlocked to a pole. I ate cheese and biscuits and listened to the birds and the crickets, whose hum increased as the sun sank away.

Heaving the rear panniers back onto the bike next morning, I found myself talking out loud.

'Only two more weeks and I can get rid of some of this stuff. Just hang in there until then, could you, Woody?'

Every time I picked up the panniers, I was assailed by the thought that the load was too heavy and that, at some point, the back end of the bike would simply collapse. How paradoxical. My daily life had become a celebration of the relative freedom from possessions that touring cycling offers, demonstrating over and again that you can travel the world with a few panniers' worth of stuff and not only have everything you need but feel positively happy about living with fewer things. In a world where over-consumption by the privileged is causing such environmental havoc, that living with less can be liberating seemed like one of the most valuable insights imaginable. What a gift to the world, from the bicycle. Yet, there I was, knowing all this, and still hauling too much weight.

And if I was on some level struggling to accept that I would be fine – better, even – if I lived with less in *these* circumstances, where I could literally feel the truth of it every time I lifted a pannier, how difficult would it be for all of us, the rich and relatively rich citizens of Western, industrialised countries, to move away from consumerism in our 'normal' lives? The largely unspoken assumption that our happiness and well-being is rooted in economic wealth and possessions is constantly being stoked.

I bungeed the tent onto the top of the panniers and wheeled Woody out from behind the sweetcorn hiding place, back onto the road. The road swung south-east, tracking parallel to the coastline, though some miles from it. I pedalled it for days. Sometimes the 'desert' was carpeted with a creeping plant with white bell-shaped flowers, like a convolvulus. Sometimes there were shrubby bushes and small trees, often thorny. At other times, the bushes receded and there were little pyramids of sand, a foot or so high, with greenery on top, looking like eyebrows. I often glimpsed the movement of creatures I could scarcely see against the ground, recalling Wade Davis' description in *El Río* of 'doves the colour of the earth'. Once, when stopped to eat a five-mile biscuit, I realised that hundreds of small birds were flying high above me, heading north-west. I could just hear their distant calls as they passed.

I found myself grinning with a sort of disbelief. I was on the road, cycling through a small desert in northern Peru! Peru was unlike anywhere I had ever been before. It was all so different, and strangely hard to grasp. Huge spaces and huge vistas, the kind of expansiveness that triggers a sense of release, as if your spirit has been unknowingly constrained by closer horizons. A long, straight road through a sandy, low landscape under a pale, hazy-blue sky, a solitary tuk-tuk coming the other way. Then, for miles and miles, the road empty of traffic, the usual noise of engines and constant horn-hooting simply gone. Here was something that could help free a person from the desires for things, I thought. Simply being out on the road in the big, wide, amazing world, and having that world constantly arrest your attention with its strangeness and beauty.

Approaching the desert's edge, there were more settlements. Half-finished houses, clusters of thin metal poles sticking up out of concrete into the sky like skinny aerials. Men making bricks. Then a group of road workers all dressed in fluorescent orange, with full face-masks and sunglasses over the top. I waved, and they all waved back. Trucks ground slowly by, some with their rear ends decorated with painted depictions of Jesus and lions and, once, a polar bear on its hind legs hugging a woman in a turquoise bikini on a beach with a palm tree.

A pale green truck overtook me, stacked high with cabbages and topped with boxes of vivid red tomatoes. Then a huge, articulated, double-truck packed full of chickens. These were a common sight, the birds alive, crammed together in boxes as if they were, well, dead cabbages, not living, sentient creatures. *I apologise for eating so many eggs. I only ever eat free-range eggs at home, I swear.*

The roadkill included large coffee-coloured, patterned snakes and a desert cat, the only one I saw, patches of ginger fur behind her ears, mouth open in a snarl, flies and small beetles already exploring the pale, flecked tan-and-grey fur, half her body missing. Later, I saw two Sechuran foxes, alive and scavenging opposite an electricity station. Then a donkey with a cart so covered in greenery you could see only its head protruding from the vegetation.

All day, I found myself mentally scoping for potential camp spots, seeing places where a tent could be hidden that I wouldn't have noticed before. Behind those pylon legs. In among that scrub. Being worried about camping already seemed ridiculous.

The road delivered me back to the ocean edge at Pacasmayo. I walked along a pier and up to a viewpoint, an enormous white statue of Jesus, arms outstretched, looking out to sea over small, colourful fishing boats hauled up on the shingle. The town looked vaguely Moroccan, with palm trees among the desert-coloured brick houses. There were stalls selling jewellery along the pier. I picked up a small thumb ring, silver with three leopards chasing around it, beautifully etched in a dark green stone.

'The leopard is a symbol of strength,' the stallholder said, as I wavered.

Another thing I wanted but didn't need – though I could certainly use some strength. The leopards came with me, joining the small wooden armadillo that lived in my handlebar box and Carlos the cotton-top keyring. I was gradually surrounding myself with animal guardians: under the circumstances, it seemed apt.

I passed unscathed through Paiján, a town with a dubious reputation – touring cyclists were regularly robbed there, I'd been warned – and at the next coastal stop, walked along the beach for a while. A few surfers played in the waves, all-grey seagulls swooping above them. I watched them in delight as they flew low and free over the water. What a contrast between these birds and the ones I'd watched struggling and fearful in the bird-ringing nets.

Crabs ran across the sand among little piles of creatures that looked like large woodlice, many upside down. They wriggled their legs if I touched them, but even the right way up they didn't move away or, indeed, do anything, as if afflicted by some weird but peaceful malady. I was full of happiness.

12. FRIENDS IN HIGH PLACES

Never doubt that a small group of thoughtful, committed citizens can change the world. Indeed, it is the only thing that ever has.

MARGARET MEAD

I had spent a couple of days in the town of Trujillo, close to the coast, at the first ever Casa de Ciclistas. I left inspired by what the Casas stood for: the values of hospitality regardless of income and a welcome for anyone who shared the love of the bicycle as a way of travelling.

Woody and I climbed solidly for nearly an hour. Then it was back to immense, sandy-grey coloured flatness, followed by miles of sugarcane. Trucks stuffed with these tall, spiky plants ground by, bristling like giant porcupines on wheels. In the background were gigantic sand dunes and diminutive hills. It was hard to tell which were which: sand dunes, small hills or small hills partially draped with sand dunes, like snow on a windward mountain face.

The next day, my last on that flat section of the Pan-American Highway, miles of fenced-off land sported rows of small, dark green-leaved fruit trees that had somehow been coaxed from the desert. White signs read '*Prohibido el Ingreso*', 'No Entry', the message illustrated in black with the silhouetted shape of a man aiming a gun. Fruit farms were not potential camping spots, then. About ten miles later, a road sign read 'Tanguche 21km'; 'Chavimochic 46km'. My turn-off. The dirt road it pointed to aimed straight through the grey dust to a distant line of angular hill/sand dune hybrids. Behind them, small mountains. And somewhere beyond the mountains, the Andes proper.

The landscape gradually shifted from sandy-grey to orange, the orange occasionally streaked with creamy whites so that the overall effect was like the pelt of an animal; a copper-coloured squirrel, maybe, with a pale-furred chest. The road – which Lucho, the Casa de Ciclistas' host, had described as bliss – had a wonderful, wild feel to it. There was scarcely a sign of humanity bar the ubiquitous power lines; it headed straight for the hills and was fabulously free from traffic. The surface, though, left me wondering when he'd last ridden it. I strongly suspected not since the floods. About as far from bliss imaginable, it was loose, draped with occasional sand dunes that could only be pushed through, and so rutted and corrugated that the collar bone I had broken 40 years previously began to ache again.

It was well beyond sundown by the time I arrived at the small town in which Lucho had said there was a petrol station with a room that cyclists could use. I'd seen nowhere I could camp. The single pump stood in darkness. I pushed Woody tentatively across the forecourt. An open doorway with a light behind it came into view. Inside, a large man and a small, timid cat. 'Hello,' I said, in my faltering Spanish, 'I've come from Trujillo and a friend there said it might be possible to have a room for the night here? Please?'

'Trujillo, eh,' the man said. And then nodded. He took me across to a shed. Inside the shed was a room with two single, metal-framed beds and nothing else. He put sheets and a blanket on one of them.

'Toilets?' I asked.

He walked me back behind two trucks to a room with a line of toilets and sinks. All functional, all clean. When I asked him the price, he asked if I had money to pay.

'In truth, yes,' I said. The man nodded and said, 'OK, ten soles,' – about £2. Somewhat battered from the long day on the rough road, I had rarely been so delighted to arrive somewhere, nor so grateful for a bed in a shed. I was under the sheets by nine, with stars and a view of the river through a gap in the wall, and slept long and hard despite the regularly arriving, horn-blowing buses.

I woke late, startled to find I'd been asleep for eleven hours. Before I left, I took some fruit juice across to the man's office, where he appeared to be about to work all day, despite having just worked all night. He cracked a huge smile at the meagre gift. I thanked him again, shook hands and cycled away. We'd exchanged no more than a few words, yet he'd exuded kindness and decency. I had been utterly certain I was safe there.

The next few days were mostly back on asphalt, a substance I was becoming embarrassingly fond of for an environmentalist. The morning mist was burned off by hot sun to reveal vivid blue skies above an astonishingly arid landscape. Not far from the small town I'd spent the night in, a line of flat-rooved brick houses stood out against the dry hills, their roadside walls painted lime or terracotta or cobalt blue, each with a single window and a single door. Then rough shacks, and a white church set back from the road. Then only hills and the river. It was a canyon road, heading east direct to the high Andes and running alongside the Río Santa, a fast, grey, glacier-fed river touched with dark green. Sandy, high-sided, loose rocky hillsides edged closer to the road and ever steeper, the dryness of the sharply angled slopes relieved only by the occasional shrub or an outbreak of tall cacti standing like meerkats on the look-out. There were small bridges as the road looped back and forth across the river, layers of subtle colours in the cliffsides – pale tan, ash, occasionally ochre – and one extraordinary chunk of rock that arched up and away from the road like a giant stone rainbow, etched in dun and grey stripes.

I climbed for days. The road was disorienting. It had started at sea level and Huaraz was at 10,000 feet. I kept expecting a monster climb, but the road often appeared to be flat, which it couldn't have been. I was gaining height and rarely making more than about four miles per hour, unreasonably slow on the flat, even for me. Then, as the canyon narrowed, the road started diving through a series of tunnels. These, especially the longer ones, were also disorienting. Something to do with the sudden change from bright sunshine to darkness and the loss of horizon, it was hard to maintain a straight course while cycling through them. Veering gently from side to side, keeping a sharp ear open for traffic, I lost count of the tunnels after the first 27.

Eventually, I came out of the canyon into a wider valley. As the road inched higher, thicker strips of vegetation diluted the overwhelming aridness of the slopes. The tan and grey mountains were patched with small fields and terraces: I started to see people hoeing fields or ploughing with stocky-looking horses. There were women with hats, a bit like the ones I'd seen in Ecuador, but higher and often white rather than black. There were pigs and piglets and brown, thick-fleeced sheep. Then, as the road began to swing south, my heart leapt as the astonishing white peaks of the Cordillera Blanca came into view, towering above the farmed slopes, immense, yet almost ghostly. *I can't believe I'm here.*

Huaraz comes from the Quecha word '*Waraq*', which means sunrise or dawn. An ancient city, founded before the Inca Empire, it was conquered by the Spanish in 1574. During the Peruvian wars of independence, the sprawling township came together to support the liberating army with food and guns, to the extent that Simón Bolívar bestowed on it the title 'Noble and Generous City'. Devastated by earthquake-precipitated floods in 1941, and again in 1970, Huaraz received so much assistance from outside Peru it was named a 'Capital of International Friendship'. Farming and tourism are the region's main source of income, and Huaraz' current manifestation is as a fabulous mixture of food markets; outdoor gear shops; mountain biking, climbing and trekking outfits; excellent cafés and Peruvian chaos, all against the stunning backdrop of the high, spiky, white mountains of the Cordillera Blanca.

For me, after months of cycling solo, reaching Huaraz was a relief. It had been a destination with a date attached to it I hadn't been sure I could make, putting pressure on my daily mileage for weeks. It also marked the beginning of some serious socialising.

Or it would have done if the toothache hadn't returned with a vengeance. The socialising began well, with a happy evening swapping stories with Tara and Aidan, whose last day in Huaraz coincided with my arrival. Over pisco sours, they told me that the scariest aspect of Tara's dog bite incident – a bite so deep into her calf muscle it had forced a two-week rest from cycling – had been the difficulty of finding a clinic for the necessary rabies vaccination. At one point, they'd ended up in a laboratory full of hysterically barking dogs, with a needle-brandishing vet poised to give Tara a shot there and then, claiming that the dog shot wasn't much different to the human one. It had been a huge relief when they'd finally found a human clinic.

I spent the rest of that night in the hostel, El Jacal, nursing a swollen face and eating aspirin. Soon, Lee – still recovering from the Tour Divide race, not helped by a serious bout of travel sickness on the bus journey from Lima – arrived with another friend, Ferga, the reception area of the hostel suddenly full of bicycles and friends hugging each other. Shortly afterwards, Chris rang from a lost taxi nearby, bike box and panniers bursting out of the car's boot – a surprise, as he was a day early, having somehow confused his time zones.

Of all my mates in the world, Lee and Ferga were up there with the people I most wanted to spend time with, as, of course, was Chris. This wonderful reunion was somewhat marred both by the toothache and a stomach upset that followed it. The dodgy stomach eventually afflicted all of us but was especially ironic in my case as I'd scarcely been ill at all so far and now seemed to have been brought down by Westernised fare in touristy Huaraz.

Before the stomach bugs had fully taken hold, Ferga, who spoke fluent Spanish, talked me into finding a dentist and acted as translator. The tooth was loose and the space around it was infected. No, the dentist didn't think that living on antibiotics for the next six months while cycling thousands of miles was a good idea. The tooth needed to come out. Yes, he could do it. How about there and then? Ferga literally held my hand as five leg-shakingly painful injections confirmed my worst childhood dentist terrors. On the plus side, I didn't feel the extraction at all.

It was some days until we were all more or less in fit shape to head into the hills.

I had nurtured the hope that after 3,000 mostly mountainous miles I would, for the first time ever in our long friendship, be able to at least almost keep pace with Lee and Ferga – much faster and better off-road cyclists than me. Post-infected face and stomach, there was no chance. We settled on a different plan. Chris and I, load lightened by leaving some of my kit in the hostel, would set off to cycle the multi-day Huascarán Circuit, as detailed in Neil and Harriet Pike's wonderful guidebook.[1] Lee and Ferga would take a longer route, and we'd aim to coincide in the town of Chacas, a little less than halfway round.

When Chris and I finally cycled out of Huaraz, I was delighted to be back on the road. Despite that I'd been cycling for months, cycling with friends and without a laptop or project visits to plan seemed like a holiday. It *was* a holiday. And this place and this cycling was something else.

Our route would take us around Huascarán, the highest mountain in Peru, at 22,205 feet, through Quecha-speaking villages and some of the highest farmland in the world. Pedalling north back the way I'd ridden in, we were flanked to the west by the Cordillera Negra, the black mountains, and to the east by the Cordillera Blanca, named for their permanent, white snow caps and glaciers. With each curve of the road, increasingly spectacular mountain vistas beckoned us on.

Soon we swung east, straight towards them. Women in tall hats, skirts and fabulously colourful shawls hauled huge piles of sugarcane strapped to their backs in oversized papooses. Pigs, sheep and cattle grazed alongside the road and the patchwork of farmed squares and rectangles ran far up the steep slopes. Beyond them, brown mountains. And beyond them, the tantalising white peaks of the Huascarán range, shimmering and alluring as a siren. Camped close to the river after a short day's ride, I relished the chance just to be there.

For Chris, the slow progress and multiple stops as we headed up the valley towards the first pass were not so much an opportunity to enjoy a calmer pace and time in a wilder environment as a necessary altitude-coping strategy. If you cycle slowly from sea level to Huaraz over the course of three or four days, you will end up gaining an almost textbook amount of altitude per day. Arriving there straight from Lima by bus takes you directly to 10,000 feet, often with less-than-ideal consequences. Having acclimatised to Huaraz, the first few days of climbing higher left Chris with a headache, an irritating cough and nausea when he stood up. Classic symptoms of mild altitude sickness. I called an early halt, stuck him in the tent and, while he slept fitfully, spent the afternoon and evening sat on a rock by the river, listening to birdsong and gorging my eyes on the partial view of the dark mountain flank that sloped up in the distance to the beginning of a glimmering glacier. As the sun faded, stars emerged through the tall, slender eucalyptus, slowly at first and then in their millions, sparkling chaotically behind the silhouetted mountain beyond. Occasionally, trucks on the road would briefly illuminate the treetops with their arching headlights. Then darkness again. When I eventually climbed into my warm sleeping bag, I fell asleep to the sound of the river and the image of stars glittering among the dark trees.

The next morning, icy cold until the sun reached the tent, we took things slowly, eating porridge and listening to the birds. Two lads in school uniform walked past and waved, heading upriver.

Back on the road, the climb to Punta Olimpica, one of the highest road tunnels in the world at 15,544 feet, arrived quite suddenly, a dramatic series of road loops showing as a pale-grey line, zigzagging straight up the dark slope face that reared ahead of us like a wall. The road was lined with wild lupins, vivid purple against the mountainsides, the shimmering white of the highest peaks heart-lurchingly beautiful beyond. It was a good ten-mile climb, and it took us hours. It was the kind of climb I loved, visually dramatic, hard without being cruel. I fell into a rhythm and powered – slowly – up the snaking loops, revelling in the return of some strength. *I can do this, and I love this. I can do this, and I love this.* At the top, surrounded by white peaks, I was exultant, flinging my arms wide towards the spiky summits, in that top-of-the-world feeling. Just ahead of us was the entrance to the tunnel. Chris was mostly just relieved to have made it without a total implosion of his head. The descent, flying down on tarmac through the mountains was glorious. We waved at a Japanese cyclist heading up, very late, on a crazily laden bike, and shot down past what was left of the glaciers, back into the patchwork of farms and tiny villages.

On the far side of Punta Olimpica is the small and beautiful town of Chacas. Chris and I had been sitting in the lovely square when Lee and Ferga rolled in. It was a heart-warming moment, seeing our friends ride round the corner. Hugs all around. Then ice creams and chat. I was hungry for their company and for conversations, about everything from disc brakes to Scottish politics, environmentalism in Latin America and the best ways to deal with Peruvian dogs. And I was hungry for banter; even when it involved Lee repeatedly pointing out how much more weight I was hauling than she and Ferga with their lightweight 'bike-packer' approach.

The next few days were fabulous for me, harder for Chris. We dropped down from Chacas to the river, then up through stands of eucalyptus and mixed farmland – maize, potatoes, grazing animals. Then onto a dirt road that switchbacked upwards through the brown and green patched slopes, the first glimpses of the big white ones reappearing, alluring as ever. There were tiny villages, properly remote, with women in traditional dress, sheep and pigs in harnesses on tethers by the roadside, small herds of cattle, donkeys and the occasional aggressive dog.

Among the steady climbs were short, steep sections that set my heartrate thumping. I am, in essence, lazy, and have perfected the art of cycling long hours while barely getting out of breath. The steepness combined with my lack of aptitude on rough surfaces rendered this skill useless. Yet I was relishing the sense of the world opening out in front of me. How amazing that a bicycle can take you off-road into places like this!

The final big climb on that route, Portachuelo de Llanganuco, is almost as high as Punta Olimpica, though the road is emphatically not tarmac. Chris decided to tackle it by bus: I left a day ahead of him, through the small town of Yanama, a lovely ride in sunshine eventually looking back down the valley to the town below. I passed women spinning yarn onto small spools while walking along the road. Occasionally, they called out a stream of Quecha as I rode by, the tone of which was hard to read.

At the next tiny hamlet, a group of Europeans gathered at a trail head were being briefed by a trek leader and behatted, local women were leading donkeys stacked with luggage. Beyond them, piles of multicoloured sweetcorn heads were drying in the sun, the traditional deep yellows set off by pale cream and grey heads; others a rich brown as if already toasted or such a deep, luscious red that the kernels looked like pomegranate seeds. By the roadside, lines of straw and mud-brick rectangles hardened in the heat.

As the day wore on, the villages and farms fell away below me. An occasional combi van went by in a miasma of dust, but otherwise, the track was empty bar me and Woody, winding slowly, slowly upwards. The Pikes had suggested a discrete camping spot which I somehow overshot. I reached the first set of switchbacks, the land sloping steeply up or down either side of the road. By the time I finally found a flat spot near a small lake, it was close to dusk. There was a tent already there. I set mine up on a dried-up cattle wallow below it, out of site of the resident tent and the road, but in sight of the mountains. Then I walked over to say hello. There was a shout.

'*¡Hola Kati!*'

It was my Bulgarian friends, whom I'd last met in Riobamba nearly two months ago. With the whole of Peru to camp in, we had somehow coincided in the crook of a switchback in the high Andes. Full of smiles and warmth, they had just come over the pass from the other side. The road surface was terrible, they said, but the scenery was stunning.

'It's busy, though,' they remarked, with wry grins.

It sounded like a different world over the pass, with tourist buses and a wedding at the lake with a blue-with-cold bride having her makeup done and someone who had hiked up carrying an inflatable unicorn.

Then we fell silent, arrested by the sight of the clouds streaming off Chopicalqui, a classic, pointed mountain top, one of the highest peaks in the Cordillera Blanca. The sun was still on the ridge that dropped sharply away behind it, turning it silver. It was heartbreakingly beautiful, and easy to comprehend why these mountains were viewed not just as home to spirits but spirits themselves: the powerful Inka Apus, both guardians and predators. Mario Vargas Llosa wrote about them in his dark and fabulous novel *Death in the Andes*; and the Apus are still honoured and feared across the mountainous regions of Peru, revered, supplicated, appeased by offerings.

The next morning, eating breakfast outside while wearing every piece of clothing I had, I watched the clouds build up around the summit. There was cloud in the valley behind me, too, where the sun should have been.

The Bulgarians came down with a mug of hot, sweet, ginger tea. When I took the mug back, they were packing to go, a cheerfully slow process. Their set-up made me look like a Lee lightweight, bikes loaded to the gunwales with an eccentric assortment of items, including spare tyres, a yellow metal Alaskan car numberplate, a Bulgarian flag and tubs of dulce de leche. They found the idea of doing huge mileages per day, or racing north to south, quite hilarious. Their recent meanderings on small roads and tracks in the mountains had, they said, taken them into hamlets where no one had ever met a tourist before. On a darker note, in northern Peru they had seen decapitated mountains, their tops taken off for mining.

When they'd finally reattached everything onto their bikes, I waved them off and, giving up on waiting for the sun, packed my still-wet tent. The rest of the climb to the top of the pass was brutal, slow and wonderful. I had been heartbroken to see the clouds re-gather. After days of clear skies, it seemed deeply unfair that I would cross this pass, with its legendary views, on the one day that was overcast. But the grey sky made the mountains look like a silver etching, fine detail of rocks and fissures all visible, silver on black and slate grey; then shading down the steep slopes into coppery brown with rivers of grey scree. I stopped for the twentieth time to catch my breath and take it all in. Then more zigzagging road thick with gravel.

Suddenly popping out through a narrow gap in the rocks that rammed up against the road, I was there. The top of the pass.

'WOW!' I yelled out loud, reaching for more sophisticated vocabulary and finding only wordless awe.

There was an outrageous mountain vista on the other side, even in cloud. Madly serrated ridges at crazy angles, knife-thin, stretched jaggedly up to the receding glaciers and the still-white peaks beyond. An absolutely crazy zigzag road fell down the far side, switching and turning to the valley below.

Completely bewitched by the view, it was a while before I noticed it was also windy and cold. Dropping down a few bends of the road to find some shelter, I ate lunch sitting on the back of a parked bulldozer. I had been there perhaps twenty minutes before I realised there was someone in it. The descent, as I'd been warned, was rough, small boulders strewn among the thick, rutted gravel. I was profoundly grateful not to be riding up it, which I think might have broken my heart. Woody and I slithered and skidded slowly downwards, the icy grey landscape gradually gaining colour, until we were back among the purple lupins and a yellow-flowered bush, frequented by hummingbirds.

Chris was waiting at the corner that marked the turn to the campsite. Walking down the track together, a sign welcomed us to Huascarán National Park.

'And welcome to your luxury accommodation for the night,' Chris said. He had bagged a grassy spot by the river, next to a stand of small, gnarled trees.

'These are the famous Polylepis trees,' he said.

The trees, with their fabulously twisted, tangled trunks and rich, foxy-orange, flaky bark, are fast-growing and tough; and they flourish at high altitudes, providing an important habitat for birds and other animals. The park, in theory, was a stronghold for them.

The trees were only one of Huascarán National Park's attractions. Most of the Cordillera Blanca lies within it – the world's highest tropical mountain range, with 27 peaks over 20,000 feet, and over 600 glaciers – and more than ten species of mammal wander its 340,000 hectares, including spectacled bears; pampas cats, mountain cats and puma; vicuña; white-tailed deer; viscachas;* long-tailed weasels; Andean foxes and hog-nosed skunks. There are over 700 species of plants and at least 120 species of bird, ranging from the giant hummingbird to the Andean condor.

UNESCO recognised the park as a Biosphere Reserve in 1977 and a World Heritage Site in 1985 for its rich human history, which includes pre-Colombian archaeological remains and cave paintings. But the protection that these designations should offer wasn't really working. There were conflicts between long-resident farming communities and the biodiversity aims of the park,† including over the Polylepis trees, a popular source of timber and firewood and badly impacted by grazing animals. On top of that, despite legislation banning mining in all of Peru's national parks, concessions had been granted and illegal mining was common. Hydropower projects involving dams posed another threat. And of course, park boundaries and designations were useless in the face of climate change that was driving the glaciers into retreat – with all the negative consequences for farming, human communities and wildlife that that entailed.

It all seemed to confirm the conclusions I'd reached on the journey so far. Nature reserves and other protected areas

* Rodents that look a bit like rabbits.

† Faced by national parks the world over, including in the UK.

were vital to protecting and restoring biodiversity. So were the suite of conservation approaches that included protection and regeneration of habitats, community engagement, education, research, tackling poverty, creating sustainable incomes and calling for good governance.

But so long as we humans routinely met our needs in ways that put such immense pressures on other species and natural systems, they would never be enough. We needed to tackle the things that made conservation and reserves necessary in the first place. Unsustainable forms of farming, mining, fishing and other industries; our gigantic, climate change-inducing use of fossil fuels; the economic system that demanded and legitimised it, and the worldviews and values that underpinned it all.

In my mind's eye I could almost see two strategies running side by side like roads or, yes, cycle lanes.

The national parks, protected areas and 'conventional' nature conservation strategies and projects of the kind I'd been visiting and was still learning about.

And then, well, transforming pretty much everything else.

The next morning, we woke to find that we'd camped on a spot used by combi drivers to deliver men working on something that involved long, heavy black pipes on the other side of the river. And squarely on the walking route to a famous lake. We skulked in the tent as hikers began to trickle by in dribs and drabs, then busloads. On the plus side, we'd evidently been checked out by cattle in the night, and brown and cream birds with a fast, amusing run like sandpipers on a beach were soon working the cow pats around the tent, coming right up to the doorway and occasionally peaking in. Numerous dark, silver-patched butterflies twinkled at the stream edge. Then a team of small, sharp, slim insectivore-beaked birds moved in, flipping the cow pats for the grubs underneath, and hunting the butterflies (who mostly got away).

Hauling our bikes to the road some time later, the still-rough descent dropped us into a different world. A lone, orange Polylepis tree stood by an astonishingly turquoise lake, brilliant in full sun. Then a traffic jam of tourist buses churning dust as thick as smoke.

We were back in Huaraz in time for Peruvian Independence Day. In the El Jacal hostel kitchen – they had saved us a room despite the pressure of bookings and our uncertain return date – we cooked a huge pasta and veggie sludge, with a black sweetcorn starter and quantities of red wine, plus Peruvian flags. Then we walked to the main square, where a band was playing music from the USA, sung in English, which seemed a slightly strange way to celebrate national independence, even if from the Spanish.

A few days later, Chris and I escorted Lee and Ferga to the bus station after a final meal together. I had spent most of the day in the well-known gringo hangout, California Café, a gem of a place with staff we were by now on first-name terms with, and wonderful food and coffee. Adopting a café as a temporary office had become a favoured part of my travelling routine, and California Café was my regular workplace in Huaraz.

On that day, the work had included an interview with a Dutch journalist from the *Huaraz Telegraph*. He told me that despite the fact that drinking water and agriculture in the region were utterly dependent on the glaciers, and that many climate change experts were predicting the glaciers would be gone in under 30 years, no one in the Peruvian government – local or national – seemed to care much about it.

'We are at the heart of the problem here,' he said. 'But the mayor of Huaraz is more interested in a new plaza and shopping malls than climate change action. Even though the loss of glaciers will have – is already having – devastating effects.'

This was not exactly a phenomenon unique to Huaraz, Peru or even South America. What is it about us humans that we can have robust information about environmental impacts we are causing, that will be catastrophic, and that we know how to stop – and yet choose not to face this and act, but engage with distractions instead? On a bad day, the prioritisation of shopping malls over retreating glaciers seems like a summary of the human condition.

Chris and I still had a couple of weeks. The original plan had been to do a second ride beginning and ending in Huaraz but, on a meltdown day surrounded by maps in California Café, the implications of my extra month in Colombia, and of spending six weeks with Chris basically riding in circles, finally dawned on me. I was running out of time. About halfway through the journey in terms of months, most of the mileage was still ahead of me. We changed the plan. We would head back into the mountains – but on a roughly south-easterly trajectory, instead of a loop. That way, Chris could still see some more of the Andes, while I could make progress in the direction of Patagonia. We would both get more time in some of the most astonishing mountains in the world, together.

Meanwhile, I was ducking a dilemma. While I was doing everything I could to keep down the carbon footprint of my journey, Chris was flying back and forth to spend time with me. My environmental impact or my relationship? It was unanswerable. The carbon cost of Chris' flights was gigantic. But as we fell into step towards the hostel to start packing, the conversations of the past weeks picking up again as we went, it was impossible to regret he was there.

13. THINK LIKE A MOUNTAIN

The cowman who cleans his range of wolves does not realize that he is taking over the wolf's job of trimming the herd to fit the range. He has not learned to think like a mountain. Hence we have dustbowls, and rivers washing the future into the sea.

ALDO LEOPOLD, *A SAND COUNTY ALMANAC*

'No way!' I called to Chris. 'It's for real!'

We stood with our bikes, looking at a huge sign at the head of the road we were about to turn onto. '*LA RUTA DEL CAMBIO CLIMÁTICO*' it read in dark green lettering. Lee and Ferga had ridden this way before joining us in Chacas. We'd assumed their reports of a road called 'The Climate Change Route' were a wind-up.

We'd left Huaraz after numerous rounds of goodbyes at California Café and El Jacal, having pretty much become part of the family in both places. A day or so later, as the road became gravel not long after the turn-off, my happiness from having enough stuff to have options vs unhappiness from weight–load ratio took a radical swing towards the latter. The difference between how the long-suffering Woody felt to ride on gravel now in comparison to when his load had been lightened on the Huaraz loop was prodigious. We sank into the gravel like a boulder into bog. Woody appeared as stoical and unperturbed as ever, but my knees complained bitterly, and I struggled to keep up any kind of pace. There was no evading it: I was carrying too much stuff for off-road cycling at any kind of altitude.

We made slow progress up through a wide valley between a huge vista of rounded, brown, dry, decimated-looking hills under a blue sky and a brilliant sun. A cool wind blew at our sides. Chris, now cycling more strongly than me, was not just sanguine about my lack of speed but delighted to move slowly through this new landscape. Occasional, small, semi-circular stone shelters and shieling-like summer dwellings dotted the otherwise featureless, brown and tan slopes. We'd seen only a few cattle, but these hills were clearly heavily grazed.

A considerable quantity of traffic was headed for Pastoruri, a tourist destination somewhere ahead of us. Waiting to let a combi go by at a narrow point in the road, we saw that yellow, dandelion-like flowers were pressed in among the short, dry grass. And then that hundreds of tiny moths were moving among the flowers, in turn hunted by a flotilla of brown and white birds – none of which we would have noticed if we hadn't stopped.

Around the next corner we were brought to a halt again. A single, enormous, dark finger pointed upwards from the tan-coloured hillside. Another corner, and the hillside was covered in them. Puya raimondii, the biggest bromeliads in the world. Growing up to ten metres high, the long flower-head emerges from a rosette of cactus-like, yellow and silver, barbarously hooked leaves. From a distance, the leafy base of the plant – itself taller than me* – looks round and shimmery, so that the effect is of dark pillars growing out of gigantic Christmas tree baubles. They are known as queen of the Andes in English, *titanka* in Quecha or *puya de Raymondi* in Spanish after the 19th-century Italian scientist Antonio Raimondi, who immigrated to Peru and made

* Not that that's setting the bar very high: I'm about 5 ft 1 in.

extensive botanical expeditions, announcing the plants as new to Western science (though undoubtedly not new to the locals) in 1874. We called them Raymonds.

A relative of the pineapple – I'd heard them described as 'pineapple meets triffid' – Raymonds can live up to 100 years before they flower, once only. Thousands of greenish-white flowers cover the entire stem, producing somewhere in the order of 6 million seeds. They die shortly afterwards.

Before we'd reached the Raymonds, the landscape seemed almost familiar, like an overgrazed sweep of rounded hills in the Dales or the Pennines. Now we had advanced far enough up the valley for the distant range of dark, craggy mountains to come into view and, beyond them, glacier remnants. Add in the dark and sparkling Raymonds across entire hillsides and the landscape had become, to our eyes, utterly extraordinary.

I was ready for a break by the time we passed a sign welcoming us back into the Huascarán National Park. Coot-like birds circled on a small lake behind it, and a bunch of little shore waders fast-trotted at the sandy edge among mountain caracara – a black and white scavenging member of the falcon family. At the ranger's station, we paid our tiny park entrance fee and walked over to a wooden building with yellow painted archways, the park's climate change information centre. Inside, we found information boards – and a glacier map. A white line like a thick contour snaked over the 3D model of the mountain topography, marking where the edge of the glaciers had been in 1970. There was an ominous gap between that line and the white paint that indicated where the glacier remnants were now.

Peru is set to become one of the countries most affected by climate change of any in the world, with droughts, floods and severe weather events all predicted to worsen, and the shrinking glaciers disastrous for farming, wildlife and fresh water supplies countrywide. Peru's Andean biodiversity is especially at risk. As the climate heats up, many species who can move, do, heading to cooler climes or uphill. If you already live high on a mountain, you've nowhere left to go. Mountain species all over the world have been called the 'canaries of climate change' for this reason.

The information boards set out all these threats. They showed how a selection of species in the park were threatened by the interconnected impacts of climate change and habitat degradation. Among the at-risk species: mountain lion and vicuña, the national animal of Peru. And the Raymonds, classified by the IUCN as endangered.

But there was nothing that suggested what any of us could do about it. And most of the visitors to the park didn't even see the boards. We were there for over an hour, the only people in the building.

That evening we camped a short distance from the road, hidden by a small rise. Below us was a plain, with a river running through it. The plain was scattered with stone enclosures, tiny huts with thatched rooves and ant-sized cattle. Locals walked up and waited on the road to be picked up by a combi. Various folk wandered by the next morning, wishing us good day.

That day the cycling was hard. There was a lot of loose gravel, and many steepish inclines. Where the loose gravel and inclines coincided, my heartrate would rocket into the red zone, my breathing so hard and desperate it connected with my bowels – a truly unpleasant sensation. On these slopes, my mind kept anticipating the need to stop before I was actually forced to. I did a lot of pushing. And a lot of singing, even if, lacking breath, much of this was internal. The higher we got the stronger my sense of being in the right place, despite the toil of shoving a heavily laden bike uphill. I would look at the towering walls of crazily jagged peaks and be moved beyond words by their beauty. And by the sense of being utterly small; a sense not oppressive but liberating.

Chris, who has a background in geology, would look at the mountains and mutter 'protalus rampart' seeing, not just beauty, but a creation story, crystals changing their basic essence so that ice can glide; glaciers nosing their way through loose rock across the dizzying expanse of geological time. When Chris talked of the forces that had created these peaks, a whole new dimension of immensity came into play, a timescale in which the effort and discomfort of a single, small, upright mammal did not even register.

And then the timescale would switch again, and I'd be brought back into the way we typically understand time's passage – on a scale calibrated to our own lifespan – and think what a strange and precious thing it is to have the option of putting yourself in the way of a high mountain pass, for what amounts to fun.

And the people living here, what might the mountains mean for them? Had this changed over time? People had lived up here for 2,000 years. It must always have been a tough place to live and to farm, even without the glaciers' retreat, though it had also offered sanctuary from the Spanish conquistadors. If I'd had the language skills to ask the people waiting on the road whether they found the mountains beautiful or harsh, a physical manifestation of geological time, an obstacle to be overcome, a sanctuary or a prompt for thoughts about their own transience, how might they have replied?

That night we camped just beyond the turn-off to Pastoruri, in the crook of the road that curved round towards the next pass. A combi went by as we set up the tent, a gigantic amount of luggage stacked on its roof, plus a live sheep. Small birds like wheatears called from the rocks and tiny lilac flowers were scattered among the tan-coloured grass. Below us, horses, hobbled to prevent them wandering too far, grazed the short pasture, clearly adept at moving despite having their forelegs bound together. An astonishingly jagged line of mountains ringed the far horizon, changing left to right from tobacco-brown, to ice-capped and shining. We were at about 16,000 feet. It felt high, and wild. We sat and watched the sunset, orange and turquoise in bands, with orange clouds. Small, dark clouds drifted slowly across the orange ones, then everything turned grey. Then grey and silver as an almost full moon rose slowly and silently into the darkened sky. *How amazing that a few bits of rip-stop nylon and some down make it possible to stop here, to sleep safe and warm in a place like this.*

There was a hard frost in the night, and ice inside the tent in the morning. The pans we'd left soaking outside were frozen solid, sporks still in them. Then the sun reached us, and the temperature was transformed. We sat outside the gently dripping tent, defrosting ourselves, eating and watching the world go by. The horses had moved further away. The combi drove back past, sheepless.

'We should go to Pastoruri since we're so close,' I suggested.

'Hrrmh,' Chris replied. 'You go. I'd rather mooch about here. And anyway, I don't think I can face it.'

He had a point. Pastoruri was a glacier that, over the last few decades, had drastically reduced in size. I nodded.

The Pastoruri visitor centre was a few kilometres away, uphill, but a joy with only one, almost empty, pannier. A couple of tourist coaches overtook me, which should have prepared me. It didn't. Rounding a gritty corner, I was astonished to see a huge car park already half-full of buses and a small village-worth of terracotta-rooved buildings, looking as out of place among the barren, grey landscape as if they had been dropped there from another planet. Lines of stalls sold Peruvian hats and beautifully patterned scarves alongside bland, made-in-China varieties designed for quick, cheap warmth. There was also practically a food mall. Beyond the buildings, a concrete walkway snaked up the dry slopes towards a wall of steep crags, loose rock streaming down from the rugged buttresses. Groups of tourists – mostly Peruvian I guessed from their accents – milled around the stalls, on the walkway and alongside the walkway, where guides led a long line of riders on small horses. Almost everyone was swollen and bulky with huge coats, hats, gloves and sunglasses. Many were wielding selfie sticks and not a few looked as if they found walking anywhere uncomfortable, let alone at 16,500 feet.

I locked Woody to a post and joined the relatively small stream of people heading upwards. At the high point, a lake the colour of dark-blue ink lay at the foot of the wall of buttresses. A slender stripe of ice ran along the very top of the rocky summits, as if someone had outlined them in white marker pen. At the far side of the lake, a high bank of glacial ice crouched alongside several strangely shaped chunks of former glacier, disconnected from each other and from the ice on the mountain tops. These chunks were so deeply scarred they had an almost greyish hue, the criss-cross lines across their surfaces telling something of the story of their long, intimate relationship with the rocky mountain faces they had descended and shaped. Yet they also looked out of place, fractured, diminished, like some sort of wild, dirty-white animal in a zoo, constantly being stared at.

By the time I was walking back down again, hundreds of people were walking up. There were long queues for the horses and the car park was packed with coaches. Was it a good thing that thousands of Peruvians were witnessing their shrinking glaciers? Signs explained how the glaciers had receded over the past 30 years, and that their predicted disappearance threatened water supplies. But while this was connected with climate change, nothing explained what caused climate change itself. One sign asked viewers to take care of 'the only home we have' but gave no clues as to how. And the information was conveyed in a strangely calm tone. You wouldn't take from the signs that Peru, like the rest of the world, was facing climate and ecological emergencies that threatened the future – and present – of life on earth; and you could easily ride a pony to the glacier and not even read them. The visitors to Pastoruri had been sold a spectacle when what they were looking at was a catastrophe that would severely affect them all.

Back at the tent, we decided to stay another night. We were in the most beautiful place imaginable, and quite likely would never return to it. Why move on if we didn't have to? My sense of urgency in relation to southerly progression was, on that day in that place, outweighed by a powerful desire to linger.

That evening we watched the sun set again. There were no stripes this time, just a dark, salmony-orange fading into deep, turquoise blue. A fox trotted over the horizon at a moment we both happened to be watching, her silhouette dark and clear against a sky still rich in colour.

In the night, Chris woke, gasping and crying out. Something about the nature of the night coldness and the automatic reaching for breath that can kick in when you sleep at altitude had triggered a reliving of the kayaking accident. We sat up in the icy tent and talked. The near miss had left a lot of darkness. I stayed awake a long time after Chris went back to sleep. It had been so very close.

The bikes were coated in frost when we woke next morning. We packed up slowly, waiting for the tent to thaw and dry, glad to have a little more time in this extraordinary spot. The combi van went by and gave a friendly hoot. Heaving the bikes back up to the road, the first pass was less than a mile around the corner. Then we dropped into a huge, wide valley, scattered with tiny huts and shielings. A teenage indigenous girl wearing Western clothes waited at the roadside. What was her life like? Would she live it out here or move away? I wondered whether she saw the jagged mountains that reared up around her valley as beautiful or a barrier to a different existence.

The final pass, a little over 16,000 feet, revealed the most astonishing views yet, with layer after layer of outrageously spiky, white peaks behind jagged, brown peaks, receding into the far distance. The mountains closer to us showed strange formations in the rock and contorted fold lines as clearly visible as if they were contours of a re-entrant drawn on a map. Chris, still cycling strongly, had faded just before the pass. I was exuberant, in love with this place, these vistas, this altitude, this mode of being. These mountains had been thrown up by a collision between tectonic plates: an ocean plate and a continental one. Now, looking at range upon range of them, at mountains we typically perceive as archetypically solid, permanent entities, I grappled with the mind-twistingly disorienting thought that the immense, slow-motion forces that created them were still underway; that these mountains were moving.

We spent a final night camped high and wild before dropping back down onto tarmac. It felt strange to be leaving the mountains for 'normal' roads with asphalt and trucks. The valley was suddenly filled with the scars, pipes and buildings of the Huanzala open-pit zinc and copper mine. And at one point, a set of dinosaur footprints: three toes on a foot half as long as my body, standing out like a plaster of Paris cast from a vertical slab of rock right by the road. The rock would have been horizontal sand when the creature walked this way roughly 250 million years previously. Chris said you could tell from the fossilised ripples that there had either been water running across a shallow riverbed or a tide going out when the prints were left.

We arrived in Tingo Chico just before dark a day or so later. Dervla Murphy, the writer and pioneering traveller, had also spent a night there on the Peruvian journey she made in the early 1980s with her daughter and a mule. She had not been overly impressed.

'[T]his sleazy newish settlement,' she wrote, never one to mince her words, 'hoped to become an important river port; now it has an air of irremediable failure and seems to be inhabited chiefly by drunks, prostitutes and thin dogs.'[1]

The next morning, we stocked up on food and, for the next few days, rode long hours on gravel, through tiny, high settlements. Photos of guinea pigs featured on café menus.

There were outbreaks of mild hostility – the first I'd experienced – with shouts of '*gringos!*' and occasional laughter. The road climbed and fell, climbed and fell through the brown hills, past steep, farmed terraces and adobe houses with tin rooves, sometimes patched with thatch or bits of plastic.

Finally, we dropped down for real, wrangling with an increasingly mad descent. The road reverted to washboard, with gravel frozen/thawed/frozen into steep speedbumps, cavernous potholes, loose gravel, loose thick gravel and banked-up gravel. Occasionally, I would hit a rock, or a pothole faster than I'd meant to and, almost bucked off as Woody lurched, skidded and jumped, began to feel I was riding a horse rather than a bike. Twenty-five miles later, we were back in eucalyptus country. Thirty-five miles later, we were still going down, the air warming fast and thick with oxygen and menthol.

We arrived on the outskirts of Huánuco just before dusk, bone-shaken by the rough descent, and were thrust straight into mayhem. The Huánuco Carnival appeared to have started two days early. The town was absolutely mobbed. The approach road was closed and the detour was packed with tuk-tuks, buses, pedestrians, motorbikes and roadside vendors selling shoes/hats/trinkets/food. Pushing the bikes, we finally reached the central plaza, which was in full swing with a stage, music and crowds. By this time, it was dark. We could just see the beautiful, old, custard-yellow, Grand Hotel Huánuco through the dancing mass of people. Finally reaching it, we heaved the bikes straight inside, to the consternation of a sternly disapproving man on the front desk. Eventually rescued by a friendlier colleague, the bikes were whisked off by two porters while two more hauled our grimy luggage upstairs, clearly amused by our equally grimy state.

'*¿Senora?*'

The door was opened onto a vast room with a balcony looking across the plaza, giving great views – and sounds – of the full-on party outside. It was the perfect place to land before Chris headed back to Lima, and home.

14. HEAVY METALS

Those who want change tend not to be in power. Those who hold power tend not to want to change.

TIM JACKSON, *POST GROWTH: LIFE AFTER CAPITALISM*

I'd travelled with Chris on the bus to Lima, and we'd said goodbye at the airport. Then I went to the immigration office in a less-than-successful attempt to extend another about-to-expire visa. *Oh, how I love confusing and apparently pointless bureaucracy conducted by humourless people in vast disorienting buildings in a language I still barely understand.* Extension-less and on the brink of being in the country illegally, the journey back to Huánuco and the Grand Hotel took all day, the road switch-backing steeply over layer upon layer of desiccated tan and brown mountains, tiny settlements clinging to them, for hundreds of miles. Even on the bus it was exhausting. It was probably just as well I didn't know that I'd soon be making the bus trip to Lima again.

Back on the road after dragging myself away from the Grand Hotel, I returned to the world of mildly dubious *hospedajes* and sketchy roadside camp spots. I was finding it harder than I'd expected to readjust to cycling alone. And my feet were hot. In addition to Chris, I had said goodbye to a hefty chunk of luggage, after a massive round of stuff-editing. The casualties had included my much-loved

SPD sandals. They were sorely missed on the long, sweltering climbs, though the temperature at night still plummeted.

For the next couple of days, the road climbed steadily upwards among the brown mountains. Small shacks clung to the slopes at bizarre angles. *Living from the land here must be a million miles from romantic*, I thought. It would surely be close to terrifying, in fact, to have your survival be completely dependent on the food you could grow. If things went wrong, there was no option of nipping to Tesco's instead. For how many people across Peru – across the world – was this true? How would their already precarious lives fare as climate change exacerbated all their challenges; as water from glaciers dried up, as ecosystems started to collapse?

I arrived in Cerro de Pasco in the middle of the afternoon after a slow start, waiting for my frost-coated tent to thaw. The camp spot had seemed at the sketchy end of the sketchy–ideal spectrum, but dusk was falling and the tent, while visible from the road in daylight, would vanish in the dark.

Once the morning sun had reached me, I crawled out and sat in my Therm-a-Rest chair – a happy survivor of the stuff cull – writing up the journal while the tent dried. I was on a small cliff edge, with a damp-looking valley below, a clear, brown stream wandering slowly through it. There was a flock of large, dark birds I would have taken to be vultures, except that they seemed to be grazing on a patch of pale grass. In the sky, occasional swallows dived past a pair of lapwing-shaped birds who flew upwards, then tumbled down in a complex courtship sky dance.

Beyond the valley were rounded, dark mountains, textured with dry grass and outbreaks of grey rock. I could see one roof and a few squares of cultivated land on the far slopes. Otherwise, there was, unusually, little sign of people. Behind me was the road, trucks and buses roaring by, the noise slicing through the blue sky. But what I faced was the river, the birds, the mountains. *Biodiversity is right here, on this roadside with the traffic.*

Cerro de Pasco, at 14,210 feet, is one of the highest cities in the world. At the top of the final climb, an immense, sharply angled, yellow metal archway welcomed me to ***CAPITAL MINERA***, 'the mining capital', in thick, black letters. I cycled through the archway, a portal to another world. No, the same world, just an aspect of it I wouldn't usually see. Beyond the archway, acres of grim, grey, rectangular houses. There was a steep descent complete with bicycle-killing open manholes, then another climb up through more grim buildings to the area marked 'city centre' on the map. The whole place felt bleak. And, despite the sun, it was cold. Very, very cold. I resolved to leave as soon as I possibly could.

An hour after I'd found a grim-looking *hospedaje*, Wilmer hooted outside as promised. Wilmer was a local activist who had offered to tell me about Cerro de Pasco's mine. He drove us first to a restaurant. The restaurant was cold; we could see our breath hovering above our food. Wilmer, dark hair, round, high cheek-boned face, big smile, was easy company and patient with my Spanish. Over a plate of avocado salad, a mountain of chips and a hot pineapple juice, he drew diagrams.

'The mine is in the middle of the town,' he said, echoing what Margarita had said, and pointing to a circle in his diagram, ringed with houses.

I assumed he was speaking figuratively but, even if I hadn't, nothing would have prepared me for what I saw next. The remaining chips in a plastic bag for later, we left the restaurant and drove to the mine. It was, literally, in the middle of the town. Standing on a raised piece of land, I stared, aghast and completely lost for words, at a vast, grey crater, perhaps a quarter of a mile deep. It had stepped sides and an access track that zigzagged its way to the depths below. On the far side, a white dot halfway down the track was actually a truck. The whole thing was ringed with razor wire. Beyond the wire, houses hunched right up to the edge of the crater, just as in Wilmer's diagram.

Cerro de Pasco had been a mining capital for hundreds of years, though the gaping hole in the city centre was relatively recent. In the 15th and 16th centuries, silver was said to have wept from the rocks, 'filling Spanish galleons'. In the early 1900s, Cerro de Pasco was Peru's second-largest city, and 'fancy carriages and European consuls graced its streets'. By 1903, the world's highest railway had been completed through the high Andes. With access greatly improved, the North American 'Cerro de Pasco Corporation' bought their way in, and J.P. Morgan and the Vanderbilt family, among others, made fortunes from copper and the remaining silver. Until the 1950s, miners 'dug out ore the old-fashioned way, through tunnels'. But then the company switched to open-pit mining and discovered rich seams of lead and zinc right under the town.[1] The mine was now operated by Volcan Compañia Minera, a Peruvian company with links to the commodities giant, Swiss-based Glencore.

'The worst impacts are on our children,' Wilmer said, as I looked out over the grim chasm in shock and disbelief.

Lead and other heavy metals had been found at such high levels in the blood of so many residents that there had been a proposal to translocate the entire city. But not all of the roughly 60,000 inhabitants wanted to leave. The US Centers for Disease Control and Prevention has identified ten or more micrograms of lead per decilitre of blood as cause for concern. In 2012, children with lower levels of lead had been found to suffer harmful, irreversible effects and as a result this figure was revised downwards. The current view is that there is *no* safe level of lead in a young person's blood. It affects brain development, and many of Cerro de Pasco's kids have suffered impaired learning and permanently lowered IQs. In the worst cases, it can cause anaemia, convulsions, organ dysfunction and death.

'We don't have the expertise in Peru to treat this,' Wilmer said, quietly. 'That's if it's even treatable at all.'

Wilmer came to collect me again the next day. He had invited me to join him on a trip involving, I thought, a drive to *un bosque* – a forest. There would be two artists, him being one of them, and school children. I had only the sketchiest grasp of what it was about, but it sounded intriguing, positive and fun – an antidote to the mine. Despite having arrived in Cerro de Pasco resolving to leave as fast as possible, I'd said yes.

We drove out of town past a small mountain of mining waste – 'the building next to it is the hospital' Wilmer said, in passing – and onto a huge plain of pale tan and sand-coloured grass. There were herds of cream-coloured llama – big and fluffy like oversized sheep – and smaller, brown and off-white alpacas. We arrived at an area covered in rocks. The Bosque de Piedras de Huayllay, it turns out, is a stone forest, considered to be one of the seven wonders of Peru. It covers several thousand acres, bristling with huge rock formations, many standing tall and in strange shapes, named after what they are most taken to represent. There were gigantic humans – The Walker and The Thinker – and a variety of animals – The Dog, The Turtle, The Snail, The Penguin.

The artists, Wilmer and a man called Jacob, set up their easels in front of a rock known as The Upside-Down Sombrero and started to paint. I was treated to a guided walk on a trail with a young, attentive guide who knew about the plants and animals as well as the rocks, drawing my attention to tiny blue flowers among patches of star-shaped moss, or to llamas grazing near a rock that resembled a gigantic, towering serpent. By the time we returned, Wilmer and Jacob's paintings were both well underway.

Various groups of children clustered around them or sat on rocks working on their own paintings. A group of mums sat to one side with a couple of teachers and a local mayor.

I wandered around the groups, looked at paintings and joined the mums, drinking hot tea from a flask. When the paintings were finished, the children's pictures were judged, photos were taken and prizes were handed out. Wilmer spoke powerfully about awakening the artist in every child and about how important artists were to society. This was clearly heartfelt – I knew that he used paintings to help explain and make vivid the impacts of the mine.

Later, over huge plates of Cuban eggs – eggs, rice and plantain – we talked about the day. The restaurant was still cold. Everywhere we went in Cerro de Pasco was cold, in a bone-reaching kind of way that was somehow intensified by the grimness of the mine. But I was buoyed up by the day, by being with people doing something creative and positive. And I was looking forward to escaping from Cerro de Pasco the next day.

'Would you help us make a short film about your journey and the biodiversity impacts of the mine?' Wilmer asked, when we drove to his office that evening.

At the office, a centre for mine-related activism, Wilmer added a whole new dimension to the issue of the mine.

'There is no clean drinking water anywhere in the city,' he said. 'Water has to be brought in by tankers. There is often not enough.'

And then he said something about pollution on a nearby lake, known for its wildlife.

'That's where we want to make the film,' he said. Of course I said yes.

The next day, Wilmer picked me up in a truck. I had agreed to the trip to the lake on condition that Woody came too, and that I fetched up at different accommodation at the end of the day. As we drove out of town again, past the hospital by the waste dump, past a heavily polluted former lake – whose name was 'the lake of gulls' – now dried-up and streaked vivid orange without a bird in sight, I sat in the truck among the cameras and tripods and felt something shift in me. This was not OK. How had we become so diminished as human beings that we accepted this monstrous mine and its impacts as a sad necessity? Not that everyone *did* accept it. But the people who did not accept it – and who wanted to change it – were not the people with power.

The mining companies had so much power they could operate in any way they pleased: punch a chasm in the heart of a residential area; maim and kill young people with the entirely predictable health consequences of lead poisoning; pollute and render undrinkable an entire city's water-source; poison whole watersheds. And the power of the mining companies extended into governments, who made decisions contrary to the welfare of their people and ecosystems because of it. As my dad had put it in an email about the mine, it was 'unfettered capitalism' in which awful things were 'justified' because they bought in big money, and sick kids and trashed ecosystems were treated as acceptable externalities.

How could anyone hope to find ways to protect the diversity of life on our planet – the diversity that this journey was teaching me is even more wonderful and valuable than I had ever imagined – without grappling with this monster? It was the same conclusion I'd reached after visiting the oil-exploited rainforest and the unprotected, 'protected' national park.

The rest of the day was bizarre, a surreal manifestation of the good, the bad and the ugly. On one side of Lago Punrún were stunning, vivid blues and turquoises, covered in birds, the spectacular dark shapes of the high Andes in the background. I was filmed walking slowly along the beautiful shoreline, pushing Woody and looking wistfully out across the water. On the other, a grim cluster of mining-related buildings hunched up against the grey and orange sludge that was the lake. Just offshore, the hard rectangles of a fish farm floated on the grimy surface, possibly the most insane location for such a farm in the world.

'The water here is dangerously polluted,' Wilmer told me, not that he really needed to point this out.

And of course, the height of the lake meant that any contamination would feed straight into the watercourse, spreading out and poisoning the water and the soil as it filtered down into rivers – including the Amazon-feeding Huallaga and Mantaro – streams, villages and ecosystems below.

'It means that the *campesinos* can't use the land for farming,' Wilmer said. 'And animals are poisoned, both wild and domestic.'

I cycled down a steep, switchback track to the lake edge, a drone buzzing above me. And then sat on a rock surrounded by the bleakness of mining fallout, with Wilmer asking carefully worded questions about my journey and its biodiversity focus in Spanish and me answering in English. Where did your journey start? Where have you been before here? Why is biodiversity important? What did you think when you saw the mine and this lake? What would you say to the president of Peru if you met him?

'I would say that I am shocked,' I said. 'I think of Peru as a country with strong values. Of course, Peru needs to generate an income. But the costs of this mine – to human health, to biodiversity, to water – are just too high. These things must be valued above profit, not the other way around. The potential to bring in income via ecotourism instead is enormous. And there would be so many other positive benefits too. Mining may even be part of the future, but only if it can operate without causing these kind of impacts on people and nature.'

And so on. The extent to which my answers actually related to Wilmer's questions was not clear.[2] But sitting on that rock, talking slowly about the lead-mine and its impacts, I felt full of a sort of calm fury that seemed to lend me a temporary clarity.

The mine was a catastrophe for people and nature, the most vivid, visual example of unconstrained, brutal, ruthless capitalism I had ever seen or imagined – almost like a caricature. Except that it was for real.

The next day, I left Cerro de Pasco in convoy. Wilmer, his colleague Elli and a lad called Roy cycled with me behind a car with two other lads filming us out of the back of it. A police car brought up the rear. Occasionally, the filming car would shoot ahead and someone would jump out and release a camera drone. People waved at our odd procession as we cycled and drove slowly up and out of the town.

As we crested the last steep rise, Wilmer shouted, 'Don't stop at the summit!' and we swept down a fabulous, long descent on the other side, the drone chasing after us. We stopped for goodbyes. Elli was clearly moved, telling me she had suddenly understood 'the freedom/bike thing'. Then the drone buzzed back up into the blue sky for a final shot of me riding off alone, the others watching my receding back.

The next day, hungry for beauty and a positive story, I left my panniers at the friendly, though cold, Neby's Hostel in Junín and cycled north on a minor road back towards Lago Junín. Lago Junín is an important birdwatching destination and the hostel, apparently in honour of this, had rows of stuffed birds in glass cabinets along the corridor, alongside a row of hats. It also had a kindly owner who knocked on my door in the late evening and handed me two hot-water bottles. Junín lies at an altitude of 13,500 feet, but this didn't seem to fully explain the temperature. It was as if the deep coldness of Cerro de Pasco emanated out across the area around it. Utterly unresponsive to layers or blankets, it was like no other cold I'd ever encountered.

Within minutes of leaving the town, heading for the lake felt like a good decision. I pedalled past flocks of llama and a turquoise church, slowly warming up. There were birds by the side of the road and on the tan-coloured slopes, including a small, dark bird with a bright orange patch on its back and many curlew-sized, black birds with brown heads. Then, as the fields became increasingly marshy, flamingos. The flamingos, both beautiful and a little ungainly, were almost pure white with pink flushes that showed only when they lifted their wings.

Lago Junín, or Lago de Chinchaycocha, is enormous – roughly 200 square miles. It lies on the upper reaches of the Río Mantaro, one of the three so-called 'true' sources of the Amazon. Among the wildlife are two species of birds found only on Lago Junín; the Junín flightless grebe – depicted against a wavey lake on the Reserva Nacional de Junín signs – and the Junín rail. Reaching the point where the lake proper came close to the road, I sat on a wall and looked out. Flocks of small brown birds hunted among tall reeds on the shoreline. In the distance, flamingos were making a sort of grunting noise as they fished and a large, dark-chocolate bird with a white tail dived at spectacular speed into the water.

Prior to the interview with Wilmer, I had thought hard about how to answer the question, 'Why should we conserve biodiversity?' and had ended up taking the ecosystem services approach, talking about how biodiversity constitutes flourishing ecosystems – which give us clean air, fresh water, fertile soil and other vital things. Now, looking out across the still wild mini-lagoons and the huge lake, listening to different bird calls having cycled past flamingos, the inadequacy of the 'ecosystem services' answer seemed as insultingly obvious as it had in the Amazon. The idea that nature is our life support system was a critical part of the story, of course. But somehow, just by existing, the flamingos alone demonstrated that it could not possibly be all of it.

That evening, buoyed by the day, the images of birds and reeds and water still vivid, I retreated early to bed with the hot-water bottles. I wanted to find out more about the lake, and the animals and plants that lived there. Among other things, I found some research for a postgraduate thesis online. It talked eloquently of the many plant species found in the high Andean wetlands of the lake, including plants that spend their lives fully submerged and plants that float, rootless, on the lake surface. It talked about the richness of the birdlife, both endemic and migratory. It mentioned that the Junín grebe was classified as critically endangered, and hence particularly vulnerable to changes in habitat. And then, using impeccably referenced research data, it talked about how seriously the wildlife on and around the lake had declined in recent decades and how badly polluted the lake was – from lead and other heavy metals.

As I left Neby's Hostel the next day, the kind owner told me that a German family travelling by bicycle with a young daughter had also stayed there. Chris and I had met a German family travelling by bike not far from the dinosaur footprints. It had to be the same people. The image of a delicate network of tenuous yet somehow wonderful links between cyclists and other travellers from all over the world, on the move across time and space, stayed with me as I cycled out of Junín, past Amerindian women sitting on road corners, dressed in hats, shawls and huge skirts, selling small sheafs of oats and hay as they had been when Dervla Murphy walked through in 1979.

The road ran close to rail tracks, climbing and falling across the high wide plains, though I never saw a train. Cattle, horses, llama and a small herd of vicuñas grazed the huge brown hills. Towards the end of the day, it fell into an unexpectedly spectacular, switch-backed descent, dropping down into eucalyptus stands, through terraced hillsides and ending in fields of commercially grown flowers and the town of Tarma. Tarma, a large sprawl of houses reaching up the slopes around it, had a colourful centre and seemed to specialise in fabulous cake confections. Trayfuls of cupcakes topped with mounds of cream and chocolate-traced cherries sat beguilingly in shop windows. It was wonderful to wander about without feeling that coldness was literally sapping my life energy.

The climb back up through the eucalyptus belt and into the mountains took most of the next day. I passed a woman washing her hair in a bucket of water by the road's edge and stopped not long after for lunch of bread, soft cheese, avocado and tomato sitting just off the tarmac. As I ate, I was waved at by a man with four dogs rounding up thickly curled brown sheep, the same colour as the land.

I arrived at Huancayo some days later, cycling uphill through heavy traffic to a hotel in the centre. It was a Friday, and I wasn't sure how long my stay would be. Ivonne from Acción Ecológica had given me the contact details of a man called Marco Arana, an anti-mining campaigner. For some reason, I had understood that he was based in Cusco, and that he was an activist there. It had taken a considerable amount of confused email exchange before I figured out he was based in Lima. And that he was a congressman. Congressman Arana, via Margarita – his PA – professed interest in my journey and its aims. I professed interest in his anti-mining stance. It was agreed we should meet. Then there was a long silence. Fairly late on the Friday I arrived in Huancayo, an email arrived. It read: 'Congressman Arana looks forward to meeting you at 12.30pm on Monday ...'

Huancayo is a mountainous city, at an altitude of a little over 10,000 feet. Lima, of course, is on the Pacific Ocean, with a couple of hundred miles separating the two. It would be pushing it, to say the least, to try and cycle to Lima in two days – and then it would take me many more days to climb back up again. I resigned myself to another long bus ride.*

By the time I had sorted bus tickets, somewhere to leave Woody and a cheap place to stay the night in Lima, I'd also managed to find out a lot more about the congressman. When it came to commitment to environmental and social justice, Congressman Arana was the real deal. He'd grown up in Cajamarca, a region and a city in Peru where the last Inca emperor, Atahualpa, had been kidnapped and held ransom by the Spaniard, Francisco Pizarro. Atahualpa's followers had the paid the ransom – cavernous roomfuls of gold – but Pizarro had executed him anyway. In Cajamarca, Peru, gold and conflict seemed to be ever-recurring themes. And like the Cajamarca I'd visited in Colombia, Peruvian Cajamarca had been dominated by gold-mining, and the threat of further gold-mining, for decades. Arana, the same age as me, had grown up with the aftermath all around him, starkly visible in the decimated landscapes and polluted water courses, as well as the poverty of the people who lived there.

Arana had studied sociology and a bit of philosophy (I liked him already). His master's thesis was the first ever written in Peru on socio-environmental conflicts. And then he trained to become a priest.

During his training, he'd been invited to make a documentary about mining and its impacts. What he witnessed – adults and children working deep underground without helmets, shoes, decent food or the education that might have given them different options – mirrored the inequities he'd grown up with. From then on, his story was one of activism. Arana set up soup kitchens, and nutrition and basic health

---

* A bus-justifying 'biodiversion' to use the excellent term a friend had coined for me.

courses. He founded a college aimed at giving young people with limited resources access to good secondary education. He formed EcoVida, the first ever ecological organisation in Peru and, later, an organisation called Grufides, focused on human and ecological communities affected by extractivist industries. And he became increasingly engaged with challenging the expropriation of *campesino*-owned farmland for the gigantic open-pit Yanacocha mine, operated jointly by a US multinational, a Peruvian company and the International Finance Corporation of the World Bank.

By the time Stephanie Boyd made her award-winning documentary, *The Devil Operation*, Arana was a key figure standing up against the outrageously powerful mining companies, mediating between the farmers and the police and campaigning for justice when local activists were killed in protests. It was a now-familiar story, complete with a mercury spill that poisoned 900 people.[3]

The Devil Operation focused on the events of a particularly brutal era, when the proposal to extend the Yanacocha mine to Cerro Quilish, a mountain that is both spiritually important and critical to Cajamarca water supplies, caused public outcry. Activists, including Arana, were systematically spied on by FORZA, the Yanachocha 'security firm', and sent death threats – given added potency by the gunning down of one of the activists, shot on his farm in broad daylight in front of his neighbour.[4]

The more I read, the more of these stories I found – including that of Maxima Acuña, 'The Badass Peruvian Grandmother Who is Standing Up to Big Mining', who was beaten up so badly when she resisted being forcibly evicted from her own land that she was hospitalised.[5] By the time I reached Lima, my research had left me both fizzing with outrage and utterly amazed that Arana was still alive.

The ornate, white, colonial-style congress building, the word *Congreso* in huge letters under the red and white Peruvian flag, could hardly have provided a starker contrast with the mining-scarred mountains and poverty-ridden *pueblos* I'd seen from the bus. Complete with gigantic statue of a man on a rearing horse, everything about it spoke of power and the kind of elegance that proclaims wealth and status.

It was also the kind of place that is tricky to enter. I was stopped at the entrance, given a number, dispatched to a queue, asked to take out my laptop and tell them the brand (which I didn't know) and given a phone. On the phone, I talked with someone who turned out to be Margarita, who duly came down to find me, just as I was being reluctantly persuaded to exchange my passport for an identity tag. Finally, we were allowed into a blue-painted, open-air inner courtyard, Margarita exuding easy-going warmth and friendliness and dispensing huge hugs. I sat in an office anteroom, chatting with Carmen and Jorge, two young activists who spoke some English and would help translate. Then the door opened, and Congressman Arana joined us, leaning against a table while he chatted. It was all very relaxed and informal.

Focused, attentive, charismatic, Arana asked questions about my ride and the route I had taken. As soon as I could, I shifted the questions to him and his work. He spoke very little English: we muddled through with me communicating via a weird mix of terrible Spanish and English translated by either Carmen or Jorge; or Arana understanding my English but replying in Spanish which was often left untranslated and which I only half understood. Opening up a laptop, he showed me images of the impacts of open mines on Google Earth – vast areas utterly denuded of any vegetation or other life apart from humans in dumper trucks – and of the impacts of gold-mining on rivers in the Amazon. Then, perhaps even more shocking, a map of mining concessions that had already been granted in Peru, an entire country with scarcely a mine-free area, the promise of devastation in all corners of the land.

'These are huge, huge challenges,' he said, over lunch in a restaurant nearby – our party had been ushered in and swept upstairs to a private room. 'We know the solutions. But they are not popular.'

Arana argued that Peru urgently needed diversification, so that mining became less important economically and could be reduced. Textiles, including the fabulously colourful, traditional indigenous designs, could play a much more important role in the Peruvian economy, he argued. As could selective, high-value, sustainable fishing. And of course, tourism, which Arana maintained could be developed in ways that made more of the extraordinary heritage of Peru – its archaeology, cultures, history, wildlife and landscapes – as well as Peruvian cuisine that showcased the range of vegetables and fruits that were widely grown but rarely appeared in restaurants.† He talked of the need to increase taxes for the mining that remained, and to have much more control over how mines and mining companies operated.

'I'm not alone in this,' he said, with a wry smile. 'A whole eighteen out of 130 congressmen and women think that there is another, better model of development than the one based on extractive industries and neo-liberalism. But at least there are eighteen of us ...'

He talked of corruption at all levels, and how much of a problem that was; and of the need for protected areas actually to be protected. He said he believed that the solutions included a strong focus on giving property titles to the indigenous communities, and that the power of the

† I heartily agreed. In a country with 6,000 species of potatoes, in restaurants, I was typically only ever offered chips.

indigenous movements was one of the things that gave him hope.

'As well as young people. Like these two,' he gestured at Carmen and Jorge, 'and the growing climate change movement around the world.'

Then, with a grimace, he explained that Peru currently had no climate change laws at all. The predictions about the impact of glacier loss on water supplies were terrifying. But there was no water shortage at the moment, or at least, not in the main cities. And so, politicians were simply not interested.

'We are working on a legal framework for tackling climate change right now,' he said. 'Getting it implemented will be the difficult bit.'

Then he was gone, back to the office, leaving me with Carmen and Jorge to finish up. A press officer asked questions about my journey and took photos.

Then I chatted with Carmen. 'The press position him as a terrorist,' she said, 'because he stands up for the indigenous, and for small farmers against big development.'

I nodded. A few months ago, I would have assumed this was a fairly wild exaggeration. Now I knew it was probably an understatement.

The congressman came through to say goodbye just before I left. There were more photos and hugs all round. Walking back past the gleaming white congress building, I had a pang of regret. I had been invited to come back in a few days' time, with Woody, to meet the entire congress and tell them about my journey and its aims. The logistics made this nigh-on impossible (and anyway, what *would* I have worn?) and I had declined. At that moment, I wondered whether I should have somehow made this astonishing invitation work. The encounter with Congressman Arana had been so very powerful. In a general climate of widespread cynicism about politicians and their values, in conjunction with the urgent need for so many changes that cannot be made without them, it had been downright inspiring to meet such an ethically driven, brilliant, passionate, committed person in politics.

As I walked back towards the bus station, I was already thinking of all the things I wished I'd asked Arana and hadn't, and about how much I would like to meet him again. The unlikeliness of this ever happening made the meeting both more precious and more poignant.

15. GUANO, DOUGHNUTS AND CONDORS

You never change things by fighting the existing reality. To change something, build a new model that makes the existing model obsolete.

BUCKMINSTER FULLER, QUOTED BY KATE RAWORTH

The Doughnut model of economics: the space in which we can meet the needs of all within the means of the planet.

KATE RAWORTH, *DOUGHNUT ECONOMICS*

Leaving Huancayo after the detour to Lima, I found myself thinking about Chifa restaurants. I was now seeing them on almost every street. They served food described as Chinese/Peruvian fusion,* and the ancestors of the Chinese–Peruvians, or *tusán*, who ran them were the many thousands of Chinese and other Asian immigrants shipped to Latin America in the 19th century. As the slave trade ended, they had been brought in to work on the sugar plantations, ostensibly replacing slave labour with paid labour, but on such low wages and living in such poor conditions they were effectively slaves themselves. Chinese labour had also been

* Essentially, rice with bits in.

contracted to build the railway from Lima to La Oroya and Huancayo, which had transformed the fortunes of the Cerro de Pasco mine owners. And it had played a key role in the so-called 'Guano Era' of stability and affluence in mid-19th-century Peru, when the country repaid debts accumulated during independence with revenue from guano mining.

Guano, the accumulated excrement of seabirds and bats, is packed with nitrogen, phosphate and potassium. An excellent fertiliser, it can also be used to create explosives. One of the most important sources of it is the Guanay cormorant, found on islands and remote headlands on the Pacific coast of Peru and northern Chile. Indigenous peoples of the region have long known of guano's importance and the Inca valued the cormorants so highly that any interference with them was punishable by death. The birds flourished and the guano was used sustainably for centuries. Then came the Spanish. They mined the guano with ruthless ferocity, using black slaves and then, after slavery was outlawed in Peru, 'Asian coolies', all working in appalling conditions. With mind-twistingly short-sighted stupidity, the scale and manner of the mining caused severe habitat degradation and millions upon millions of seabirds – and their highly valued shit – were lost. The Guanay cormorant is now listed as near threatened by the IUCN. Sometimes it seemed to me that the whole story of how different cultures treat nature, and each other, could probably be summarised in the story of that bird.

Cerro de Pasco had left me stunned. I felt as if that grim chasm of a mine had seared an indelible mark and left me utterly convinced of the need for profound, systemic changes.

But what did that actually mean? With the mine falling ever further away behind me, I was on the hunt for ways forward: if not for solutions, exactly, for positive trajectories and guiding lights.

One of these was *Doughnut Economics*, which I went back to reading in the evenings. I wondered whether one way of summarising the challenge was in terms of values. Values in the sense of the things a Martian anthropologist would observe that we humans consider important or take as goals – that we value – were they to study us.[†] They would observe that values in this sense are not only found in individual human hearts and minds but are embedded in systems and organisations too; and that they are highly influential. They would note that the values currently dominant in human economic and political systems – or those in the industrialised, Western world – are profit and growth, material and financial gain. That the humans in these systems are, as the Yanomami shaman Davi Kopenawa describes us, 'The People of Merchandise'.[1]

One night I realised that in *Doughnut Economics* Kate Raworth was arguing, in effect, that we needed to change the operating values of our economic systems. That we needed a system, or systems, that values meeting the needs of all people and staying within our planetary boundaries – articulated in terms of biodiversity, climate change, freshwater use and so on – and that values these things above all else. Or as a group called the Wellbeing Economy Alliance[‡] argues, systems that have the primary goals of human and environmental well-being. Economics in service of people and nature, not people and nature in service of economics. It was currently all the wrong way around: turn it around and everything changes.

† As opposed to things we've already judged to be good, such as honesty or integrity.

‡ With the wonderful acronym WEALL. See: https://weall.org/about.

From Huancayo, it was a four-day ride to Ayacucho, along (and up) a remote and beautiful valley, on a road whose edges frequently disappeared down a precipitous drop to the Río Mantaro, en route for the Amazon.

I spent the whole of the first morning climbing; enjoying both the climb and the feeling of being strong after a few days off the bike. The hillsides were patched with rectangular plots and criss-crossed with old plough lines. There were brown fields covered in squares of red plastic, with people harvesting something onto them. Women hauled sacks of camomile over their shoulders.

Then I was over the top and into a crazy descent. The hills were suddenly less rounded, more chaotic and completely different colours, the rich caramel, camel, tans, oranges and browns switching abruptly to greens and greys. There were repeated stretches where the road ran right under unstable rocky cliffs. Road signs warned of dangerous landslides, and the road itself was pitted with alarmingly large craters and scars from previous rockfall. Sometimes, the rockfall was still on the road.

There was not much I could do about this danger. Nor about another one – a *Rough Guides* online post warned of buses being held up by bandits between Huancavelica, slightly to the south-east of my route, and Ayacucho. This was mostly happening, it said, at night. I would be especially vigilant about not riding at night and hope that bandits would not think that a person on a bike was worth bothering with. Chris' suggested remedy for most problems – taking a bus – clearly wouldn't help. Yet I couldn't quite get my mind to take either risk seriously, to believe they could happen to me.[§]

Later, having made it to the small town of Izcuchaca before the rain, I met another cyclist. Alvaro from Spain was

---

[§] They didn't, but I didn't know they wouldn't at the time.

heading north from Ushuaia, Argentina. He wasn't bothered about the bandits either, but then again, he seemed unfazeable in general, and told hilarious stories about scrapes he'd gotten into.

Alvaro was travelling on a low budget and that night was sleeping in the train station.

'More often it's the fire station,' he said. 'They usually have beds and Wi-Fi. Or the town hall. But there isn't a fire station here, and the town hall is locked, and no one seems to have a key.'

I was only the second solo female cyclist he had met. He told me the wind was so strong in Patagonia he had taken to cycling at night when it was slightly less ferocious. For him, travelling north, the Patagonian winds had been sidewinds or – *oh how good to hear this* – headwinds. He also told me that under no circumstances should I miss Bolivia, which I was contemplating to save time, and that, in any case, after the climb up from Cusco, Bolivia was entirely flat, and so crossing it would scarcely add any time at all.

For the next few days, I rode above or alongside the Mantaro, sometimes a pale, khaki colour, sometimes a greener jade; sometimes in a beautiful valley, at others a steep-sided canyon. There were cacti with gorgeous, canary-yellow flowers sprouting straight from the thick, oval, spiney leaves; and small villages, shot through with colour and poverty.

A huge hydro-dam was a visual affirmation of something Congressman Arana had said – that hydropower should be adapted to have less impact on the river. On one side of the gigantic concrete structure was a full, green river. On the other, a diminished trickle. A sign read: 'No unauthorised photographs, on pain of arrest,' annoying me sufficiently to take one.

Other signs issued orders in contradiction to the ecological impacts that the dam would already have had. 'Conserve the plants!' 'Conserve the animals!' And one admonished: 'The river is not a dustbin. Keep it clean,' with an angry-looking fish holding up a placard.

My ride through the canyon became ever more spectacular. The scrubby, scattered mountains on the far side of the river kept keep changing colour, with outbreaks of rosy-pink reverting back to the dun/camel/toffee/custard palette. Outrageous, upward-sweeping rock curves were festooned with further signs warning of dangerous geology.

Gradually, the road widened out and the canyon fell away. It was hot. Strange to think that at home, early September would mean the beginning of autumn. Here I felt completely disconnected from seasons and disoriented by southern hemisphere climate realities in general. I still struggled to get my head around the idea that heading south meant heading for colder climes, not hotter ones.

Late that afternoon there was a massive crack of thunder and immediate lightning that scattered the dogs as I came into Esmerelda. I stopped to ask some lads about places to stay and they waved me to shelter just as the rain became torrential. Standing with them in the doorway of a small café, they asked all sorts of questions and shook their heads at the thought of cycling alone for seven months.

Later, back in the café, I read Dervla Murphy while drinking herb tea strongly flavoured with washing-up liquid and navigating around the distinctively chicken foot-shaped item at the bottom of my 'yes, of course, vegetarian' soup. Dervla Murphy was about to arrive in Huanta, tomorrow's destination, delighted by the colours in the rocks and mountains. I loved the thought that her feet and hooves, and my wheels, had sometimes travelled the exact same roads. By the time I had finished the rice and very fried eggs that followed the soup, she had reached Ayacucho, failed to find any food for her mule and reluctantly moved on. She found the city beautiful and full of cathedrals.

Nearly 40 years later, that was still a pretty good description, though strictly speaking, the cathedrals were churches – 33 of them, one for every year of Jesus' life. I found a café with a first-floor terrace view across the central square and ate my way through piles of vegetables and fruit – I had been craving broccoli for days – in between pizza and coffee. I had not been troubled by bandits, but I was troubled by the route ahead – or at least, aware that it needed some thought. From the comfort of my café sanctuary, I studied the maps and plotted.

The challenge was two-fold. First, the route from Ayacucho to Cusco ran across some fantastically folded and complex mountains, the contours clustering together with eyebrow-raising density. On the bike, this would translate to multiple climbs up to around 13,000 feet followed by heartbreaking descents to 6,500 feet and up again. It looked to be one of the hardest sections of the ride so far. Second, I was about to have company. Lizzie was a former student of outdoor studies at the University of Cumbria – where I used to teach – who had made a short film about building Woody. Now she had won a grant to come to Peru to film me riding him. I knew that Lizzie was both fit and tough. But she had limited experience of cycle touring and none at any kind of altitude. She would arrive with no acclimatisation at all. Ayacucho is at 9,000 feet, so the first part of the acclimatisation plan was a no-brainer – we would hang out in the church- and café-rich town for a few days. The interesting bit would be once we were underway – 13,000 feet isn't dangerously high, but to go straight to that altitude could give a person a cracking headache. To be on the safe side, once accustomed to Ayachucho's altitude, Lizzie shouldn't really be gaining more than about 1,600 feet a day. Since the road would start climbing almost immediately, that could make for some very short rides. On the other hand, given that she'd be filming, we'd probably be doing pretty short days in terms of distance, anyway.

And that was the other part of the challenge. Since Chris' departure, I had fallen happily back into travelling solo, moving at my own pace, making decisions that affected only me, spending my evenings working and reading. I was not just good at being alone – I positively enjoyed it. I would need to figure out how to adjust, not only to having another person with me, but someone I didn't know very well. I had agreed to her joining me, and it would be brilliant to have some decent film footage. But I also knew I would find the constant company and the film-related stopping and starting difficult. And that none of this was Lizzie's problem (or fault), but mine.

Reluctantly giving up my terrace-view seat, I went for a wander around. The square was lined with arched, white and terracotta buildings among the churches, while streets of yellow buildings with ornate, black metal balconies led away from the square at all angles. Quechuan women sold tea on the pavements, pouring hot water from a flask over fresh herbs in a polystyrene cup. Others sold chocolate and sweets, and some were clearly sleeping on the streets, small piles of blankets stacked behind their wares.

As well as poverty, Ayacucho, for all its beauty, was a city with horror in its not-so-distant past. It was one of the worst affected areas during the time of the Sendero Luminoso – Shining Path – terrorist movement, which began in 1980, a year after Dervla Murphy had walked through with her daughter. Led by Abimael Guzmán, aka 'Chairman Gonzalo', a Marx and Mao-influenced professor of philosophy at the university in Ayacucho, some of Sendero Luminoso's most brutal attacks were in the Ayacucho region. The Peruvian novelist Mario Vargas Llosa wrote about this era in *Death in the Andes* – one of the darkest, strangest, most impactful books I'd ever read[¶] – brilliantly capturing the movement's transition from one motivated by the not unreasonable

¶ With an utterly unexpected ending.

conviction that the political system needed to be overthrown, to one that perpetuated brutal murders and cruel 'trials'. The peasants that Sendero Luminoso claimed to represent turned against them; the military were called in and inflicted atrocities of their own and by the time Guzman was captured and jailed for life, approximately 70,000 people had died – about three-quarters of whom were Quechas. The warring parties were, without exception, white or *mestizo*.

In the end, the ride to Cusco was more spectacular than it was difficult and the challenges, for me at least, were more psychological than physical. Lizzie, riding a heavy, borrowed bike loaded with filming kit as well as camping gear, would still manage to shoot ahead to take shots of me labouring uphill or, occasionally, flying down them. Nevertheless, between the relentless, repeated climbing, the need to gain altitude slowly and the inevitable pauses for filming, our daily distances were emphatically on the short side. The stress of this was relieved by the fact that I had predicted it and had made no other commitments for nearly three weeks. All we had to do was make it to Cusco. Yet I was soon finding travelling as part of a pair more difficult than being on my own.

We had ridden out of Ayacucho and into a landscape smelling of powerfully of eucalyptus, always stronger after rain. Women carried bundles of herbs or fodder strung across their backs in vividly coloured, Peruvian-patterned papooses. In one village, women with long black plaits under black top hats led pigs on tethers.

For a while, the land was convoluted, red soil and soft orange stone sculpted into pillars and buttresses. Deep canyons dropped down sheer to small rivers, with patches of farmland at crazy angles. Towns were few and far between, and we camped in odd corners, waking to birdsong. For me, the time spent finding a discrete spot and then spending the evening in and around my small tent was now part of the way I wanted to live on this journey. Outside all day, being inside at night had begun to feel less like a source of enhanced comfort and more like being cut off. In the tent, I could still hear the sounds of other life around me, or the wind, or even the traffic. I was learning to trust my intuitions and judgement about suitable spots and, tucked away under canvas, I always felt safe. For Lizzie, the somewhat bizarre location of some of these often just-off-the-road spots, left her feeling she was putting her life in the hands of a person with a distinctly blasé attitude to personal safety and risk.

In the tiny town of Ocros, we came inside again, spending a night in a cheap room among blue and pink buildings lined on the sides of a small square. As we left, cycling a slow lap around the square past the saffron-coloured church, a burst of laughter brought us to a halt, followed by a huge hug from a Quechuan woman amused by Lizzie's filming antics. Spectacular switchbacks took us slowly up to an immense, high plateau, the kind that makes you want to stay high for ever and never come down, like a kid up a tree. The plateau was grazed by llamas and flown over by the black and white caracaras, or carrion hawks, fierce enough to take on vultures despite their smaller size.

The filming, despite occasionally turning unintentionally slapstick, focused our attention on where we were, and the story I was trying to tell. Two encounters stood out. The first was with a woman called Genevieve, who came to say hello when we set up our tents on a layby near her house. She returned with her family and hot potatoes. As we sat outside, eating together in the fast-deepening dusk, she told us how much she wanted to travel. She wanted to go to India, she said, and other countries that were beautiful and interesting. Genevieve, who we had earlier seen working the land with a heavy-looking implement resembling a pickaxe, had never been to Cusco, little more than a hundred miles away. How likely was it that she would ever leave Peru?

Later, mulling over the conversation from the fabulous warmth of my down sleeping bag while thunder and lightning cracked around outside, I was almost overwhelmed by the sense of my own privilege.

The second was with condors. It had been one of those evenings, arriving in a small town, when everything was first joyful, then resistant. The joyful bit started when a man flagged us down on the outskirts and offered us a cold beer. Lizzie declined, but I accepted with gratitude. He was a little tipsy, as I soon was too, the beer hitting my dehydrated system with added efficacy, but the situation was utterly uncomplicated. He simply saw us arriving and thought we might like a drink. In my case he was more than right – it was up there with the most welcome cold beers ever.

The resistance came when we then tried to find somewhere to spend the night. The first place I asked about a room clearly had several, but the woman who came to the door didn't much like the look of me (visibly tired, hair and face wild with dried perspiration and dust and now smelling of beer as well as road sweat – who could blame her) and claimed they were full. The second (and only other) place had rooms, but the person who had the key to them worked late and wasn't yet back. We would have to wait. We sat on a bench with our loaded bikes propped up next to us, and we waited. It got dark and cold. Eventually, the keyholder returned, and things fell slowly into place. Eating and warming up again in a small café/restaurant, the woman who ran it told us about a friend of her son who took people into the mountains to see condors. Another friend had a car and could, if we liked, arrange the whole thing. And so it transpired that we took the day off cycling and were driven higher into the mountains up a track that wouldn't be considered compatible with your average, clapped-out, two-wheel drive, family saloon car anywhere outside Peru.[**]

[**] Or perhaps, thinking of the moto-taxis in Colombia, outside South America.

There are places in Peru, we were told, where condors are fed, luring them where they can be viewed from close-by. This was not one of those places. After leaving the car at a registration point and paying a small fee, we walked, for some considerable time.

The condors, when we eventually glimpsed them, conjured every superlative you could ever think of in relation to a large, airborne bird. We saw two of them, hanging in the vivid blue sky and then gliding at high speed across it with little more than a single flap of their gigantic, fingery-tipped wings. Even from a distance, and even after hours of hiking and waiting, the birds had a more-than-physical power – power to uplift, to inspire. It was impossible to see them without feeling better about the world. Despite all the clichés about condors and South America, I found watching them unexpectedly moving. It seemed that condors affected most people this way.

'These birds are part of the future for Peru,' said the deeply tanned, smiling-eyed ranger we met back at the small registration office.

Two towering, 20,000-feet, spiky, white peaks had emerged from the cloud as we walked back down. The place was beautiful beyond words.

'There are increasing numbers of condor-watching experiences on offer in Peru,' the ranger said.

Between the birds and the sheer beauty of the place, it was hard not be excited at what this meant. It was the kind of ecotourism with the potential to bring in significant income that Congressman Arana had talked about as critical to freeing Peru from its dependency on damaging extractivist industries. It was doughnut-compatible economics in action, wing-born.

The next morning, still feeling buoyant about the condors, we had breakfast at a roadside café, watching the small-town world go by. Women with pink/green/blue/red/orange-striped shawls wrapped around babies or goods for sale. Estate cars packed with entire families, school kids in the back with the tailgate open, their legs hanging down to the road. There were hugs when we left and shouted goodbyes and *bien viajes* as we cycled out of the town. It seemed to be an especially friendly part of Peru. Even the dogs.[††]

The next day, Cusco appeared over a hill, a bowl of brown buildings suddenly beyond and below, the immense white mountains just in sight behind them, and an advert for a guinea-pig farm on a large sign in the foreground. Cycling down into the increasingly busy outskirts, I felt a familiar mix of emotions. I was undeniably looking forward to a few days off the bike, some food that was neither eggs and rice nor eggs and chips, a generous quantity of pisco sours, access to a laundry and a good shower. But there was always a sadness, too, at leaving the mountains, leaving the road, coming fully back inside.

Not that we would be in Cusco for all that long. I had set up three visits for the coming days. None were located on my north–south trajectory: we would visit them via motorised transport.

This was especially useful in relation to the first visit. The Manú Learning Centre was a good day and half away to the north-east by bus and boat, across the stark, dry mountains of the Cordillera Vilcanota and down to the Río Madre de Dios. The centre, run by the Crees Foundation, sits a short distance along the river from the small town of Atlaya, on the edge of the Manú National Park and Biosphere Reserve in the remote Peruvian Amazon – the largest and one of the most biodiverse, life-rich biosphere reserves anywhere in the world. I couldn't wait to be back in the rainforest.

[††] I was intrigued by the thought that there were dog cultures as well as people ones. If the dogs were aggressive in one place, they were usually aggressive for several miles around. Ditto friendly dogs.

16. OLD FORESTS, NEW ROADS

The human race is challenged more than ever before to demonstrate our mastery, not over nature, but of ourselves.
RACHEL CARSON, *SILENT SPRING*

Butterflies surrounded us. They were on the foliage, on our heads, on the butterfly identification books and all over the white gauze trap that hung like a ghostly, squat-ended cylinder from a nearby tree.

'They can't resist the smell of putrefied fish,' Diego explained. 'So we use it as bait.'

I studied a butterfly that had landed on my hand, lifting it up until it was close enough for me to see the dust on its wings, a tiger-stripe pattern in blues leading out from the slim body to a bright band of eggshell and turquoise, edged with inky-dark wing tips. The creature's long, slender antennae quivered gently from between its composite eyes.

'May I?' asked Diego.

He picked the butterfly up between finger and thumb and took it off to be identified. Another landed on my finger almost immediately. It seemed extraordinary that such apparently delicate creatures could be trapped and handled without harm. Not to mention that they were attracted to rotten, stinking fish.

Manú National Park covers an area of about 170,000 hectares that is roughly the size of Fiji, Kuwait or, that frequently used indicator of comparative size, Wales. It was declared a national park in 1973, when Peru was headed by the military leader turned left-wing dictator, General Juan Velasco.

In 1977, two years after Velasco had been deposed, UNESCO declared Manú National Park a Biosphere Reserve and, in 1987, a World Heritage Site. The ecosystems within the reserve range from mountain grassland with peaks as high as 13,000 feet, to cloud forests on the lower slopes and rainforest in the Amazon basin. It is home to over 1,000 species of birds – roughly double the number found across the UK – untold numbers of insects, about 15,000 species of plants, fifteen primates, eight wildcats and over 200 mammals, many of whose names I had never heard before this journey. Tayra. Collared Peccary. Jaguarundi. Pacarana. As well as giant armadillos, puma, sloth, spectacled bears, various monkeys – including the pigmy marmoset, the smallest monkey in the world – wild guinea pigs and tapirs. Plus, according to the large varnished wooden information board we saw as we entered the reserve, 300 *hormigas* (ants); 27 species of *guacamayos* (macaws); and 650 *escarabajos* (beetles). And an astonishing 1,307 species of butterflies.

Butterflies are an indicator species. They are super sensitive to the conditions around them, which means that they can be used to monitor or even predict change in the ecosystem they are part of. It also means they can be used as a sort of proxy for biodiversity in general, though this is more controversial.

'If you have many different species of butterflies in a particular ecosystem – a rainforest, for example – it's a reasonable assumption there will be high species diversity in general,' Diego had explained at the briefing before our early-morning, wellie-clad walk into the forest to the butterfly trapping areas.

Most of the briefing concerned appropriate safety responses to animal encounters. The wellington boots were compulsory, a measure to reduce the likelihood of snakebites to the ankles.

'We use frogs and other amphibians as indicator species, too,' Bethan – who had helped organise our trip – had told me on the minibus journey.

I'd been gritty-eyed after a horribly early start but was unable to sleep in the bus, not least because I didn't want to miss a moment of the scenery. We crossed a stark, dry, pale khaki-coloured mountain range with one spectacular long ridgeline. Then we dropped down into cloud-forest lushness for an overnight stay in a small lodge, before descending to rainforest and the Río Madre de Dios. To my surprise, I had found myself yearning for the bike; to be on the road under the big blue sky and on my own again. Having said that, the suddenly enhanced company was wonderful. Bethan, who worked for the Crees Foundation, was currently on holiday and travelling with her dad, Chris, and brother, Bryn; and we would meet colleagues of hers at the centre. It promised to be a sociable time. At a brief stop at a viewpoint, standing in the hot, dry sun looking out over Atalaya and the Madre de Dios where we would catch the boat, I had a moment of sheer, piercing happiness.

Back in the minibus, I sat next to Bethan again. She had studied for a degree in social media communication before getting a job with the World Land Trust, an organisation that works with local partners to protect biodiversity, by creating reserves and providing permanent protection for habitats and wildlife in some of the most threatened ecosystems around the world. Captivated by the stories of conservationists in the field, she had travelled across Central and South America visiting conservation projects and collecting these stories for over a year. After that, she'd taken on the job of social media communications for the Crees Foundation, spending as much time as possible in Peru and, especially, in Manú.

'Crees does a range of different kinds of work,' Bethan explained, 'including scientific research. We have a mix of interns, volunteers and "normal" ecotourists visiting and helping with the research. It's "citizen science" in action. The research we're doing currently is one of the most important studies we've ever been involved with – looking at the recovery rates of biodiversity in regenerating rainforests.'

If you cut a rainforest down, how fast can it grow back? And how diverse will this new rainforest be compared to the old one? While this kind of study had often been carried out in areas that were still being impacted by people, the Crees research looked at the best-case scenario. Where the damaged forest was fully protected, the results were staggering. The forested land they were studying at the Manú Learning Centre had been farmland only 30 years previously. In the late 19th and early 20th centuries, swathes of the forest had been cut down and cacao, sugar cane and cotton planted instead. Fast forward 30 years since the farming had stopped and the land protected, and the researchers found that up to 87 per cent of the biodiversity in the forest had returned.[1] Admittedly, this was an average figure,[*] but it was still a hugely exciting result, showing that damaged and even completely cleared rainforests can recover, given a chance – and that *almost all* of the biodiversity in those forests can return. Given the amount of rainforest worldwide that is no longer primary – about two-thirds of the world's tropical forests

[*] Averaged in relation to land that had only been selectively logged, as well as land that had been either partially cleared or totally cleared. In the latter case, the rate at which biodiversity returned, and how comprehensively, would likely be lower.

have been subject to human impacts – let alone the amount of forest that is still being degraded or destroyed, it was a piece of research that gave tremendous cause for hope.

This particular forest had a long history of humans impacting on it less than favourably. In the 1800s, the rubber boom caused ecological and social havoc.

The rubber boom ended abruptly in the early 1920s. Then, in the 1950s, the government had decided to bring in settlers.

'The Incas had failed to colonise this area,' Bethan said, 'which tells you something about how tough it is here. But the government of the day thought it was a good idea to populate it by relocating people. It was mostly driven by the idea that settling people here would help prevent Brazil encroaching on Peruvian territory.'

The settlers, brought in from other parts of Peru, had been given a piece of forested land and left to their own devices. There was no provision for healthcare or education, and many didn't know how to farm. With few choices, the people of the area – known, somewhat confusingly, as colonisers – slowly but surely destroyed or degraded the rainforest with unsustainable slash-and-burn agriculture, with illegal logging and with gold-mining. How else could they feed their families?

'And that's another important part of the work we're doing,' Bethan said. 'This is an amazing part of the world. It's culturally rich as well as massively biodiverse.' The languages alone confirmed this – the first language of a little under 50 per cent of the people who lived in the area were indigenous ones, including Quechan, Aymaran and Arawakan languages† among numerous others. There were still uncontacted peoples in the heart of the reserve.

† There are about 46 languages in the Quechan family of languages alone.

'Yet the people here live in poverty,' Bethan continued. 'Tackling poverty is a human rights priority, of course. But it's also a conservation priority. The only way these forests are going to survive is if people can make a sustainable living here. We have to demonstrate that sustainable, forest-compatible livelihoods are also good livelihoods. Not just good, better than the alternatives. That's where Reynaldo comes in.'

Reynaldo had arrived as a young man in the forest, having spent his whole life in the mountains until then.

'Many of us came here at the same time,' he explained, in Spanish, with Bethan translating.

We had left the sunny clearing with the cluster of bamboo, wood and thatch Manú Learning Centre buildings and travelled by boat to the small town of Salvación to meet him. The long boat sat low on the brown river as we motored upstream.

'Initially, we cut down the forest to grow food crops,' he said. 'When the soil became exhausted, we moved on and cut down more forest. Eventually, I realised we were destroying the forest we depended on.'

Since then, Reynaldo had dedicated himself to figuring out how to grow crops without damaging the forest. And to planting trees. By the time I met him, he had planted around 30,000, grown from seed, each tree nurtured and protected as it grew.

Reynaldo now worked for Crees as their 'productive enterprise manager', teaching his community to farm sustainably. They did this by using agroforestry, a mix of crops that can be grown among the trees – a huge turn away from monocultures such as bananas grown on land razed of trees – as well as organic food production in the community gardens and allotments on already-cleared land close to the town.

Later, I went with Lizzie and a Crees employee called Christina to Reynaldo's home to interview him in his farm gardens. When we had finished, he gave us a tour. There were various lush-looking fruit and veg plots, bursting with produce, everything grown without any use of artificial fertilisers, pesticides or herbicides.

And then there was a bare compound, thick with genetically modified, overly fast-growing, white chickens. Already so heavy they could barely stand, and too heavy to walk, they crouched, panting, in the heat.

Almost all the (vast) quantity of chicken eaten in Peru comes from these genetically modified hens, I learned later. In relation to industrial-scale agriculture, nothing surprised me any longer about how animals were treated. Suffering is built into the system. But I was still regularly caught off guard by the number of small-scale, sustainable, ethically driven farms – like this one – that seemed to have a genuine blind spot in relation to animal welfare. It was as if the quality of life of the animals simply didn't matter. The guinea pigs were another case in point – in small, barren cages, they rushed to the far end of them when we approached, clearly terrified, calling out the '*cuy, cuy*' noises that gives them their Spanish name.

I wanted to focus on all the many positives about food production here and ignore it. But I couldn't. The hens and the guinea pigs were leading stressed, uncomfortable, highly restricted lives. And it was so unnecessary. It would be relatively easy to farm them differently. A bit more space, shelter, shade,[‡] enrichment – it wouldn't take much to improve their conditions.

Beyond that, in relation to the hens, it would involve abandoning the genetically modified varieties in favour of birds that grew at a natural pace, and that were able to walk, scratch, dustbathe and do all the things that hens naturally

‡ Domestic hens are descended from jungle fowl and love to be among trees.

do. People would still be able to eat chicken affordably – if slightly less often. And the animals would be leading much better lives. Even here, where the poverty was profound and the need for protein urgent, it surely couldn't be OK to prioritise human needs at *any* cost, to keep these animals in any conditions at all. Human needs were not just trumping animal welfare, they were deemed to make it irrelevant. It was that sense of entitlement again, albeit in a very different context.

Back at the centre, evenings had become sociable times. Bryn, initially quiet but with much to say of interest once drawn out, designed 'pod' office space as his day job, but was about to move across to the post-disaster humanitarian world and design shelters instead. Chris was a former GP who had travelled to Australia via the United States in a Volkswagen van with one-year-old Bethan when he'd first qualified as a doctor. He told hilarious stories about their adventures along the way and fast became a friend and ally on the pisco-drinking front – when Huret, whose many jobs at Crees included that of bartender, felt like making them.

Among the other visitors was Adam. Adam was at Manú with his partner and home-educated son.

'We call it de-education,' Adam explained, grinning. 'He's only ten and already in recovery from Western civilisation.'

Adam was looking into the long-term viability of tourism in Manú.

'Ecotourism is critical to the success of places like this,' Adam told me one evening over dinner.

'And to Peru as a whole, come to that,' he added.

'But isn't there a huge contradiction if the model of "ecotourism" depends on encouraging long-haul flying?' I asked.

This question troubled me a lot.

'Well, he said. 'I'd like to see a country bold enough to tax all incoming, long-haul flights an extra $50, say, to be invested back into environmental conservation. A sort of biodiversity/ecosystem offset.'

But even without that, Adam, like Andreas and Gabrielle, believed the negative environmental impact of flying people in for a well-constructed ecotourist experience was outweighed by the benefits.

'The key is to make sure that as much as possible of the money it brings into the area, stays in the area,' Adam said, 'and doesn't "leak" back to Lima or, worse, outside Peru.'

Then the conversation shifted to communication, with Bethan telling us that, when she was working with local children, she would ask them if they knew how many frogs lived in their rainforest, compared with the UK.

'The UK has seven native species,' Bethan said. 'Seven compared with 155.[§] It can give the kids here a real sense of pride. I tell them the UK is frog-poor, while they are frog-rich.'

And from there, talk turned to the evidence we'd seen of bigger animals in the area. Copybara – the largest rodent in the world, so large that small monkeys have been known to use them as taxis – had left prints on the grey-sand beach by the river. And to my delight, I had come across the distinctive tracks of puma on a muddy bank, conjuring an image of the big cat walking slowly down to drink at the river.

My favourite experience, though, was the dawn chorus bird census walk – not just the fabulous upwelling of sound, but Diego's ability to listen to the forest cacophony and unravel each song, then identify the bird that was singing it.

§ The highest number anywhere in the world according to Mongabay: https://news.mongabay.com/2014/01/287-amphibian-and-reptile-species-in-peruvian-park-sets-world-record-photos/

A wall of sound was transformed into distinct calls, uttered by an astonishing array of birds.

Later, we sat outside, not talking much. A full moon rose slowly above the silhouetted trees. Stars caught in the branches while below the air was thick with fireflies and moths with incandescent eyes.

Behind everything that Crees was engaged with lay an ominous, complex dilemma. A new road. The road was to run along the eastern edge of the national park and into the forest, heading for Brazil. Some of it had already been built, illegally, under the auspices of a pro-road local mayor – before permission for it had been granted by the government.

Diego's views on the road were, like most people I asked, mixed. On the one hand, many of the communities who lived along its proposed route wanted it. They were currently isolated and cut off, slow and unreliable river transport their only access to healthcare or the chance to trade goods.

'Obviously the communities here are entitled to what the road means in terms of better access,' Diego said.

But the section of the road that had already been built was about four times wider than the five-metre width recommended for roads built with the purpose of connecting communities.

'It's being built with large trucks in mind,' Diego said. 'The pro-road mayor is also a mining man.'

The road would make it much easier for illegal loggers to get logs out, and for mining to take place. Both were big drivers of biodiversity loss here as elsewhere. Then there would be the increase in traffic, and the 'fishbone' effect, where lots of little roads branch off the big one. The environmental impacts of the road on the forest and river were potentially immense. There was also the possibility that this road would be joined up with the road on the Brazilian side. If that happened, Diego said, the impacts could be catastrophic.

Lizzie and I headed back to Cusco by minibus. Bethan, Bryn and Chris had left the centre before us, heading the other direction by boat, to go further upriver. I would have loved to go with them, experiencing, as always, a great, sad reluctance to leave the forest and return to the city.[¶]

As we wound our way back up the narrow dirt road through the cloud-forested mountainside, manoeuvring past other minibuses, the thought of large numbers of huge trucks grinding towards the rainforest gave me serious grief. How would this environment cope? All the positive work, and all the optimism in relation to ecotourism and agroforestry, suddenly seemed to be held in a precarious balance against the negative consequences of big roads, and everything that came with them.

¶ Though not to leave the chiggers, who gifted me a legacy of bites that itched, ferociously, for days.

17. *BUEN VIVIR*

But whether we travel with the nomadic Penan in the forests of Borneo, a Vodoun acolyte in Haiti, a curandero *in the high Andes of Peru, a Tamashek* caravanseri *in the red sands of the Sahara, or a yak herder on the slopes of Chomolungma, all these peoples teach us that there are other options, other possibilities, other ways of thinking and interacting with the earth. This is an idea that can only fill us with hope.*

WADE DAVIS, *THE WAYFINDERS*

The silver Toyota truck hurtled out of Cusco, past the Jesus statue on the hill and onto the road to Písac, gaining speed as we roared by terraces built for farming by the Incas some 500 years previously, and up into the tan-coloured mountains. In the back, I hung on to a door handle with one hand and a leaflet that was (mostly) in English with the other. It read 'Asociación ANDES. Together towards *Sumak Kawsay*.'

The driver, Kike Granados, had worked for ANDES – Association for Nature and Sustainable Development – for fifteen years. ANDES worked with the Potato Park, which was where we were headed.

After a breakfast stop in Písac, famous for its Inca ruins and vibrant market, we came to a halt at a large sign. A colourful design showed two high mountains with a large sun rising behind them and a river flowing down towards strips of land being worked by two figures with long implements something like hoes. Below it were the words, *Parque de la Papa. Comunidad Pampallacta. Banco de Semilla.* Potato Park. Pampallacta Community. Seed Bank.

Behind the sign, which was protected with its own small, tiled roof, a group of Quechen men stood by a scatter of thatched buildings. We all shook hands. Neither Lizzie nor I really knew where we were. I had made contact with someone from ANDES who had arranged our visit to the Potato Park. When we arrived at the ANDES office, early in the morning on the agreed day, my contact wasn't there, and no one seemed to know quite who we were. They'd taken us along anyway. Now that we were at the park, we had no idea what the plan was. It was one of those occasions where the way the day unfolded was outside our control – and all we could do was relax, give in and see what happened.

The park, we learned, was a huge area: 9,000-plus hectares, with six different communities living within it. The seed bank that the Pampallacta community hosted was in a dark, barn-like room with Potato Park posters and photographs on the wall at the far end, and rows and rows of multicoloured, plastic banana boxes on wooden-slatted shelving across the rest of it. In the boxes were small potatoes in string or paper bags, all labelled: the seed potatoes for hundreds of different species.

Back outside, Lizzie and I hung about trying to figure out what was what, as the Quechan men and our group – Kike and two other ANDES colleagues – moved stones from a patch of ground just by the road and then started to dig it over with wooden implements called *chakitaqllas*; the implements, I realised, that were pictured on the sign. The long handle of the *chakitaqlla* ended in a long, palette-knife shaped steel blade. Above the blade, two shorter pieces of wood were lashed on at right angles. After a while, I understood that it would be OK to join in. I studied the technique – one hand on the top of the long handle for leverage and control and a hand and a foot on the shorter wooden handles. Then I dug. This was a rapidly warming activity, which was useful, as it was hailing and, not having realised that the Potato Park was higher than Cusco,* I hadn't brought enough layers.

After a while, we drifted up to the main building and hung out under the large porch while the hail turned to rain.

I was about to give up on anything else happening, and use the time to do some journaling, when one of the Quechen men, Nasario Quispe Ameo, opened a drawer in a cupboard in the hallway. He beckoned us over to look. In it were the traditional clothes he wore when tourist groups visited. There was a pointy hat in crazy colours with long tassels, and a poncho with a classic squares-and-diamonds design in pinks and dark burgundy, with a red, yellow, blue, purple and green fringe. Nasario could see that, the digging warmth having worn off, I was cold. He gestured that I should put the poncho on. It was heavy, and the coldness bounced off its colours like water off a waterproof pannier.

Then he laid out a piece of black and white patterned cloth on the porch floor. From another drawer, he bought out bag after bag of different kinds of potatoes, beans and sweetcorn, and set them out on the cloth, talking in slow, careful Spanish all the while, about half of which I understood. Soon the cloth was piled up with produce. I had seen some of these varieties in markets but, brought together, the diversity was astonishing. Alongside the familiar buttery-yellow sweetcorn were ears that were cream and dark grey; ears that were a deep orangey-red with saffron swirls; ears that were black and maroon. And there were potatoes from inky violets to rich creamy-yellow, patterned like snail

* The park is at an altitude of around 13,000 feet. Cusco is at about 11,000 feet.

shells, from smooth and round to almost carrot-shaped, and as knobbly as an artichoke. Nasario cut a few in half. In one, the flesh was milky-white with a vivid flush of cherry pink. Another was bright purple.

After that, lunch was declared. We all sat in a darkish room on benches or upturned boxes, eating halved avocados bought from the market followed by mounds of salted corn, colourful potatoes and hot beans, everything set out on cloths unfolded in the middle of the floor. Someone handed round a huge bag of sweet bread rolls. Gorgeous music – Peruvian creole – played from an elderly tape deck in the background. The men's faces, the music, the easy chat between the others, the food: a pang of appreciation shot through the fog of tiredness, and I felt suddenly alive and deeply grateful to be in that room.

The Potato Park is an 'Indigenous Biocultural Heritage Area'. The aims of ANDES and the park are to use indigenous knowledge to help protect biodiversity, especially in relation to domesticated crops; to use this diversity to help climate-proof critical food supplies; and to do so while protecting the cultural diversity and traditions of the area, of which there are an abundance. Over 200 languages are spoken across the Andes. In addition, there are two biodiversity hotspots, two important 'centres of origin of major cultivated species',[1] and twenty of South America's 72 World Heritage Sites.

The region is also, as I'd already learned, unusually threatened. As with many mountain environments and cultures, the impacts of climate change have arrived in the Andes sooner and with more ferocity than in lower regions. They are already witnessing changes in weather patterns – including an increase in severe weather events and weather that is utterly unpredictable – plus crop failure, arrival of pests, loss of glacial meltwater and greater risk of mud avalanches. Many people in the region live from the land, directly dependent on the ecosystems and the basics of life they provide, such as water and food – the so-called natural resource base. In the Andes, human-caused impacts on forests and rivers and grasslands, on biodiversity and climate – on our planet's ecological boundaries – have long been not just understood, but experienced.

So far so familiar. But ANDES argues that the Andean mountain cultures are not only among the people in the world most vulnerable to environmental change – they are also one of the greatest sources of wisdom about how to tackle it. Local and indigenous knowledge, traditions, values, livelihoods and cultural practices are potentially game-changing in relation to achieving the transformations we all need.

For a start, these are cultures that have long known that diversity within wild ecosystems is key to their resilience – a central tenet of Western ecological science. ANDES and the Potato Park argue that it's key in relation to domesticated crops as well. If you have two species of potato and the climate becomes hotter, or drier, or wetter, you may be unable to carry on growing either of them. But if you have 2,000 species, there's a good chance that some of them will have what it takes to cope with the new conditions, or even to flourish.

Because these are cultures that have already had to figure out strategies for dealing with climate and other environmental change, they are decades ahead of us in terms of knowing which crop species can best grow where.

And it wasn't just about discrete pieces of knowledge, such as which potatoes can cope with climate change. It was also about the way this knowledge, and the resulting produce, was exchanged. Across South America, indigenous coastal and mountain communities have long traded seafood and farm produce via barter markets and plant-swapping – and these traditions are proving to be really helpful in the face of rapid environmental change. They had effectively arisen thanks to steepness. Because of it, people had lived close to one another for centuries, yet in ecosystems and climates that were very different. The social structures that emerged facilitated regular exchange of the different things that could be grown from the high mountains down to the sea, built on the core values of respect and reciprocity among human communities – and between human communities and wider nature.

Compare this with the 'modern' globalised food system,[†] immensely vulnerable to the environmental impacts it itself is causing, and it seems quite astonishing that we[‡] have any lingering notion of superiority at all.

In short, the potato seeds we'd seen in the potato seed bank, in conjunction with what these people already knew about which varieties grew best in different conditions and their collaborative, exchange-based, reciprocity-valuing social structures, added up to a unique mix of knowledge, practices and values that could be seriously helpful in dealing with the onslaught of climate and environmental change we face. Including one of the biggest environmental change-related challenges of all: how to keep feeding ourselves.

Yet, indigenous peoples are, to this day, still treated as second-class citizens, across South America and the world. Worse, long since the conquistadors left, they have suffered extreme injustices, ranging from forced sterilisations of over 200,000 Quecha and Aymara women during the presidency of Alberto Fujimori,[§] to the ongoing persecution

[†] The 'food system' includes everything from production to distribution to dealing with waste.

[‡] 'We' as in Western, industrialised societies.

[§] Fujimori was the president of Peru from 1990–2000.

of indigenous 'land-defenders'.[2] And indigenous peoples' knowledge is too often still treated as backward or simply dismissed, with Western experts advising in a hierarchical, top-down way, rooted in the assumption that Western knowledge is superior.

ANDES' approach to knowledge, in contrast, was 'horizontal' rather than 'vertical'. Indigenous wisdom was valued and respected on equal terms with any other kind of knowledge, including that derived from Western science.

Back on the porch, digesting my lunch next to Nasario's arresting display of vegetable diversity, I wondered what it said about my own assumptions that I'd initially been surprised to find a source of hope in potatoes.

As potent as that from the Manú research about biodiversity recovery, hope lay in the social context around the potatoes, too. Hope that this knowledge would help Andean cultures sustain food systems that are resilient: that have the capacity to recover from climate and other shocks, and that can adapt to the changes ahead. Hope that it could help other cultures, including so-called modern ones, move in this direction as well. And hope that they could do all this while resisting the cultural impacts of globalisation, protecting and celebrating local traditions and approaches.

Not that all customs are beyond reproach simply by being local. One featured in the leaflet involved a prospective wife being required to prove her worthiness by peeling one of the knobblier potatoes to the satisfaction of her future mother-in-law. Nasario, watching me read, gave me a quizzical, 'do you like what you're reading?' look. Potato-peeling challenge notwithstanding, I gave an enthusiastic, thumbs-up reply. Sitting on the porch of the Potato Park building, cloaked in Nasario's multicoloured poncho and reading the ANDES leaflet, I had the sudden, powerful sensation that I could see beyond extractivism and everything that came with it to a different and vastly more positive world.

What lay behind the work of ANDES and the Potato Park was both an alternative worldview – or 'cosmovision' as it's often called in South America – and the radically different approach to 'development' that I'd already begun to encounter. *Buen vivir*, or *vivir bien* in Spanish, is frequently translated as 'good living' or 'living well'; though some argue that 'plentiful life', 'harmonious life' or even 'to know how to live' comes closer.

There is no single model of *buen vivir*, and it's constantly evolving. It has its roots in indigenous worldviews, including *sumak kawsay*, the cosmovision of the Quechua peoples. Similar belief systems are articulated by the Mapuche in Chile and Argentina, and the Aymara of Bolivia. Yet, it is relevant far beyond the Andes and '[i]t certainly doesn't require a return to some sort of indigenous, pre-Colombian past'.¶

Contemporary critiques of capitalism and (conventionally understood) 'development', as well as some varieties of Western environmentalism and feminism increasingly feed into it too.

What *buen vivir* perspectives all have in common is an attempt to work out, in theory and in practice, profoundly different approaches to the questions about what it means for people to live well, and for a country to develop and progress. They offer 'systemic alternatives' – alternatives at the level of economic, social, political systems, worldviews and values. Alternatives to colonial habitation as a way of being in the world.

Within this diversity, a common theme is the importance of community. From *buen vivir* perspectives, the individualism of the West makes no sense. Well-being is only possible within a community, and individuals can't be understood

¶ According to Eduardo Gudynas, a leading *buen vivir* scholar and activist.

in isolation from the communities they are part of. And the understanding of what a community *is* is very different too. There's no distinction between 'modern' and 'non-modern' peoples, or between humans and other-than-humans, people and nature.

All members of the community are valued for themselves: the idea that humans could own the earth's resources or put a monetary value on nature via a calculation of the worth of ecosystem services – of 'natural capital' – is nonsensical anathema.

From *buen vivir* perspectives, the notion of trying to enhance human well-being and improve society by adopting the resource-hungry, growth-dependent economic systems and high-consumption values of the West is ridiculous. Progress, according to this worldview, is not the acquisition of more money or more possessions – held to be impossible because nature, or *Pachamama*, has limits – but greater harmony or equilibrium. Harmony between people; between people and nature; between all of this and the cosmos. And the values of reciprocity, respect and complementarity will be key in achieving this, rather than capitalist competition.

Perhaps most important is the assertion that consuming less does not mean a reduction in quality of life. Quality of life requires a basic level of consumption, of course, but beyond that, quality of life is to do with relationships, experiences, community, harmony. Which means that living with less – and from an environmental perspective this is absolutely necessary for many in the West – doesn't have to be understood as a sacrifice, but a gain: the insights from cycle touring and *buen vivir* coming together in a fabulous smorgasbord of positive, potentially world-saving ideas.

Nasario beckoned again. I left my seat on the porch and went into a sort of greenhouse. There I helped him put compost into pots for future potato seeds. We chatted. Nasario had travelled widely, presenting the work of the Potato Park, and I found his Spanish easier to understand than most. He told me that potatoes originated in parts of Peru and Bolivia, and that archaeological evidence suggested they'd been eaten here for thousands of years. They'd powered the Inca Empire and arrived in Europe** towards the end of the 16th century, possibly with returning Spanish conquistadors who'd used them as food on the journey and then planted the leftovers. Now, there were three Peruvian seed banks where specimens of every variety were kept. That way, if there was a catastrophe in one of the banks, there would still be seeds elsewhere. He told me that roughly 1,400 varieties of potato were grown in the park, and at least 4,000 across the whole of Peru. And that they were not just collating their existing knowledge about these and other food plants, but building on it, experimenting with different varieties – including wild ones – to test for resilience to changes in rainfall and temperature, and enhancing their knowledge of which species grew best at different altitudes.

Nasario fell silent. After a while he looked straight at me.

'Climate change is terrible,' he said.

It had already brought torrential rains, huge hailstorms, extreme temperatures – hot and cold – he said, and this was making life very difficult for farmers and wildlife alike.

I asked him what he thought about the Peruvian government, and whether he thought mining was compatible with respect for *Pachamama*.

'The government does nothing,' he replied. 'And no.'

Gradually, people began packing up and moving back to the truck. When we finally went to leave, all the Quechen

---

** Where they were largely considered food for the working class, and for armies.

men climbed into the open back, jumping out at different houses and villages on the way down the mountain. In the delicious warmth inside the cab, I felt for how cold it must have been out there.

Machu Picchu. All superlatives and clichés apply. It is simply stunning, the dramatic mountain setting as much as the astonishing architecture. In the classic view of it, neat, organised-looking clusters of (now roofless) rectangular, stone buildings cover the top of a flat, grassy area, from which steep-sided terraces – on which food was grown for the inhabitants – drop away at a dizzying angle. A cone-shaped, green mountain towers behind it, falling away in a long, green ridge, the trees and ferns of the sub-tropical, moist forest somehow clinging to the sheer sides. And behind that, range after range of high, forested mountains stretch to the far horizon. The drama and beauty of the human-made constructions are matched – exceeded – by the drama and beauty of the mountains, and yet seem profoundly in place.

The stories attached to it are equally arresting. Machu Picchu was built in the 1450s at the time of two great Inca rulers, Pachacutec Inca Yupanqui and Túpac Inca Yupanqui. Some say that Pachacutec ordered its construction as a royal estate for himself, after a military victory. Others, including our guide, said that it was created as a sort of university research centre, inhabited by an elite who studied philosophy, religion and the Inca cosmovision.

Either way, it was 'tangible evidence of the urban Inca Empire at the peak of its power and achievement – a citadel of cut stone fit together without mortar so tightly that its cracks still can't be penetrated by a knife blade'.[3]

And somehow, it was kept hidden from the Spanish conquistadors, despite that Cusco, which the Spanish invaded in 1533, is only 50 miles away. The Spanish never found it.

It is thought that about 750 people would have lived there before it was abandoned, only about 100 years after it was built. It remained largely unknown for centuries until it was 'discovered' by the North American academic Hiram Bingham, thanks to a tip-off by a local farmer.

The gold that was assumed to have been there has never been found. There are theories aplenty about who might have stolen it – including Bingham himself. Bingham, a lecturer in South American history at Yale University (and possible inspiration for Indiana Jones), had reached the area by travelling on the Río Urubamba that runs through the Sacred Valley at Machu Picchu's feet. The site, when he found it, was almost entirely grown over. Returning on numerous occasions with the support of *National Geographic*, Bingham is credited with turning Machu Picchu into one of South America's most renowned tourist attractions. The switchback road that snakes up the mountain from the river to the site entrance is called Carretera Hiram Bingham – the Hiram Bingham Highway.

Bingham spent much of the rest of his life traipsing through forest in search of other 'lost' Incan cities. Many think that there's at least one more, the fabled lost city of the jungle, somewhere out there in the still-vast forest.

In addition to their worship of the sun, Inti, and the Apus, the trilogy of condor, puma and snake was especially important in the Inca belief system. It's this trilogy that gives grounds for believing there is another city. From above, Machu Picchu is said to be shaped like a condor, while Cusco, the centre of the Inca Empire, has the form of a puma. Is there a snake city? Almost certainly: but it has never been found. According to our guide, wherever it is, it is guarded to this day by Inca-descended warriors. And just possibly, the lost gold was not stolen by Bingham, or anyone else, but relocated there.

But the story I most loved was told, again by our guide, about the Inca who ruled at Machu Picchu. They were, he said, known to be brilliant farmers as well as engineer–architects, as the terraces attested.[††] These people, rich beyond imagining, with precious metals that were to finance untold Spanish skirmishes, *valued seeds more highly than gold.* Imagine. Imagine if we had embraced those values, that wisdom, in the West. How very different our priorities and systems would be. How different our current environmental challenges would be. It's just conceivable that we might not even have any.

Back in Cusco, Lizzie and I did a final round of filming in the tiny yard of the small flat we'd rented for her last few days. I was trying to talk about *Doughnut Economics*, and the framework it offered for thinking about ways forward, identifying the social and environmental boundaries that any development model should operate within. *Buen vivir* would definitely be at home there. While I talked, I gesticulated with a chocolate doughnut I'd bought for the purpose. It became stickier and stickier, disintegrating in the heat. On the fifth take, I gave up and ate it.

Lizzie boarded a bus for Lima and the journey back to the UK, where she would face the huge challenge of editing the many hours of footage into something short and coherent. I wandered the city, savouring my last days there, but also troubled. I had not always managed to keep to myself how difficult I'd found Lizzie's constant presence – through no fault of her own – and the huge amounts of time the filming had taken up, however much I valued it. I sensed some dark stuff beneath my own bad behaviour, and it niggled

[††] The terraces were built to ensure good drainage and soil fertility for growing food, while also protecting the mountain itself from erosion and landslides.

away at me. No wonder I found solo cycling easier. The time with Lizzie left me thinking that my ability to travel alone, perfectly happily, might be rooted in weakness and character flaws, as much as in strength.

One day, I went inside the cathedral, an extraordinarily sumptuous mix of wood, gold and silver. The Spanish built it on an Inca site, dismembering their buildings and using the stones. Back on the steps, I sat imagining the Spanish army marching over the huge brown mountains to Cusco, and wreaking havoc on the people and culture they found there. Cusco had been the capital of the Inca Empire from the 13th to the 16th century, when Pizarro arrived and sacked the place. The story of the Spanish in Cusco was, as ever, a story of brutality and of an incredible displacement of one culture by another: new god, new values, new worldview, new mythology, new customs, new practices. And yet the cathedral was full of '*mescla*' – mixed-upness – such as Catholic altars inset with mirrors, which the Inca believed allowed you to see your own soul. The Inca also liked glittery-ness, perhaps because of Inti, their sun god; and the Spanish had added more mirrors, as well as representations of potato flowers and llamas in the highly decorated wooden alcoves, to help get the Inca on board with their new religion.

For three centuries, the Inca had overseen an extraordinary empire, the largest in pre-Colombian America, despite not having steel or iron, the wheel, draft animals or the written word. They operated largely without money and without markets, their economy described variously as 'feudal, slave, [and] socialist ... socialist paradise or socialist tyranny'.[4] They also had no hunger. Now, women in traditional dress with lamas or small lambs tried to sell photos, and a man in full Inca get-up stood by the Inca stone wall on the way to the plaza. As I walked, people came up to me offering massages for twenty soles.

PART 3

Bolivia, Chile and Argentina

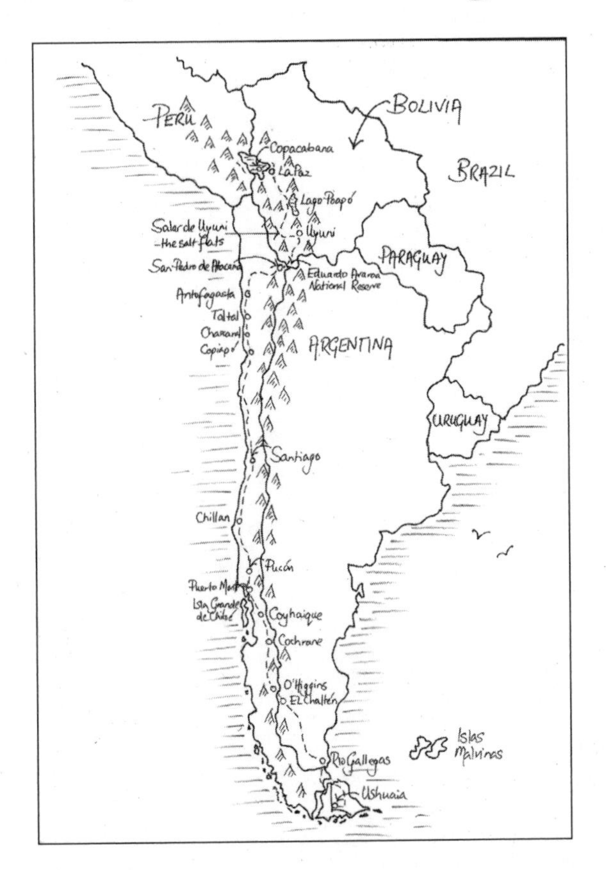

18. EDGELANDS

No hay nada que no se pueda resolver con una taza de café.
CAFÉ CAPPUCCINO, CUSCO

BOLIVIA

Happy Planet Index score: 45
Rank: 75th out of 152 countries

The message was written in white chalk on a small blackboard, below the words 'Café Cappuccino' and above the words '2nd Floor'. The translation: 'There is nothing that can't be resolved with a cup of café.'

I climbed the stairs, found a window seat and ordered a strong latte. I was grappling with a dilemma and hoped the sign was right.

Time was closing in on me again. It was perplexing. Despite having doubled the number of months I'd thought I'd need to ride the length of South America in a leisurely manner, deadlines kept appearing over the horizon. Deadlines that were either hard to meet or verging on impossible. This time there were two: I had arranged to be in Pucón, a town roughly three-quarters of the way down the length of Chile, by mid-December, to meet up with Chris again for Christmas. And beyond that, I wanted to reach the southern tip of Patagonia by mid-February, just before their winter started for real and with enough time to turn around and get back to Valparaíso to catch my cargo ship home. There were only three months to go until deadline one. And at least 3,000 miles. The fastest route would be to drop down from Cusco to the coast, a good 400 miles away, and then leg it south across flat northern Chile. Even then, and even with upping my daily mileage and not stopping much, the timing was now tight.

The dilemma arose thanks to the tantalisingly close proximity of the Bolivian border. And thanks to Bolivia having huge biodiversity and other environmental stories, not to mention some extraordinary landscapes and an above average quota of political interest and intrigue. But if I went to Bolivia, I would need to take a bus to catch up the mileage and make the deadlines. Which horn should I grasp? The pure, 'I rode it all' adventure horn, or the 'find the biodiversity stories' horn? This was especially tricky given that the aim of *The Life Cycle* was to use the first to raise awareness of the second.

I consulted the world via Facebook and emailed my dad and my brother. A lovely email from my dad was not long in arriving.

'You should go to Bolivia,' it read. 'The environmental stories are more important than cycling the whole way.'

Then, from my brother, the opposite.

'Cycling *almost* the whole length of South America has much less kudos,' he wrote. 'And cycling for thousands of miles on endless, flat, hot, straight roads is as much part of the challenge as the Andes.'

He was right of course. But then, so was my dad.

By the time Woody and I finally left Cusco, it was 24 October. A kindly group of staff and residents at the hostel waved us off as we bucked skittishly away down the steeply cobbled street. I hadn't cycled for over three weeks. We trundled down now-familiar streets and joined the long road out of town. Two women grappled with an enormous load of firewood stacked on the front of a cargo-adapted tricycle, one pedalling, one pushing.

The feedback from friends on Facebook had been overwhelmingly in favour of pursuing the environmental stories rather than non-bus-diluted ride purity. After much agonising, I'd decided to go for broke. I would head for Bolivia *and* I would to try and cycle the whole way and still make it to Pucón on time. This might not be possible, but I would worry about that later. For now, it felt wonderful to be back on the road. Not long out of Cusco, a full rainbow ringed the sun, as if an endorsement of my Bolivia decision.

As I drew away from the city, acres of young sweetcorn grew alongside a dark green river. There were low buildings, rooved with tin or with terracotta tiles, many daubed with the names of candidates in forthcoming elections painted in royal blue or red on white walls. One read 'REGION 2018' followed by the Inca flag, in rainbow stripes at a jaunty 45 degrees. Huge, colourful portraits of Atahualpa began to feature on building ends. Then a gigantic statue of a chocolate-coloured guinea pig, standing in heroic pose on a stone plinth. The guinea pig dwarfed the sculpture of a condor that hovered above a road sign nearby. Different cultures ate different animals, that I understood, though I didn't eat any of them myself. But the guinea pig thing seemed to have a unique weirdness, perhaps best captured in an advert for a restaurant that featured a guinea pig in a chef's outfit brandishing a cooked guinea pig on a stick.

For several days, I rode alongside a trainline, though I saw only one train. It was blue, with four carriages, travelling slowly in the other direction. The road rose, surreptitiously, towards the Abra la Raya pass ahead. It was slow going. *Come on, you know how to do this. Stop thinking 'am I nearly there?' Stop resenting the hill, the road, the wind.*

On previous long rides, albeit on skinny-tyred road bikes, I'd aimed at 75 miles as a daily average. On this journey, my daily miles kept dropping. 50. 40. 30. I'd left Cusco vowing to get it back up to 50 but was already doubting I could. *How simple it is to say, OK, I am going to crack on, do faster miles and longer days. My reality is a cliché. Easier said than done.* I was never quite sure what was making it so hard.

The pass was a little over 14,000 feet. I just missed a drenching as rain threatened, then backed off. Rolling down the other side, everything changed. Life learning from the road: *no hill goes up for ever.* Now I was on a gentle but definite descent and, at last, the miles were rolling by. There were long stretches of road that were almost straight, stretching out ahead across the huge, wide plain, with mountain ridges for hundreds of miles on every horizon.

I was crossing an extensive stretch of *puna* – high montane grassland ecosystems that exist above the treeline and below the permanent snow-line – the road rising and falling around 12,000 feet. Between the occasional small towns, a smattering of poor-looking dwellings. Stone buildings, many collapsed, hunched alongside the road beside mud and wattle shacks, often with no windows. There were few people about. Some were with their animals – sheep in such small numbers that each was tethered separately, a few cattle, many llamas. The animals were the same brown and tan colours as the mountains. *What is life like here?* I wondered. *The people who live backed away from the road in the tiny mud-brick houses, what do they wish for?* As for the ecosystem, *puna* grasslands are tough and resilient, but grazing was damaging them nevertheless.

The days swung between fabulous and grinding. Fabulous when a tailwind gifted me the illusion that my fitness was back, and I would race across the high, wide *puna* under the fierce sun with a wonderful, swishing sense of being in motion, fast, the grin-making, intoxicating payback of effort. On other days, it felt as if the universe begrudged any forward motion at all. On those days, every pedal stroke was an effort, every inch of slowly nudging forwards gained at a physical and mental cost.

I caught up with the Bulgarians at a Casa de Ciclistas in the large and busy town of Puno. The Casa was a yard behind a hotel. It was dark by the time I arrived, more than usually traffic-frazzled after days on the open plains. A friendly man on the front desk of the rather smart-looking hotel helped me through the wood-panelled lobby with the bike and into another world. Llamas were painted on the walls of the smallish space in which were various plants, a white cat and a cluster of plastic tables and chairs. Two tents had already been erected in the gaps. The Bulgarians greeted me with headtorches and big hugs.

I hadn't been expecting to cook that night, and the Bulgarians and Camilo, a Colombian they were now cycling with, shared their pasta with me. We traded stories – in a disjointed sort of way – about where we'd been since we last met.

The next day was hot and sunny. After topping a hill, the wind rose. We were getting into edge country, I thought: the edge of Lago Titicaca, the largest freshwater lake in South America and one of the highest in the world; the edge of Peru; the edge of land you could possibly make a living from. People were ploughing with cattle and sitting with their animals. Sheep, cattle, pigs. On the lake, the big, circular tops of fish farm enclosures sat silent on the grey water. What sort of fish would be fighting for space among heaving, crowded bodies under the surface in those grim cages? Trout? Salmon? Given the astonishing journeys that wild salmon make throughout their lives, confining these animals in tiny, packed cages seemed especially brutal; all the more poignant for the immense, watery space around them.

Modern humans have not had a harmonious relationship with Lago Titicaca.* In 2012, the Global Nature Fund nominated it 'Threatened Lake of the Year'. Climate change has continued to shorten the rainy seasons since then and this, plus the shrinking of glaciers that feed the lake's tributaries, have caused the lake's water levels to fall.† The lake's rich biodiversity‡ is threatened by the introduction of new species and by water pollution as cities around the watershed grow, outpacing their waste and sewage treatment infrastructure in the process. It was a sorry story, I thought, looking at the grey, choppy water.

I left the Bulgarians when they stopped to skin and cook a dead rabbit for lunch. I had no problem with eating road-kill, but the border was ahead, and it only functioned during certain daytime hours. When I reached it – a couple of kilometres past a village and, of course, up a hill – there was, to my amazement, no queue. The official I spoke to, a large, world-weary but kindly man, studied my overdue passport stamp, shook his head, told me – as expected – I had to pay a fine, wrote me a chit and instructed me to go back to the village and settle it at the bank. I legged it back down the road, found the bank in a lovely square of custard and maroon-painted buildings and joined the long queue. A

---

* The name comes from the Aymara, early lakeside dwellers whose word *tití* has an unusual range of meanings, including 'Andean mountain cat' and 'lead-coloured'.

† The water levels have always fallen and risen, but since 2000, they've only fallen.

‡ The lake is home to more than 530 species, including huge populations of water birds, frogs, snails, bivalves, reeds, other aquatic plants and of course fish, of which 90 per cent are endemic.

lengthy wait later, I paid the 200 soles fine which, at around £35, was considerably less than I'd feared.

Then I legged it back up the hill to the border. There was now a queue there too. When I finally reached the front of it, my payment receipt was scrutinised. I needed multiple copies of it, apparently, as well as of certain pages in my passport. There was someone who offered photocopying services across the road. Photocopying duly completed, I returned to the border queue, where my passport was examined for so long that I began to wonder whether I was showing up on their computer as some sort of troublemaker/activist/environmentalist.

Finally, to my not inconsiderable relief, I was stamped and released. Free to enter Bolivia! By this time, the others had all turned up, Camilo and the Bulgarians having raced to the bank in a tuk-tuk to pay their fine before the border closed. We cycled together up and over the hill to the Bolivian side. This piece of officialdom proved relatively straightforward for me – but horrible for them, with the announcement that they needed to pay $100 each, something to do with Bulgaria not being a Schengen country. This, given their extremely low budget, was a fairly disastrous amount of money. Eventually, I left them debating whether they would enter Bolivia after all, or cycle around it in a massive huff, and rode off in the fast-gathering dark towards Copacabana, about six miles away, still just visible over the lake.

I was showering in an old, welcoming and cheap Quechen family-run hotel when the Bulgarians and Camilo arrived. I'd told them where I planned to stay and promised to buy them all pizza if they came, though they'd said they would camp closer to the border if they crossed it at all. We headed to a pizzeria, which joy of joys, also had cold beer. They were thoroughly fed up with the border bill, but food and, of course, beer – paid for thanks to the credit card my dad had insisted I take 'for emergencies and luxuries only' – could only help. As the mood relaxed, we talked about whether, if they had the choice, they would travel with more money.

'Not this time. We've learned so much, and the bad times are always followed by good,' was the Bulgarian consensus. Followed by, 'but next time, yes, definitely!'

Various conversations with them, and others, had convinced me that travelling on a minimal budget often altered things in positive ways, making you creative, facilitating different sorts of interaction with people. But I wouldn't have changed the amount of money I had either (not exactly lavish, but considerably more than they had). Camilo, mouth full of salami pizza, managed to snort and grin simultaneously. He had no doubt at all that he'd like to be travelling with more money and, for example, check into a hotel rather than camp in the plaza, which was their plan. In the end, they camped in the courtyard of the pizza place, which turned out to also be a hostel. When I left them there, hauling their sleeping bags and mats out from pannier bags, there were big hugs all round. They would leave tomorrow as they were aiming for Santiago by Christmas and were feeling the mileage pressure, a feeling I knew well. Their route would take in the salt flats, as would mine, so it was possible we would meet again. It was good to say goodbye on a positive note.

I woke early after a restless night and spent a couple of hours checking emails and Instagram in bed. Finally dragging myself out from under the blankets, I went down to breakfast just after nine, only to find that Bolivia was an hour ahead of Peru and that breakfast, which finished at ten, was over. *They might have mentioned this at the border.* A kind man on reception rustled me up some food and coffee anyway.

I spent the rest of that day alternating between the usual 'new country tasks' and 'somewhere with Wi-Fi and other facilities' tasks – phone-sorting, money, laundry, emails – and sightseeing. Bolivia had felt positive from the minute I crossed into it, and different from the edge-land of Peru in a way hard to define. Somehow livelier. Even in the short, dark miles after the border, I had sensed that the landscape was becoming more varied. In Copacabana, I relaxed unashamedly into tourist mode. There was a large, white cathedral with Moorish-looking tiles on the outside and a riot of gold on the inside. There were cafés with salads and coffee, stalls in the streets selling food and electronics and hats and fabulously colourful blankets. The cashpoint at the bank delivered me a wadge of Bolivianos without demur. Then I walked to a beach on the lake where horse-rides and small cafés selling fresh peach juice were on offer.

I left the next day, heading for La Paz. The road climbed steadily but relentlessly for the first nine miles up into the tan and brown hills. At the top I caught glimpses of the glinting grey lake water below, and a range of white mountains, hazy in the far distance. Somewhere on the lake were the Islands of the Moon and the Sun – where the Inka sun god, Inti, was born – though it was too misty to see them.

Small towns came and went, some with nothing in them bar half-built houses and the occasional gated entrance to a tiny shop. The houses were mostly an orangey-toned brick, the few with plaster painted with the colours and logos of phone companies.

After a while, I began to pass groups of people sitting outside playing musical instruments. They wore heavy black cloaks and, the women, tall bowler hats. Among the instruments I recognised drums and flutes, but it was hard to discern either a tune or a rhythm. In fact, to my ears, it sounded like a racket. Some sort of festival? Normal Wednesday evening practice?

I spent another night in a hotel, feeling unsure about wild camping. Was my ability to pull out a credit card and pay to be off the road and safe part of the problem, consolidating and confirming, in its own small way, the chasm between rich and poor? Or part of the solution, bringing at least some money into the local economy? I couldn't figure it out, though my mind felt excited and alive. My body felt tired, though. A mirror in the bathroom revealed hundreds of new wrinkles, my skin sagging and crinkling across my sun-exposed face and arms. My lips had cracked open again and I woke in the night with my bottom lip stuck to my top lip. In the morning, there was blood on the pillow and sheets. *Che would not have cared about cracked lips. Nor be sleeping on hotel pillows, come to that.*

Back on the road, the Cordillera Real stretched across the horizon, a huge, silent row of white peaks above the dry Altiplano;[§] above the roadworks, the shacks and the animals. By now the dull grey edge of Titicaca was about half a mile away, no longer scenic. A sign read 'Don't Contaminate the Lake', while piles of dumped rubbish, mostly plastic, slouched precariously on steep slopes above the water.

As I got closer to La Paz, there were more people playing their strange music on drums and flutes, and cars with flowers wound around their wing mirrors. The musicians waved and returned my *holas* as I went by. The road was rising steadily. Bracing myself for a monster climb, I pedalled by a 'motorway only' entrance booth and was well onto the half-built road beyond it before I realised the climb was a descent. I dived down it, through thankfully light traffic – the single narrow lane was not navigable by both me and, for example, a bus. I switched between racing to keep ahead of wider vehicles and pulling off onto the section of road that was still being built, grateful it existed even if it were

§ High plateau.

like riding over a boulder field. *This road would be absolutely horrible going up*, I thought. In the brief moments that I could look further than my front wheel and the traffic, I saw cable cars above and a huge white mountain ahead. Then I plunged into the massive brick-coloured sprawl that is La Paz.

19. THE CRITICAL PERSPECTIVE OF REBELLIOUS SPIRITS

Natural disasters in Bolivia have been getting worse with the passage of time. It's brought about by a system: the capitalist system, the unbridled industrialization of the resources of the Planet Earth.

Capitalism has only hurt Latin America.

President Evo Morales

I had booked a hotel that seemed to be in the centre of things, and easy to find. But just for two days. After that, I'd find somewhere cheaper.

Woody was established in a ground floor room which also housed spare chairs and tables, and a picture of Mother Mary. Someone genuflected in front of it while I was lifting off the panniers.

'The bike can live there,' the receptionist had said when I checked in. 'Totally safe.'

She told me that the music I'd been witnessing was part of a festival of souls.

'Yesterday, people's houses would have been full of the things that the departed most liked,' she said, in excellent

English. 'They would spend time with the souls all day and night, and today, with the musicians, accompany them to the cemetery.'

There would be celebrations across the whole of Bolivia, she explained. And yesterday was a national day off – which probably explained why I'd encountered relatively little traffic on the way in. I mentally thanked the dead for inadvertently helping to keep me alive.

The hotel was several floors high. There were murals on every door and wall; Bolivian scenes, mostly towns and mountains, and colourful balconies and stair rails painted in orange and green. Breakfast – which was sumptuous – was on the top floor, with views across the densely packed, many-shaped rooves of the city to the big white mountain. This was Illimani, the highest peak in the Cordillera Real at 21,122 feet, and the second highest peak in Bolivia. My room also had a rooftop vista, plus a fabulously hot shower and good Wi-Fi. I already knew I wasn't going to leave.

My inbox seemed like another affirmation of the decision to come to Bolivia. Following leads from friends, I'd made contact with various environmentalists and now, as the messages downloaded, I saw they included numerous invitations to meet and talk. The first was from Ely. I'd been put in touch with Ely by the Swedish ambassador to Bolivia, who I'd been put in touch with by the lawyer and activist Polly Higgins, whose campaign to have the UN recognise ecocide as a crime against humanity had resonated well in Bolivia, a country famed, like Ecuador, for its Earth Rights legislation.* I was to meet Ely in front of the St Francisco Church and then have lunch.

I walked there, past street vendors and car chaos, the distinction between market and street constantly eroded as

* The Bolivian Law of the Rights of Mother Earth was passed in 2010 and revised in 2012.

sellers spilled off the pavements and onto the road with their wares. Bunches of herbs, paint, nuts and raisins, stockings, meals. Between the stalls, sculptures and artwork leaned against railings and street corners, appearing unexpectedly wherever I looked. Then a long street of tourist tour organisers, most of whom were offering mountain bike trips on 'death road' – exactly the kind of road I was doing my best to avoid. In front of the church were crowds and rain. Ely was running late, and I was vague with hunger by the time she arrived. Short, energetic, beautiful, dark hair, dark eyes. About my age.

Lunch was in a nearby restaurant in a gorgeous old building that Ely told me used to be a well-known music venue, a fabulous contrast to my standard roadside bread and cheese. We had peanut soup, a peculiar pumpkin and rice dish, a delicious, slimy, chocolatey concoction and a cappuccino. Ely radiated warmth and energy. She talked more or less nonstop, and I was captivated and slightly disbelieving. How, exactly, had I had the luck to land here, in this restaurant, with this vibrant woman who held views and information about pretty much everything I wanted to know? She talked about mining and its terrible impacts, and about how it conflicted with both Earth Rights and *buen vivir*. These concepts were profoundly important, and deeply rooted in Bolivian culture, she told me. But there was a chasm between what they stood for and what was actually happening.

She became most animated when she talked about President Evo Morales. How the initial euphoria and hope at his election – the first ever indigenous South American leader, lauded for his environmental and social commitments, regaled at international climate conferences – had turned to disillusionment and even fear.

Buen vivir had been institutionalised by Evo Morales in Bolivia in 2006, and by Rafael Correa in Ecuador in 2007, she explained. It became central to the official discourse of both countries, discussed nationally and internationally. It had appeared to be central to the profoundly anti-capitalist development plans of both countries, too.

'But Monsanto also arrived in Bolivia during Morales' era,' Ely said, her face wrinkling in distaste at the very name of the multinational, famous, among other things, for having been a major producer of Agent Orange. Monsanto then became known for its production and patenting of 'terminator' seeds, genetically modified so that they only produced one year's crops, thereby ensuring that farmers returned year on year to buy more; and the so-called Roundup crops, genetically modified to be compatible with Monsanto's pricey, carcinogenic, glyphosate-based pesticide of the same name.[†]

'And Evo Morales' model of the good life now has little to do with the philosophy of *buen vivir*,' Ely continued. 'It has everything to do with money and consumerism. And power. But not for the Bolivian people as he promised us. Power for him and his elites.'

Ely seemed to know everyone in Bolivia and promised to introduce me to most of them, should I wish. It was a really good conversation, and it brought me alive. As we went to leave, though, I was swept by a wave of tiredness.

I figured it didn't matter. I could spend another day or so in La Paz – I needed the catch-up time for organising

---

[†] One of the most vocal and sustained critics of Monsanto is the Indian scientist and activist, Vandana Shiva. See her work for further information on Monsanto and related issues. The World Health Organization has referred to glyphosate, the active ingredient in Roundup, as a 'probable human carcinogen'. Monsanto are currently involved in multiple lawsuits over damage caused to neighbouring crops by their new herbicide, Dicamba. On the plus side, they no longer produce terminator seeds.

project visits in any case. And I was about to step onto a roller-coaster of meeting people.

Back in my room at dusk after a wonderful, extended detour via the Bolivian National Art Gallery, I opened the window. A full moon hung large over the city, which looked beautiful, in a cityscape sort of way. Tara and Aidan, I'd learned from Instagram, were still ahead of me, on a wilder, tougher trip. They were often off-road, camping on the salt flats and in the mountains and among the volcanos. I was spending more time on tarmac and in towns – primarily because of the project visits. Given the aims of my ride, it was worth it. But next time, I thought, I would love to be doing it more their way.

The next morning, I met with Lilian, who worked for the Alianza Gato Andino – the Andean Cat Alliance. Andean cats live in the high Andes of Peru, Bolivia, Chile and Argentina, in rocky, arid landscapes with little vegetation. Their challenges include a dramatic shift in the attitudes of the people who also live in and around these fragile habitats.

'The Andean cat used to be considered sacred, a creature that bestowed fertility, given credit for good harvests and healthy livestock,' Lilian said.

Today though, she told me, if you asked local kids what these cats ate, they would all shout, 'llamas and cattle', even though the cats are clearly far too small to harm either. Facts notwithstanding, the cats are, like all predators, widely hated; the attitude shift, in her view, largely to do with the spread of Christian evangelism and white farmers. It reminded me of attitudes to wolves I'd encountered in the USA on *The Carbon Cycle* ride.

Back in my room, I felt that I'd returned to both a sanctuary and the edge of a whirlwind. On my 'to do before you leave' list: update journal; write a 'friends and family' letter;

write a blog for the Wild Lands Trust and a short piece on the relationship between hurricanes and climate change for a magazine (this one paid the princely sum of £25); respond to invitations; set up future invitations; and of course, the inexorable, relentless, inescapable item, 'catch up with other emails'. What does it take to escape one's own email backlog? More than cycling across a distant continent, clearly. I had two hours before I was due to meet a woman called Carmen in a coffee shop.

After so long on the high Altiplano, Alexander Coffee, with its choice of lattes, high alcoves, funky music and multiple food options, was both outlandishly wonderful and a touch overwhelming. It was taking me longer each time to transition from on the edge of feral to someone who remembered how to behave in a civilised way around cakes. Carmen spotted my slightly unsure approach and greeted me with a fabulous hug.

'We're over here,' she said, ushering me safely to a table in the corner. 'And this is my friend, Okiram.'

Carmen had represented Bolivia at the most recent international climate change convention. Okiram worked for the Japanese government in La Paz. Both were passionate activists.

I told them a little about my journey, and my recent conversation with Lilian.

'It's a real problem, and a real rural/urban split,' Carmen said.

Many people in rural communities were, in her view, now genuinely scared of wild animals – partly because animals were losing their habitat and coming closer to places where they lived. Sometimes eating livestock was the only way those animals could survive.

Okiram nodded. 'And people in urban communities, as a rule, just don't get this,' she added.

In Bolivian law, they told me, it was illegal to kill any wild animal unless this was part of an indigenous tradition (and you were an indigenous person), or unless it was part of a management plan. The trouble was that the law wasn't enforced.

'There's a whole bunch of wild animals that *could* be managed in a sustainable way,' Carmen said. 'If we had a management plan. But we don't. And our government really doesn't care about wildlife.'

'But surely ...' I began, still clinging to the vestiges of my belief in Evo Morales as a green leader and Bolivia as the heartland of *Pachamama* and the champion of Earth Rights.

'Yes, yes, I know,' Carmen said. 'Do you know what we call him now? Evil Morales.'

The scandals around him could no longer be hidden, they told me, despite that there wasn't really a forum in Bolivia for debate or information exchange.

'Day by day, we are doing nothing and creating the perfect scenario for a really bad future,' Okiram said. 'Take water. Again, the trouble is we don't have a plan. What we have is lots of *Pachamama*-istic discourse, and in reality? Violent extractivism.'

It seemed that a plan was beyond urgent. In relation to fresh water, Bolivia was suffering from shrinking glaciers because of climate change; overuse and pollution of water by silver-mining companies and other extractivist industries; and yet more pollution from certain forms of farming.

'In *Bolivia*, where we've already seen riots over drinking water,' Carmen said, with an almost furious disbelief. The relationship between Bolivia and water had even, she reminded us, featured in a James Bond movie. Yet, Bolivia had not learned to see beyond profit, short-termism and misinformation.

Okiram agreed. 'Take the disappearing Lago Poopó,' she said. 'It's as much about mining as it is about climate change. But we never hear this. We call it "a high-quality bad story".'

They both thought that we needed laws to protect water cycles, not just water. And experiential education to counter the disconnection between people and the way water cycles, among other things, worked.

Okiram suddenly grinned.

'Actually, young people in Bolivia are educating themselves via social media,' she said. There's real hope in that. Also, in small initiatives. The big, left-wing NGOs don't want to accept that Evo Morales is now bad news. But the small, independent ones can be much freer, and more creative. It's important. We need to be ready with lots of new ideas for the moment when the fake stability cracks open.'

As I went to put my notebook away and head off, they gave me a serious warning about trying to exit Bolivia by cycling west across the salt flats – which I'd been contemplating and had run by them as an idea. Not only was it a major drug-runners' route, they told me, but it was probably peppered with land mines. They weren't too keen on the idea of me cycling south to the Eduardo Avaroa National Reserve, either, despite earlier insisting I should go there.

'There are flamingos. On multicoloured lakes. You will never see anything like it. And it's a really positive biodiversity story,' Okiram had said.

Their suggestion was that I should join a tourist trip and travel to the reserve with Woody on the roof of a tour bus or jeep. And then to cycle into Chile from the southern entrance of the reserve. I was definitely in favour of circumventing any likelihood of serious trouble, but figuring out which 'danger ahead' stories to take seriously, and which to take with a generous pinch of salt, was always a challenge. If I had listened to them all, I'd still be in Colombia. No. I'd still be at home.

There was one definite problem with the southern route though. And that, as usual, was time.

The next day turned into one that made me question what good things I'd done in a previous life to deserve it. Ely had invited me to join her and two of her adult children 'somewhere special' for lunch. I took a taxi to her house, some distance south.

Ely's house, in a gated community, was octagonal-shaped and beautiful, with a huge garden full of flowers, a deck with a gentle, welcome breeze and a view of the river nearby and mountains beyond. From there, Ely drove us through extraordinary landscapes. Her daughter, Valentina, on crutches with tendinitis from too much dancing, sat in the front. I sat in the back with Ely's son, José, though I was mostly glued to the window. The landscapes seemed more like moonscapes. Swathes of desiccated grey and orange swirled around huge spires of rock. The spires were increasingly being flattened for buildings.

'Expensive houses,' José said.

'This valley, between La Paz and El Alto. It used to be remote, with only indigenous people living here,' Valentina said. 'Now it's home only to those lucky enough to afford the prices.'

The lunch venue was down a track, in a garden courtyard, full of ancient, native trees and fuchsia bushes dense with brilliantly coloured flowers. The restaurant specialised in cheese. As the conversation unfolded over beer fondue, wine fondue and hibiscus juice, my amazement increased. I was spending this Sunday eating the most astonishingly delicious meal, in the most astonishingly beautiful place, with people who, it transpired, were right at the heart of Bolivian politics and activism.

Slowly, I pieced together different fragments of their story. Ely's former husband, José and Valentina's dad, was a man called Pablo Solón. He had been Bolivia's ambassador to the United Nations, known as Morales' right-hand man. He was currently in exile in Thailand, after what Ely described as 'a polite disagreement' with Morales over the new road that Morales proposed to punch through biodiversity-rich rainforest, doubly protected as indigenous territory and a national park. Pablo, she said, had been threatened with legal action over completely spurious charges.[1]

'It was meant as a warning,' Ely said. 'A warning from Morales to everyone. This is what happens if you disagree with me. Whoever you are.'

'And the Solón Foundation?' I asked, having read about it online. 'Is there a link?'

'Absolutely,' said Ely, topping up my glass.

Pablo Solón's brother had been one of the *desaparecido*, 'disappeared' by a previous dictatorship. He was thrown in jail, and then he vanished. He was presumed killed, but his body had never been released. Both his wife and father – Walter Solón – had been radicalised by the state's failure to find out what had happened, despite a court order stating that it must. The foundation, based in Walter Solón's former home, was now an art gallery, celebrating his work, and set up with the intention of 'fomenting creativity and the critical perspective of rebellious spirits'.[2]

Fundación Solón was also a sort of activist think tank or, in its own words, 'a centre for questioning and searching for alternatives based on art, analysis and activism to try and change the system'.

It had links to other international organisations, and their writings included the fabulously entitled, *Short Guide to Warming Social Justice and Cooling the Planet*. Ely had been the foundation's director until 2015; José now worked at the foundation, occasionally collaborating with his father. The entire family seemed to have been on the front lines of radical politics ever since the disappearance.

And they all shared Carmen and Okiram's immense disillusionment with Morales. They told me more about the environmental and social impacts of the road – so severe the proposal had provoked riots and civil unrest. They had been known as the TIPNIS (Isiboro Sécure Indigenous National

Park and Territory) riots, after the indigenous lands it would devastate.

They told me of Morales' links to the agribusiness giant, Cargill, responsible for vast areas of deforestation to grow soy for animal feed and described by Mighty Earth as 'the worst company in the world'.[3]

'Pablo once summed it up by saying there are now different versions of *buen vivir*,' Ely said. 'An official vision that is passable even for financial institutions like the World Bank. And another that is subversive and rebellious.'

Then they told me about Morales' support for immense mega-dams: mega-dams with mega impacts, but no potential customers.

'The dams cost a horrendous amount of money to build,' Valentina said. 'We need to export the energy they create it order to pay for them. But nobody wants it.'

Later I found a quote in an interview with Pablo Solón. The dams and the hydropower associated with them, he'd said, would:

> Inundate an area five times larger than the city of La Paz, displace more than 5,000 indigenous peoples, deforest more than 100,000 hectares and will not be profitable for the country with the current prices of electricity in Brazil.[4]

'What's driving it?' I asked.

I understood the lure of extractivist projects that were highly lucrative in the short term and that allowed a populist government to improve conditions for the poorest, improve infrastructure and so on. But the whole dam thing seemed to make no sense.

Bolivia, they explained, was, like Ecuador, heavily in debt to China. China wanted to build the dams. What China wanted to happen in Bolivia generally happened. Another aspect of it was sheer machismo.

'This dam', Ely said, 'and the road, are both essentially huge, male, mega-pride, mega-technology things. It's a fundamental part of Morales' psyche and part of the wider macho-techno worldview, sadly alive and well all over the world.'

Even so, they all thought it astonishing they were still fighting environmentally disastrous big dams that threatened people's access to water, given everything that had happened.

After the Bolivian 'Water Wars' in the early 2000s, which culminated in tens of thousands of people marching to protest against the privatisation of water and the steep rise in the price of water, privatisation was reversed and laws were passed that gave small, local farmers and indigenous people control over the water they needed. Ely had been very much involved – she'd since dedicated a huge chunk of her professional life to working on water rights, seeing them as both crucial in their own right, and a blueprint for other necessary change.

The Water Wars had, she said, helped to inspire a worldwide anti-globalisation movement and provided a model for struggles for water justice around the world. At the heart of all these struggles was the fundamental notion that water, vital for life, is part of the commons, and should be managed as such. Not privatised and sold as another commodity by multinational companies. Or dammed for hydropower that would then be exported.

I could imagine Carmen and Okiram nodding and giving a thumbs up as Ely spoke. How many other issues could be seen in this way? Fresh air, fertile soil, even biodiversity could all be understood as part of the commons, essential for all of life.

The days began to fall into a pattern. I would start working, determined to clear the decks. Then there would be a reason to go out.

One of these was the chance to join a jaguar demo outside SERNAP,[‡] the Bolivian government department responsible for national parks and other protected areas. Jaguars, already extremely rare, were regularly being killed. The culprits were often known. In another darkly bizarre manifestation of the way China unhelpfully influences events in South America, they were hunted by or for Chinese workers on the dams and the roads, who in the absence of tigers for their traditional medicines and aphrodisiacs, used jaguar parts instead. But nothing was being done about it. I joined the small but vocal group, which included Carmen and her activist aunt. The lead person was a young woman called Nohely, tall, slim, long, dark hair and a powerful, almost jaguar-like aura. There were, Nohely said, only between 2,000 and 3,000 of these extraordinary creatures left.

After about six days, I decided I had to leave. Carmen shook her head and told me I was crazy. A friend of hers, Alex, an indigenous forest guide and activist who spoke English and was a leader in indigenous rights, biodiversity protection and ecotourism was coming to La Paz, either tomorrow or the next day. I should stay to meet him. And then stay for an organised walk to a glacier to learn about water issues.

It felt like a critical decision point. I wanted to do all of these things. I could take a bigger chunk of time just for Bolivia, making the most of these leads and contacts I was lucky to be getting. And then deal with the bus consequences somewhere in Chile. But this was also a cycling story: one with a deadline in time and an endpoint in space. If there was one thing I'd – finally – learned, it was that those two things together will cause you time stress, however much you think you've planned to evade it. I decided to hold on to my commitment to cycle all, or at the very least, nearly all the way, turn down these amazing opportunities and go.

[‡] Servicio Nacional de Áreas Protegidas.

20. BIRDS AND BEES

Life lessons from the road: no road is rough for ever. Ditto smooth.

I left La Paz in a taxi. I could not face riding back up that steep, narrow, uneven, half-built and now traffic-infested hill. I had no intention of joining the dead just yet. The place the taxi dropped me off was, I calculated, 2.6 kilometres from where my routes in and out of La Paz overlapped. For the record, that's 1.6155651 miles. As I hauled Woody and the panniers out of the small van, I thought about riding back so I could restart at the exact spot I'd departed from. But that was along a crazy road, thick with trucks, culminating in a spaghetti junction. I heaved the panniers onto the bike and rode off in the other direction. *The rot starts here*, I thought. *I have just cheated for 2.6 kilometres.* Was it rot starting, or an outbreak of common sense?

As I pedalled out along a long flat road through extensive urban sprawl, I thought about Carmen talking of the difference between urban poor and rural poor. No one was growing their own food here. Long sections of the walls of unfinished buildings had been whitewashed to display painted slogans.

¡Si! Minera es Desarrollo. Evo 2025. Patrio o Muerte.

Yes! Mining is development. Evo 2025. Country or death.

On the other side of the road, acres of rubbish dumps were picked over by dogs, people and caracara birds. Then, a dramatic statue of a man and woman striding out together in parallel step towards a multicoloured, checked and wind-buffeted flag, their arms thrust forward, fists closed round short, slim poles topped with a crescent moon and sun.

I was heading south, on Bolivia's Highway 1, a huge, modern road slicing through the high, dry Altiplano. Immense spaces. A vast, washed-out blue, cloud-wisped sky. An occasional fast bus or truck – sometimes with a cargo container – hurtling by and creating a buffeting wind turbulence that left me grateful for the hard shoulder. Small settlements. Scatterings of low, scrubby vegetation on the dry, dusty earth. Birds on the wires. A block-shaped house like a child's drawing, painted Viva mobile phone company bright lime-green; thick, white, three-foot-high Viva logos on every wall.

A dozen sheep were grazing on tiny patches of almost-green vegetation just by the road, their keeper a woman who threw stones at them to deter them from wandering into the traffic. When she saw me, she smiled and came towards me. I pulled up. She was from the next *pueblito* along, she said. She asked where I was from, and where I was going. Her friendliness and interest left me flooded with happiness. I didn't know if I had made the right call to leave La Paz. But I did know I was delighted to be back on the road: this road, right here, right now.

Dust devils like miniature tornados moved across the pale-brown plains. From a distance, I couldn't tell whether it was a vehicle kicking them up, or the wind. It was usually the wind, which was fickle and changeable. One minute it was blowing hard in my face. Then it veered to a sidewind, occasionally even a tailwind. My happiness and rate of forward progress were both almost entirely in its hands. When it dropped or swung behind me, there was an immediate sense of joy, of swooshing along and sheer glad-to-be-aliveness. When it backed to a headwind, everything became hard work, glee rapidly evaporating.

Kerbside lunches of bread, cheese, tomato. A cheap room in a yard behind a small roadside restaurant. After I'd been welcomed in, the owners full of kindness, I sat in the sun, on the ground, my back against the wall. A couple of guinea pigs emerged cautiously from underneath a car. There were cared-for plants. There were also several dogs on chains, who clearly spent their entire lives there, including a fluffy yellow pup called Marita. The dogs had kennels and were evidently well fed – they were just bored out of their minds. Marita was especially friendly, desperate for company in fact. The other dogs had given in, long ago, but she was visibly struggling to come to terms with the conditions of her life, smart and alert and looking for a way out. I checked out the suicide shower* in another shed, the hot water washing away some of the tired, sun-and-wind blasted feeling I'd become used to. Then I sat with Marita for a while, playing, in so far as that was possible on the short chain. It was heartbreaking to think that this was her life, for the rest of her life. Why was it like this? I didn't know how to understand this blind spot among such obviously kind and good people.

In the morning, as the owners inspected Woody, there was a sudden outbreak of tutting. Then I was shown which way up the Bolivia flag should go – red, yellow, green, not green, yellow, red as I had it on a sticker on Woody's frame – and given such a lovely goodbye that I was teary-eyed as I cycled away.

A few days from La Paz, I reached the place from where I should have been able to glimpse Lago Poopó; a gigantic,

* A widely used term inspired by collections of loose live wires in proximity to water.

saline lake, designated under the Ramsar Convention in 2002 as an internationally important wetland, especially for waterfowl. Instead, there was dry, cracked flatland. The lake, which used to be roughly the size of Luxembourg, had disappeared towards the end of 2015 and, while it had dried out and recovered twice before in its history, this time no one was expecting its return. Climate change-related loss of glacier meltwater was often fingered as the culprit. Carmen had said the bigger issue was the enormous quantity of water extracted from the lake for irrigation and mining. The latter had taken place around the lake since the Inca started seeking out metals for their armies in the 13th century. But the scale of it had increased after the Spanish colonisation in the 16th century. And it had increased again – dramatically – in more recent times, with a corresponding upsurge in water extraction. The result had been described as ecological devastation. Many of the lake's 200 or so species of animals – birds, reptiles, fish, mammals – had simply disappeared. And it was socially catastrophic too. People had lived from and around the lake for centuries. For some communities, entire ways of life had been based on fishing.

The road ran parallel to where the edge of the distant lake would have been. As if reflecting this retreat of the living, crosses appeared by the roadside, many decorated with black tinsel. Sometimes the crosses were on or inside small, rooved structures, which also held flowers, often in a plastic Coca-Cola bottle.

At one point, a wind devil came across the road right through me, so to speak – I had to stop and hang on to my cap. Then I stood and watched it twirling along, captivated. I could see other wind devils in the distance, hurrying across the land like small tornados. Soon, the wind really picked up, forcing me sideways into the road, exaggerating the effect of passing buses and trucks, so that I was buffeted around like a leaf in the constantly changing gustiness. With the wind came an amazing light, the sky almost black on one side of the road, with thunder rumbling around it. On the other side, flawless blue. Ahead, a huge dust storm, with trucks emerging from it like small monsters.

Eventually, I reached the turn off for Quillacas, where the far shores of Lago Poopó should have been. I imagined there might be a hotel there, perhaps even a guided tour that would explain the lake's misfortunes. I strained to up my speed a bit – the town was still many miles ahead. The earth was turning a sandy-white colour, with occasional outbreaks of reddish rock, and tan, occasionally greenish, scrub. Settlements of broken buildings were set back from the road, hard to say which were deserted and which were lived in. There were cacti on the hills to the left and some rocks painted with 'Evo' graffiti. Then just huge flat spaces, with tracks leading out into the sandy scrub. I was relishing the sense that the most adventurous part of the trip was, as I rode further into these intoxicating distances, just opening up.

That day closed with a wonderful mad ride, in fading light, into another dust storm. It was like cycling in low cloud, or fog made of very fine dust, the wind veering into my face then backing to my side. The last mile was up a hill. At the top, a huge sign appeared from out of the gloom.

'Welcome to Quillacas!' it read, with a colourful depiction of spiky white mountains under a blue sky and, in the foreground, two llamas looking at a pretty, red tile-rooved, white church with a surprised expression on their faces, as if they'd just encountered it for the first time.

Behind the lovely welcome sign was a somewhat bleaker reality. The place that might once have been a hotel was a semi-derelict building at the top of a rough track. It was most definitely closed. There was no other hotel, two women I met on the street told me. I was about to head for the plaza to see if I could camp there when one said, 'I have a room, though.'

It was, literally, a room, right there off the street. The woman swept it out a bit and showed me and Woody in. I couldn't quite believe my luck. When I asked about a toilet, she pointed across the street to a rough patch of ground with broken-down walls criss-crossing it.

'There, in the dark,' she said, and left.

I spent the rest of the evening in a tiny *polleria* – chicken restaurant – eating a meal of chips, water and herb tea, and inwardly laughing at my fantasy of a nice hotel and a day off with a tour guide.

Next morning, I cycled off into a huge, deserty expanse stretching away into the far distance, the road a gigantic fork under pale blue skies. There were dried-up riverbeds, and sand cracked into strange shapes.

> *I love this. I just love this ... I have no idea if I made the right call, leaving La Paz when I did. But that decision led to me cycling here on this road now, to this moment of sheer happiness. I love these immense, dry spaces, the feeling of wildness, the hot sun, the sense of being a bit more out there. And maybe that's how it is – we can never know which other decision might have been better than the one we made. But to have a happiness upswelling like this, and to join those moments up, that's the best there is. I am so very glad I came to Bolivia.*

Bolivia was producing some of the strangest landscapes I'd ever seen. One of these was a meteorite crater, a big shallow bowl not far from the road, tinged with pink and lime green around the edges and strange mats of plants and rocks that looked burned and volcanic.

And they weren't just landscapes, but habitats, places that were lived in and full of life. I had developed the practice of asking 'whose community am I pedalling through today? Who – other than humans – lives here?' Sometimes, it was obvious. Two-tone llama. Cacti, thick as posts and increasingly in flower. A group of small mud-coloured doves on mud, their camouflage simply perfect.

Other times, there appeared to be nothing bar miles of pale, dry earth, scattered with low, sparse scrub. Spotting a bird was often the way in to seeing a living community rather than an empty space. Radiating out around the birds: insects providing food, shrubs providing shelter, detritivores dealing with the dead. A million other community members, wild, indifferent, hostile, collaborative, interactive. A living system. An ecosystem.

'Birds and Bees' would be a good chapter heading, I thought, watching a small hawk, hovering above the rutted road. The birds, my companions the whole way, creatures almost everyone loves, as Kendra had said back in Colombia, forming bridges between people and the wider ecological community – and the need to protect it. And the bees, our most famous pollinators, brilliant ambassadors for the critical yet overlooked nations of minifauna.

As it became more remote, it became easier to find places to camp out of sight. That evening, I pulled off the road and pushed a short distance into the scrub, happy to be in the tent. I woke the next day, surprised to find I'd slept for eleven hours. I'd been feeling strong on the bike and thought the tiredness had gone. I dug out a tiny mirror and saw swollen, puffy eyes. Tilting it downwards revealed that my chin had developed new cracks and fissures and was peeling; my bottom lip was bloated and sore; my face skin was a persistent pink now refusing to tan; and my hair was hanging in thick, dank strands.

Hastily putting the mirror away, I unzipped the tent flap and looked out. A small lizard was slowly emerging from one of the scrubby bushes, so well camouflaged I could only see him when he moved. He came over and angled his head up towards the bike, giving every appearance of checking it out. I crawled out of the tent. It was simply beautiful, there among the small bushes under the huge blue sky. I put water on for coffee and decided I should put being in a hurry on hold. Spend an hour being where I was, seeing who emerged.

As in endorsement, three small birds alighted on the top of a nearby bush – yellowish, green-tinged, with a mess of woody stems – and delicately ate the seed heads. A smoky-grey beetle I'd evicted from inside the tent emerged from underneath it and legged it to the nearest shrub; a few tiny red ants marched to and fro. Around us, the plain stretched away to low brown hills on most of the horizon. Everything else was flat, as far as I could see, as far as where Lago Poopó should have been.

This is worth standing up for, I thought, as I sat there in my Therm-a-Rest chair, with my small green tent and bamboo bike (whose brake cables were fraying, I'd noticed). The right of this community – these shrubs, birds, lizards, beetles, ants – to do their thing, to live their lives, every bit as entitled to be here as we are. It was also what I wanted more of – to be there among them, hanging out in wild places with whichever living things happened to be around. I needed better words, but I was beginning to get what naturalist Henry David Thoreau had in mind when he wrote that 'in wildness lies the preservation of the world'. It was necessary to be there. Good for the soul, whatever that meant, exactly. And on the flipside, utterly heartbreaking to think that we were losing these wild communities, wild places, big and small, spectacular or scrubby, all over the world.

I tried to meditate, hungry for that glimpse of freedom, the relief when your mind shuts up for a bit. When I opened my eyes, the lizard was watching me. He walked slowly towards my foot, raising his head up and down and sometimes stopping, lifting up from his front legs and looking around. He reached my foot. Then, presumably smelling something unsavoury from a lizard perspective, whizzed off.

On those days in Bolivia, I always woke excited at the thought of where that day's cycling might take me. These wilder sections were the ones I was loving the most and, that morning, from tent level, my shrubby, beautiful, unexpectedly life-rich desert had seemed immense, boundless. I loaded Woody and set off. A hundred metres later, the desert, quite literally, stopped. There was a line of wire fence, and beyond it, the land ploughed into stripes with rows of small, green plants that looked a bit like cabbages with purple centres. By the road, a big bucket of pesticide spray in a backpack. The demarcation was stark. Sand and shrubs, then lines of plough. A family eating a meal sheltered from the sun underneath a tractor.

Just before I reached the town of Salinas de Garci Mendoza, a six-foot-high artificial plant stood in a custard-coloured pot by the roadside. It appeared to be a giant version of the cabbage-like plants I'd seen earlier. Around the pot, in large green letters, were the words '*CAPITAL de la QUINUA*'. Aha. Quinoa. I'd been reading about it but hadn't realised this was the plant I'd been looking at. Quinoa had been a staple food in Bolivia for decades, perhaps centuries. Its sudden 'discovery' by the West, hailed as a superfood, had sent demand for it skyrocketing. This had brought much-needed cash into the Bolivian economy. But it had also escalated the price of quinoa so that the people who actually depended on it for nutrition could no longer afford it. Most of it was now exported. And the dramatic increase in production had meant more irrigation, more pesticide use – more impacts on fresh water, on biodiversity, on soil.

Other-than-human communities are entitled to be here, too, I thought, again, cycling slowly now. We had to find ways of coexisting.

The road was getting rougher, and more remote. Dark, salmon-pink streaks ran down the steep flanks of a distant volcano. Against that backdrop, a group of roofless, mud-brick houses and one, beautifully crafted, of stones.

Then suddenly, over the brow of the final rise, there they were. The salt flats. It was like seeing the ocean. Only white. A huge, hard, white ocean, vivid and strange against the brown mountains to the side. I dropped down into Tahua, a town that seemed half-built, with massive holes in the narrow dirt roads that carved a route through the mud-brick houses. The dirt road to the salt flats went past a cemetery, the crosses glittering with black and purple tinsel, and then a brief swathe of greenness, grazed by lamas. And then the road stopped. Two cairns marked the entrance to the salt flats. I pedalled out into the immense whiteness, as if I were pushing out to sea.

The flats were patterned with hexagons from the salt crystals, and the edges of the hexagons were slightly raised, so that they crunched as I cycled over them. On the 'road' – a line across the flats worn by generations of previous travellers – these were smoothed out, though I could still see the patterns. Surrounded by whiteness, I kept stopping, feeling compelled to take photos to try and capture something of my wonder at what I felt and saw. As soon as I stopped, I could feel the heat, burning with unusual strength. On the move, there was a breeze. It was so utterly different from anywhere I'd ever been. It was like riding in a desert; a hot, white patterned desert. I was overjoyed to be there.

Navigation can be tricky on the flats. Away from the shore, there are few landmarks, and there's an intermittent magnetic interference that can render a compass unreliable. I had planned a route to the other side using one of the few 'islands' to keep me on course. Initially out of site, Cactus Island slowly appeared as a low blur in the distance and, as I pedalled, gradually took shape, hardening into a place I could aim for. In the late afternoon, the island still some miles ahead, I left the track heading more or less west, using a compass bearing and a visual guide provided by another distant blur, blue on the far horizon roughly in line with the soon-to-be-setting sun. I rode for half a mile, and then I simply stopped, pitching my tent as the wind picked up. I'd read that it's impossible to get tent pegs into the salt flats and had brought a few small rocks to help keep the tent stationary. I tethered the windward side to Woody, for once grateful for his considerable weight. Then I watched the sun sink and the ocean change colour. It became very cold, very fast. I was drunk on immensity. And later, crawling out of my tent into the freezing night, on the sheer beauty and chaos of the crazy, southern stars, glinting at their ludicrous distance across the black night.

I reached Cactus Island the next day. It was a small oasis thick with cacti, six feet high in stands like small, spiky forests. 'There are birds here,' an English-speaking tour guide told me, 'including the Andean canary. Also insects, and viscachas.' The viscacha numbers had been rapidly dropping, and there were rumours of an Andean cat.

'On the salt flats, though,' he said, 'nothing lives. Nothing at all.'

'Not even bacteria?' I asked.

'Not even bacteria,' he said.

I could, he assured me, lick the flats without risk, should I wish to. The salt killed everything.

Back on the flats and heading into the vast white space beyond the island, I started to see jeeps. Tourist jeeps, small herds of them, travelling in long straight lines in the distance. The road beyond the island was stained dark grey where they had been. If a place had no life, I wondered, was it part of habitat-biodiversity nonetheless? And leaving aside the carbon footprint, if there was no life, did it matter that it was being driven across? Tourism here was generating income for Bolivians: income not based on mining. The impacts seemed largely aesthetic. But still. And, despite the income from tourism, the flats were threatened by the promise of vast wealth to be made from lithium mining, increasingly in demand as a component in batteries for electric vehicles. The thought of the salt flats being despoiled by jeeps and mining was gut-wrenching. Even if not from a biodiversity perspective, the survival of big, wild, relatively untouched landscapes seemed ever more important. *In wildness is the preservation … Yes, poverty must be tackled. But there must be other ways.*

The day got hotter and hotter. Chunks of cracked, burned skin were falling off my face. The cycling was relatively easy, but the flatness, the relentless sun and the sheer, huge, empty space presented its own challenges. I worked on future-tense Spanish verbs, written on a by-now rather battered piece of paper next to the map in my map case, and listened to music, and just cycled. I felt as if the beauty of the place had scorched my soul, and that now I needed to look away. I found, to my surprise, that I was not yearning for another night on the white and empty plain.

By the time I reached the western edge of the flats and the multicoloured, waving mass of international flags that are a monument to, of all things, the Dakar Rally, I was definitely tiring. There was no UK flag that I could see, though there was one from Brighton and Hove. A friendly man selling tourist goods told me I should keep going to Uyuni, though it would make for a long old day.

'There's better food and hostels there,' he said.

The road from Colchane – the nearer town – to Uyuni was great, he assured me, by which he meant asphalt and flat; criteria I could endorse.

And then a long ride through the fast-fading light, out of the ocean and back onto the road. Looking back, I could see the sun setting over the islands. There was a lasting glow above the salt flats, as if the whiteness kept generating light even after the sun had gone.

The jeep was pulled over at a police checkpoint. Five men in green uniforms surrounded Ronal, the driver, as he climbed out. There was a discussion I couldn't hear, and a showing of documents. It looked as if money was changing hands. From his body language, Ronal was not a happy man. Then one of the police put his head through the window.

'Good afternoon,' I said.

'How many years have you?' he asked.

'Many!' I said, with a big grin, assuming the question must be the policeman's idea of banter. This was a mistake.

'HOW MANY?' he repeated. He was not grinning. He was not even smiling.

'Fifty-four,' I conceded, perplexed.

'And you?' the question was aimed at my fellow passenger, a Swedish tourist called Runar.

'Forty.'

The head withdrew. There was a heated conversation, during which I was pretty certain the policeman said, 'You can't take her.'

Eventually, Ronal climbed back into the driver's seat. We drove away.

'What was that?' I asked. A silence. And then an outburst about how corrupt the Bolivian police were, how this happened all the time, how the only option was to pay them or be given a ticket that would compromise his business as a driver, even though everything was in order. I never got to the bottom of the question about my age. I was used to police in every country I'd cycled through so far either being charming – intrigued by Woody – or ignoring me completely. This was the first time on the whole journey that I'd encountered anything close to police-related trouble – ironically, while in a tourist jeep. Were the Bolivian police a different kettle of fish? Or were the jeep drivers with their (relatively) wealthy passengers considered fair game? I wondered how it might have been had I met them alone on the road on the bike.

I'd fetched up in the jeep after a day in the town of Uyuni trying to figure out my next move. Following up on Okiram's suggestion, I'd decided that visiting the coloured lakes where the flamingos lived was a must, given how significant it was from a biodiversity perspective. On top of that, it sounded staggeringly beautiful. The flamingos and the coloured lakes were to be found in the Eduardo Avaroa National Reserve, in the south-eastern corner of Bolivia. The road through it would take me to the southern edge of the country and the border with Chile. But cycling there from Uyuni would take days, and the road between Uyuni and the north of the reserve was said to be terrible. If I rode out this way, I would have to accept the need to either get a bus later or reroute my meeting point with Chris so far north that his entire holiday would become a mile-bashing, main-road mission instead of a meander around some of the most beautiful parts of Patagonia.

An alternative was to take a tourist jeep trip to the coloured lakes and then be dropped off at the southern edge of the park. I'd get an insight into how Bolivian tourism benefited from – and presented – the astonishing biodiversity of these lakes, surely a better way of cheating than taking a fast bus through great swathes of Chile. But none of the trips went to the coloured lakes alone. They all went to the salt flats first. When I went to book the trip, something in me revolted. After the soul-searing beauty of cycling alone across that astonishing landscape, and the most powerful night under the stars I'd ever experienced, the salt flats had become somehow sacrosanct. I could not face returning there in a crowded, noisy jeep and having lunch under an awning.

Finally, I'd found a compromise. One of the smaller tourist companies sometimes took a few people directly to the lakes and then to the Chilean border. I could join one of these outings. But their jeeps went straight to the lakes in the south of the reserve, whereas the vast majority of the flamingos were on Laguna Colorada, the most coloured of the lakes, in the north. Eventually, a deal was struck. I would take the jeep as far as the crossroads at the reserve's northern boundary, and the driver would decant me and Woody there, before turning off for the southern lakes. I would cycle west to the Laguna Colorada, then head south through the rest of the reserve from there. It would save less time. But it would still save some. And I'd get to see the flamingos.

Having agonised over the cheating/cycling-the-whole-way issue for so long, when I'd actually made the decision and helped lift Woody onto the roof and got into the jeep, it felt strangely irrelevant. In fact, I'd spent the first half of the jeep journey – about 140 miles in total – delighted not to be cycling. From Uyuni, the road rapidly deteriorated into thick gravel, with gargantuan ruts and sections of extreme washboard that would have been unrideable, as well as long stretches between anywhere I might have been able to source water or food. It was hilly. And of course, it was also windy, though exactly how windy was blissfully hard to tell from within the jeep.

I spent the second half thinking, 'Oh shit.'

'Is my road through the reserve better or worse than this one?' I had asked.

'Worse,' said Ronal, with an exasperated head shake. Ronal thought my plan was on the limits of sanity.

It was still only 6am when Ronal dropped me off at the crossroads after a night in a tourist hostel. It was icy cold, under a sky of such deep blue it looked unreal. The dirt road ran straight ahead, tan-coloured, marked with tyre tracks, through a wilderness of loose, dry, tan-coloured ground, towards a low line of tan-coloured mountains and volcanos, tinged with violet and cream. I could see no other form of life in any direction. I struggled to load the bike with fingers already frozen inside gloves, strapping litres of extra water in plastic drinks bottles onto the already heavy load.

Come on, Kate, get a grip. How hard can it be? Well, let's see. It could be unrideable. I could run out of

> *water or find I am physically incapable of hauling all this water and have to jettison some of it. Getting to the far end of the reserve could take days of pushing uphill, which won't be possible without water, which I will by then have run out of.*

Part of me berated myself furiously. What the heck had I got myself into? Why, why, WHY couldn't I have booked a tourist trip like a normal person? But part of me, albeit a smaller part, was looking forward to pushing this boat further out. Wilder, remoter, harder. Game on.

I rode off slowly down the hill in the direction of the out-of-sight lake, stopping occasionally to try and revive my fingers.

The true nature of the road soon started to become evident. Short sections of washboard increased to long sections. Short sections of deep sand or deep gravel increased to long sections. I made slow progress, pushing through the roughest bits or across the raised centre to test the depth on the other side. Suddenly, having heaved Woody up to a slightly higher section of track, I realised that the flamingos were right on the lake edge, now in view. I left the bike and walked across a patch of low, mustard-yellow foliage towards them, cautiously, not wanting to put them to flight. The flamingos did not seem in the least bit bothered by my approach, even when a black-headed gull came and screeched at me, hovering right above my head just to make absolutely sure everyone knew I was there.

It was real 'other nation' stuff. The flamingos, a very pale, almost white pink, tinged and striped with a stronger pink the colour of a stick of rock. The darker pink, almost magenta-coloured lake, and the contrasts – thick bands of white salt across blue and pink water. On the shore, hordes of little squabbling sandpiper-type birds, and two birds I didn't recognise at all: half black, half white. The flamingos – in my head I kept calling them flamencos – walked stiff-legged and deliberately, at once ridiculous and utterly elegant. A couple were doing that neck-rubbing thing. They were all constantly chattering and chirring away.

I wanted to hang out there in the bird nation indefinitely, but the wind was picking up and, while no longer freezing, it was still cold. I went back for Woody and rode closer to the lake for a while, constantly stopping for photos. Further along the lake, thick, long streaks of orange clashed with the magenta. I'd never seen such colours displayed so vividly in a natural setting. More flamingos. I dismounted and lay on the sandy gravel and simply gazed. The gravel was surprisingly hot. I was so very, very glad I'd come. It was one of the most astonishingly beautiful things I'd ever seen – not just the flamingos but the whole setting, and the other birds too. As for the flamingos, they were there in their hundreds, perhaps thousands. On the far side of the lake, I could see great groups of them on the water. And there was one small group in the air, wings glinting pink then flicking to white as they wheeled and turned, flying like a flock of gigantic, cerise-tinted starlings.

Eventually, reluctantly, I dragged myself away and got back on the bike. After that, the wind picked up another several notches. The road into the hills to the south got worse and worse. I was spending more time pushing. And this was still the flat section. The first climb was just ahead, and I had my eye on a curve near some rocky outcrops as a lunch spot. I didn't get close. I finally stopped where the lake road joined the road south and hunched down, almost out of the wind but not quite, eating fast, really hungry, despite a substantial breakfast of cold pizza. I was back on the road in minutes.

A jeep came towards me in a huge cloud of dirt. Ronal had said there wouldn't be any on this route, but now there were several. Two in convoy, heading for the lake, slowed, cracked open their windows, cheered me on. Their existence immediately changed both my assessment of how much risk

I was taking and my options. Another jeep stopped. There was a familiar tourist company brand sign on the door. The door opened. Ronal! He hopped out.

'You OK?' he asked.

'It's a little hard. But the flamingos are spectacular,' I said, feeling my face smile.

Ronal grinned. The previous night at the hostel, when it had become clear that I was determined to cycle, he had sat down with me and the map and marked it in biro with places I could camp. He pointed to the rock formation ahead.

'That's for tonight', he said.

It looked too close. Perhaps a little over a mile. It was only 2.30pm, though I was already done in. Ronal returned to the jeep. Then he jumped out again and handed me a plastic bag. Inside was a big round of hard cheese. And then he waved and drove off, leaving me greatly cheered by the whole encounter – and the thought of an early camp at the rocks ahead.

Three hours later, I was making desperately slow progress. The rocks seemed no closer at all. I was stopping frequently, like someone struggling with altitude, and no longer even thinking about riding. I pushed all afternoon, my shoes full of grit. Two out of three – deep gravel, headwinds, hills – is bad enough. Three out of three is a killer. But I wouldn't have said yes to a lift. I wanted at least one night there in this vast space, looking back to the Laguna Colorada from my tent. Interspersed with this resolution were a multitude of moments when the shelter of the outcrop actually seemed to be receding and I thought, *I can't do this*. And then, the trickle of jeeps having dried up, *Well, you have to. You can't camp here on the edge of the road on thick, rocky gravel in a minor gale.*

The rocky outcrop, when I finally reached it, was strangely like the lake only on a much smaller scale – an unexpected oasis of life among a high desert, albeit not one of the sandy, dune-filled variety. Here, among the huge sweeps of coffee-coloured grit and boulders, a small track led up to the outcrop of rocks that curved around a flattish spot of ground, protected from the wind. The remains of a watermelon, some loo roll and human footprints indicated it was a picnic spot. There were no humans around now, though, bar me. A viscacha sat in a sort of alcove in the rock face above the area where I'd thought to pitch the tent, greyish brown like a big rabbit, only with a long tail and slightly shorter ears. She watched me, unperturbed. A small flock of slate-blue birds landed on the flat rocks above her, and a large mouse emerged from the rocks around my feet, then skipped and jumped away. Later, the viscacha ran across the top of the rock formation with the sun behind her, moving like an ungainly squirrel, her tail backlit by the setting sun like something out of the film *Ice Age*.

I pitched the tent, made ginger tea, wrote my journal. I was already properly tired, after one day of this ride. Yet part of me was close to blissful, sat there among the rocks and desert animals with the lake still just in view back the way it had cost me so much effort to come. Woody was against a rock next to me, now virtually the same dusty colour all over. Thick, unusually sticky, tan-coloured dust, coating Woody, me, the panniers, my shoes, the map case, everything. For some unknown reason, the dust evoked a memory of scraping my knuckles on the hard, rough surface of the salt flats and being a little unnerved by how good my blood had tasted when I'd licked it off.

I woke late the next morning and got ready slowly, humouring my own prevarications – packing in inefficient spurts, oiling Woody's chain, making porridge, watching the viscacha – and then regretted the late start. It took me two hours to do the first 1.5 miles, a slowness record, pushing up a never-ending hill.

That morning set the tone for the next few days. The conditions would constantly ratchet up another notch of toughness. And another. And another. Most of the track across the vast salmon/brown, gritty landscape was unrideable in a constantly evolving variety of ways. Deep sand or gravel. Steep washboard. Rough rock. When there were rideable sections of track, the headwinds would typically kick up and make it impossible anyway; and if the wind dropped momentarily, the road would instantly become too rough to ride. I was beginning to have sense of humour failure, expending such immense, exhausting amounts of effort for so few miles. Around noon, a few jeeps would go by. Some would stop. The people inside would ask lots of questions and want to take photos and tell me I was amazing, which I began to find seriously irksome. They never asked anything practical, like, are you OK? Or, do you have enough water? Or, is there anything else you need? Like, maybe, a lift.

I took a detour down an even worse track to reach a site where I knew there were geysers and also, I thought, a *refugio*. On an especially bumpy downhill section, I had a slow-motion, clipped-in, sideways fall, landing with a hard crunch on my left arm and elbow. I lay on the dusty gravel, undamaged but a touch dazed by the sheer hardness of the ground, and whimpered. And then I got up, dug the dirt out of my cleats and carried on.

When I reached the site, there was no *refugio*. Just some stinking, sulphurous pits and steam emerging from cracked, yellow-tinged earth. Nearby, a small, round, half-derelict stone building, with walls – some bizarrely crenelated, like a sort of folly, but with no roof. Inside, a stone platform at the far end and mounds of human excrement. It was late, I needed shelter, it was still windy, and I had slept the previous two nights in all my clothes and woken with ice in the water bottles. *Just deal with it.* With a handy stick from goodness knows where, I swept a patch free of shit on the platform and put my tent there. The air was so dry the shit barely smelled. Next morning, I shared my porridge oats with two small mice who emerged from rocks close to the ones I was sitting on, while I thawed enough water for breakfast. Back on the road, another gravelly, uphill section which I mostly pushed. These days, four miles per hour seemed exhilarating.

As if in an echo of my first day, between the hardness, things would occasionally veer to the sublime. It was stunning, of course, not beautiful exactly, but wild and challenging – the enormous spaces and sudden, huge, breath-taking views over the brows of brown hills across the vast, brown desert to the strangely shaped mountains. The mountains were brown and tan and streaked with cream like the pelt of the vicuñas or of some other fabulous wild animal.

I came to another lake, turquoise and blue this time, and with flamingos, though in fewer numbers. And, about halfway across the reserve, the geysers and hot springs that did, in fact, have an accompanying *refugio* and restaurant so that, sitting inside at a table, with a plate of eggs, potato and salad, I suddenly felt I'd changed worlds.

'You should make time for the springs,' the kindly woman who served me said, and told me where to get a ticket for the hot pool. I sat in it for an hour, alone with the heat and steam, soaking my muscles with a view across the turquoise lake as faintly sulphurous smells rose around me.

And then back on the road, with the gravel and the inclines and the ever-strengthening wind. It seemed that each new day I gave up trying to ride sooner, and pushed uphill into the wind, for hour after hour.

I had taken to bribing myself with chocolate biscuits on the inclines. *Get to that rock and you can have half a biscuit.* A whole one at the summit. Finally, not long after one of these summits, I cracked. The wind was so strong I couldn't ride down the gentle, smooth, enticing slope on the other side. Then the road flattened out and the gravel instantly thickened, so that between the wind and the gravel I still couldn't ride. I stopped. But this time I threw the bike down and howled. The howl turned to violent sobs, then yells. I observed myself shouting. *I hate this fucking road.* I had never felt such frustration, working so hard, pushing myself to exhaustion and yet making so very little progress. Alongside the frustration, something close to despair. My average speed for that day so far was 1.6 mph. 1.6 mph! *I am going to be stuck here for ever in this windy gravel pit of a so-called road. I CAN'T GO ON!* Then I looked around at the empty desert around me, a single small vicuña on a distant slope and thought, *well, what are you going to do then?* I picked up the bike and went on, waving at the next jeep, as if everything was just fine. *No road is rough for ever. It is logically impossible for a hill to go up for ever. This one will also smooth out and go down eventually.*

I spent a final freezing night in a gravel pit that offered partial shelter just as I was thinking my only option on the empty plain was to spend the night wrapped in the tent that the wind would make impossible to pitch. I'd never in my life been so pleased to see a gravel pit. About halfway through the fourth day, as I sat eating a stale roll and the last of Ronal's cheese, I realised that what made the difference between being physically exhausted and being physically exhausted *and* emotionally destroyed, was optimism. I had thought, time and time and time again, *OK, this time I really have reached the top and the end of the hard bit and the road is going to change and this will all get easier.* And then, around the next bend, there would be another climb or another gravel swamp.

Accept it. On this road, the hard bits run all the way through. After that, I felt a sort of release. And after that, of course, the road finally did begin to change. My average speed rose to a giddy 4 mph. I began to catch glimpses of the southernmost lake, Laguna Blanca. Then buildings in the far distance.

I spent that evening in a tourist hostel opposite the SERNAP park building at the edge of the reserve. A group of birdwatchers arrived after sunset, laden with binoculars and cameras with huge lenses and hats with strings under their chins.

Later, with the help of a beer, I tried to make sense of what I had just done. It had taken me four days to cross the reserve. Four unreasonably hard days. In total, I must have pushed the bike for almost three of them. I was chuffed to have done it, but at what cost? It had been tough enough to leave a scar. On the plus side, the experience had shaken me at least a little loose from my obsession with time and miles, an obsession that had crept up on me and become compulsive. I was not sure I cared a hoot any more about my 140 jeep miles or possible future buses. I wasn't entirely sure what, but I'd learned something and earned something, riding – walking – this section of the journey. As the beer – and food – took effect, I was increasingly glad that, however hard, I had tackled this section by bike and not jeep-toured it.

Before I left, I went across to the SERNAP office to pay for a ticket for the reserve, which I hadn't been able to buy on the way in, on account of entering it at 5.30am. Inside were pictures of the reserve's animals and plants, and information about the Andean Cat Alliance. The man at the desk looked out of the window at Woody leaning against the wall and asked where I'd come from and how long it had taken. When I told him, four days from the Laguna Colorada crossroads, he nodded. And he wouldn't let me pay.

21. THE ATACAMA DESERT

'I have always loved the desert. One sits down on a desert sand dune, sees nothing, hears nothing. Yet through the silence something throbs, and gleams.'
ANTOINE DE SAINT-EXUPÉRY, *THE LITTLE PRINCE*

'There is no wealth but life.'
JOHN RUSKIN, *UNTO THIS LAST*

CHILE

Happy Planet Index score: 45.6
Rank: 67th out of 152

I rode to the border past vicuñas grazing, seemingly on dry earth. Jeeps raced by in small convoys, usually too fast, spitting stones and trailing dust clouds. A final section of deep gravel on an uphill stretch with a headwind, as if as a parting Bolivian gift, delivered me to a grey concrete rectangle not much larger than a bunker. '*Migracion Bolivia*' was painted in black capitals on a white patch. Was this the remotest border crossing in the whole world? Possibly. Inside, a single man sitting at a table stamped my passport. That was it.

'The Chilean frontier?' I asked.

'In San Pedro,' the man replied. And then, having taken in my dust-laden cycling gear, 'From here to San Pedro, all downhill.'

San Pedro was roughly 25 miles away, so this was good to hear.

Not long after, a large green road sign proclaimed '*República de Chile*' alongside the Chilean coat of arms. Immediately beyond the sign, the road surface abruptly transitioned from gravel to asphalt. Asphalt! Wonderfully smooth, hard asphalt. I was tempted to get off and kiss it.

I looked back at the terracotta-coloured, volcano-strewn Bolivian desert, said goodbye and pedalled into Chile. This new country quickly sent a message. I might have survived by far the toughest section of the ride so far, but Chile was not about to be a pushover, either. 'All downhill,' turned out to be the jeep-driver's version – miles of steeply undulating road which did not feel remotely downhill on an overloaded bike in a headwind with a knackered rider. Then, finally, the most welcome road sign ever. A yellow diamond with a black truck on a ludicrously steep slope, the truck's nose pointing vertiginously downwards.

For over twenty miles we plummeted, a fabulous, fast descent towards warmth and comfort. A few shrubs started to appear by the roadside. Shrubs! Then, the occasional bird. Three donkeys. Some purple lupins. I found myself singing.

The San Pedro de Atacama border control was clearly signed. A policeman stamped my passport and passed me on to a colleague, who said he only dealt with cars and trucks, things with engines, and a bicycle had no engine. It was unclear whether this was a problem. I waited a while for clarification, chatting with some tourists from the USA, bemused by my sudden re-entry into the world of conversations and friendliness. After a time, another border guard arrived, brandishing a form. The real question, he explained, was not whether I had an engine, but whether I had any vegetables. He was helpful and kind and interested in Woody. I filled in the form and reluctantly handed over a pepper, three green beans and a squashed tomato. Then the few remains of Ronal's cheese and a scrap of stale bread.

'Biosecurity,' he explained, gently, clearly observing the tell-tale signs of food-insecurity anxiety in my eyes.

I genuinely forgot about the bulb of ginger and half-packet of spaghetti at the bottom of my pannier. Would they wreak havoc on Chilean ecosystems? I somehow doubted it.

By early afternoon, I was wandering around San Pedro de Atacama, struggling to come to terms with the way in which Chilean Spanish – a different accent *and* different words – seemed to have kicked me back to language comprehension square one. On top of that, I was a touch dazed and out of it in general, the physical and emotional residues of the high-desert ride as hard to dislodge as the dust. My time in the desert had – so far – been short, and toughness is always relative, but I could almost imagine how it must feel to reach an actual oasis when you really, really needed one.

I found a hostel that had not only a bed, but a hosepipe – and a willingness to let me wash the hard-caked layers of desert dust from Woody and give him some oil-based TLC right there in their courtyard. I found a laundry and changed money and figured out that Chile was an hour ahead of Bolivia. I drank coffee and ate burritos and watched the sparrows and wrote my journal and watched hundreds of emails download and sent messages to Chris and my dad. Then, as it got dark, meaning to go back to the hostel, I was seduced by an outdoor restaurant with live music and one small table left by the fire pit. I took it, lingering in the friendly, lively atmosphere over an almost unbelievably tasty risotto and my first ever Chilean pisco sour, sharp and delicious. What a different situation from just a couple of days ago. Life is made so much more vivid by contrast.

Next day, I chatted with the hostel manager – a helpful, kind young man with a mop of turquoise hair called Dani – about the road ahead. Then, stocked up with road food and water and fortified by a big hug from Dani as much as by the provisions, I popped out of the bubble that was San Pedro. I passed a bunch of very drunk men on the outskirts and pushed away from the oasis and back out into the desert. A large, brown road sign confirmed it. '*Ruta del Desierto*'. A young lad overtook me on a mountain bike, wearing a green helmet and a board strapped to his back. Then he veered down a side-track and disappeared off into his own world. A group on horseback were heading out on another track. And then it was just me and Woody, climbing. We climbed for the best part of twenty miles, and the best part of the day, under the hot, dry sun. No gravel. The road was hard and smooth. I almost floated.

The Atacama Desert is one of the driest deserts in the world. Parts of it have not had rain for 40 years. It's not your classic, yellow-dune, scenic desert. There are long industrial sections, mined for copper, lithium and sodium nitrate – aka Chile saltpetre – the basis of Chile's wealth. Other immense stretches are just vastly, endlessly, relentlessly brown. One day my 'other-than-human life form' count was three flies, though there was evidence of past human life in the form of an artist with a dubious sense of humour. Occasional rocks near the roadside had been painted to look like a flower, a mushroom, an old boot and once, a slice of watermelon, red and luscious. *Who does that?!* Was it a local Sunday hobby, to come out here among the sand and grit and paint rocks to torment passing cyclists in the 30-plus degree heat?

In the attempt to make my rendezvous with Chris without resorting to further engine-assisted transportation, I had only made one arrangement to visit a project before Pucón, and even that I eventually cancelled. I rode, and ate, and slept. To reach Pucón on time, I needed to ride 500 miles a week, for a good three weeks. Seventy miles a day, every day.

On a road bike with less luggage that would have been a perfectly reasonable ask. But here, on Woody … Ultimately, of course, it would be in the hands of the winds. I was beginning to think that headwinds had nothing to do with direction or geography and everything to do with a given cyclist's karma. Mine was clearly bad.

Occasionally, trucks or cars would stop, and I would be handed a can of Monster Energy drink or a peach. These encounters always left me cheered and invigorated. In fact, I was thoroughly enjoying Chile so far. The friendliness was exceptional and, while being back in a 'rich' country immediately made my road life more expensive – and I would soon find myself missing the chaos and vitality of the farmers' markets and street food – the country felt somehow lighter at heart than Bolivia.

This was perhaps partly due, ironically or otherwise, to copper. Chile, one of the richest, most 'developed' of all South American countries, is built on mining wealth. The war between Chile and Bolivia in the late 19th century had largely been over mineral resources. The war left Chile rich and Bolivia land-locked and poor, cut off from both potential mineral wealth and the sea. Yet Chile's extractivist-based wealth rendered it vulnerable as well. As the miles went by, I often thought of Carlos in Ecuador, and what I had learned about the impacts of copper-mining, and what all the mining must mean for Chilean ecosystems, Chilean water, Chilean social justice, Chilean biodiversity. A different magnitude of threat than undeclared ginger.

Despite these darker trains of thought, the toughest riding was lightened by the 'at least it's not gravel' sentiment, everything thrown into relief by those last days in Bolivia. And I was loving being in this desert, in the hot, dry heat and huge spaces, however scruffy and despoiled. The sense of spaciousness was intoxicating. It called out, silently, *go as far as you like* … There were no boundaries, no fences, no limits here, bar the horizon. As I rode, the colours changed gradually around me, like slow-motion lighting playing over a vast stage, the filters switching from brown to pinkish brown, to pinkish grey, then back to camel or faded fox and brown again.

There were occasional tree skeletons, often near ruined buildings where presumably someone had once watered them. The signs of life were even more fascinating: plants that had figured out strategies for surviving here, little stalks of fleshy leaves with vivid violet flowers or occasional mats of green with pale, cream flowers, their fronds curling out like wildly displaced sea anemones in the brown desert. One evening I found hundreds of strange, small trail marks on the sand where I'd pitched my tent and woke to find them being traced by the feet of numerous strikingly striped, black and white beetles.

In some ways, the time pressure had simplified everything. My only job now was to ride as far as possible, every day. It was strangely freeing. Even when the headwinds left me working flat out to ride at 5 mph, hour after hour, it was clear what I needed to do. Just keep pedalling. In other ways, I fretted against it. There was a mine tour in the town of Calama organised by the Chilean company Codelco that I really wanted to join – but it only happened on Tuesdays, and I reached the town on a Sunday. I couldn't afford to lose two days' cycling by waiting.

Another regret was not having made time for one of the most highly rated tourist offers – a night trip from San Pedro to a desert observatory. I would have so loved to see the wild southern stars through a gigantic telescope and learn something about them, in the ideal location. The Atacama, high and empty and free of light pollution, is home to some of the most advanced telescopes in the world, including the fabulously named Very Large Telescope situated on a high desert plateau where the sky's stars shine so brightly they cast shadows.*

* Its rival, the Extremely Large Telescope, is still under construction.

Listening to the wind buffeting my tent one night, I read that Augusto Pinochet's dictatorship had converted an abandoned mine in the Atacama into a concentration camp. The conversion wasn't difficult – the conditions the miners lived in were so bad they were effectively housed as slave labour anyway. All Pinochet's men had to do was add barbed wire and a few watchtowers. The political prisoners that were 'disappeared' from these prisons and elsewhere were often dumped in the desert, sometimes in mass graves, the dead disappearing as the living also had.

For me, the Atacama came to represent the paradox that is Chile. I knew there were people still searching for the bodies of the disappeared across this immense space; so barren, hot and dry that it's been used for Mars landing simulations. Yet, I found I was savouring not just the vastness and stripped-down simplicity of the desert, but a sense of lightness, the sudden release of tension that is on some level always present as a person of privilege, travelling through a country that is undermined by poverty and cognisant of at least some of the myriad ways in which your own wealth and other's desperate lack of it are linked. I was troubled by this relief.

And at the same time, Chile's dark and horrifying history was just a few decades in the past, not very far below this bright and shiny surface. It was 1973 when the democratically elected Salvador Allende was overthrown by a US-backed coup. Pinochet's military junta ruled for seventeen brutal years, ending in 1990. By then, over 3,000 people were dead or disappeared. Tens of thousands had been tortured. And an estimated 200,000 Chileans were in exile.

One evening, I dropped steeply down through a fierce headwind to the town of Antofagasta on the coast and a night in, of all places, an Ibis hotel – complete with fabulous hot shower and a vast comfortable bed. I spent the entire following morning eating my way through the breakfast buffet (and smuggling out pastries for later).

Mostly, though, I slept in my tent. I found shelter behind small gravel pits, becoming expert at seeing potential campsites, revelling in the strange, industrial beauty of the place and going slightly feral. Sometimes the road – a key Chilean artery, Route 5 – would become almost a fenced-off motorway, and I would start to see potential campsites in unlikely settings, under junctions or behind major road signs. I always found I was, perhaps strangely, really happy in these spaces, slightly grungy, hunkered down behind a deserted building or a gravel pile or under a pylon. If there were a few birds and plants there with me, even better.

I had my first rear-wheel flat, and then a run of them, my back tyre wearing so thin that the tiny metal tyre spicules, picked up from the hard shoulder, could easily penetrate. Then, a fairly major mistake. After spending all morning eating and using the Wi-Fi, I left Antofagasta in a rush, with rather frugal quantities of food and nowhere near enough water, on a blistering hot day that turned out to be almost all uphill. An entry on the iOverlander app said 'well-stocked supermarket' about a day and a half ahead, and I planned to get more supplies there. When I reached it, there was nothing. Not even a house, let alone a supermarket. There was nothing on the map to indicate there would be. Why had I believed the iOverlander entry? I'd believed it because I wanted to, because it would have been mighty convenient if true, because I have an optimism filter on life, because things will always work out, right?

The situation was salvaged by the second truck driver I asked. He had stopped to check I was OK, and then to offer a lift, perplexed by my claim that I preferred to cycle. When I asked if he could sell me any water, he grinned and gave me a small bottle of water and a huge bottle of ginger ale.

The conversation with the first truck driver had gone something like this.

'Hello. I'm sorry to trouble you, but do you have any water I could buy from you?'

'You are pretty.'

(Internal grimace, outward smile.) 'Thank you. Do you have any water?'

'Would you like a lift?'

'That's very kind. But honestly, I prefer to cycle. Do you have any water?'

'That's crazy. I can give you a lift.'

'Thank you. I am happy on my bike. Do you have any water?'

'No.'

What kind of idiot drives across a desert with no water?

The ginger ale was delicious and surprisingly hydrating. It made excellent porridge. Even spaghetti. I gave myself a talking to anyway. *You are cycling across the Atacama, not the South Downs, for fuck sake. Even if you are on a road.* That evening, grubbing around in my panniers to find what food I had left, I realised that, beyond overly blasé, I was becoming a mad bag-lady as well. Bits of food – rice, cheese, half an onion, ginger, twists of sugar – were stashed in grubby bags inside bags, all a touch insalubrious and dusty. And insufficiently calorific. The pastries long since gone, I licked out sachets of honey taken from the Ibis, becoming increasingly distracted by hunger.

The hills became wrinkly and folded, like the skin of one of those overbred dogs that get horrible infections. Then, dropping down to Taltal on the coast, rounding a corner, I could see guano-covered rocks and, suddenly, hear seabird cries. I stopped to listen, delighted to hear these exuberant, vital noises of life. The sea is in trouble, but it is not yet dead, as it is in Cormac McCarthy's unbearably haunting *The Road*. I was flooded full of happiness. *There is still life, lots of it, vibrant and vivid.* And again, that sense of a deepening understanding of what my ride was about. *This is it. This is why.*

Taltal, Chañaral, Caldera, Copiapó. Heading south, long coastal stretches. Once or twice, a green and white sign with an arrow pointing inland and a figure running from a curling wave that read '*Via de Evacuacion Tsunami*' – evacuation route. The desert reached all the way down to the coast, only there, as well as the seabirds and sea lions, whose strange half-bray, half-bellow occasionally enlivened my day, there were stands of cacti and shrubs that thinned out, then vanished, as the road moved back inland again to the immense empty spaces.

A brief section of cheerful terracotta colour, then back to muddy brown, then an almost black-brown colour with streaks of pale sand, so that the desert looked momentarily like some kind of chocolate-and-cream bar. How human brains relish diversity; for me, at least, the toughest part of being in the desert was becoming not the heat but the sameness and the lack of visible life, transforming from a sense of freedom via spaciousness to a different kind of constraint. For day after day, I gazed at blue sky, brown hills and a plain in between. It seemed astonishing that I was headed for a region of Chile called the 'Lake District'; that there could be somewhere with lakes and trees in the same country as this desert.

At some point, Route 5 turned into a motorway proper, complete with unmistakable, illustrated 'No Cycling' – or tractor-driving – signs at every junction. I decided to keep going.

One evening, on a non-motorway section, when I was peering down a side road trying to discern whether I could discretely pitch a tent there, a truck pulled over. The driver got out and, with a big smile, handed me a hi-vis gilet.

'You must wear this,' he said, gesturing to the road and the traffic.

It was gigantic but weighed nothing, and I wore it religiously from then on, using my bum-bag strap to belt it around me. It went around almost twice. I cycled on, touched by the concern from a complete stranger. Before he'd approached me, I had, unusually, been struggling to find somewhere to camp. Not long after, as if he'd brought luck as well as visibility, I found a spot practically under the road but completely out of sight. For much of each day, I looked at the world through a 'possible campsites' lens. It significantly altered what I noticed. *There is a lesson in there somewhere*, I thought. If we looked at the world through an 'opportunities to live more sustainably' lens, would that similarly change what we saw?

As the long, hot days stacked up, sustaining the pace became harder. Physically, I was fine. I was tired but not trashed. I was thin – probably thinner than I'd ever been in my adult life, my cycling shorts loose where the elasticated ends used to grip my thighs – but still strong. Mentally, though, I was increasingly fed up with working relentlessly hard for relatively slow progress. I had been cycling south, pretty much as fast as I could, without days off, for two weeks – after weeks of hard cycling before that – and Santiago was still hundreds of miles away, let alone Pucón. The 'better than cycling on thick gravel' thought was beginning to lose its power.

The truth was, I wasn't going make it on time. Yet when I thought about getting a bus, I still got a loud, internal *NOOOOOOOOOOOO!*

After yet another bout of one the most frustrating things I know – sitting on a bicycle heading down a hill but still struggling to move forwards – I realised the frustration was pointless. Hours later, still crawling into the wind, I finally accepted it. *I am not going to be doing 70 miles a day for the next couple of weeks.*

I emailed Chris. After all my agonising, Chris, positive as always, was unphased.

'Stop stressing!' he advised. 'Let's agree we'll meet on the road somewhere south of Santiago. However far you get. We can ride together to Pucón from wherever that is.'

Chris was sanguine, but I was still determined to get as far south as I possibly could. If he were honest, he'd admit to being a less-than-fully enthusiastic cyclist. But he was willing to spend his holidays cycling anyway. At least we were riding into Patagonia: I was keen not to transform this into an extended slog on Route 5.

Gradually, the desert softened. The relentless brown faded to tan. The cacti ventured further inland. One day, greenish shrubs started to appear on the faded hills alongside the cacti. Then, a patch of long, dry, sand-coloured grassy stuff, sprinkled and speckled with cacti and shrubs so that it looked spotty, like the fur of some fabulous cat. Not long after that, there was a sudden explosion of colour on a stretch of disturbed ground where all sorts of flowering plants had taken hold – vivid pinks and mats of small, pale blue, trumpet-shaped flowers and hundreds and hundreds of lilac-coloured flowers, like tiny translucent tulips.

It had been unusually wet earlier in the year and that, I was told, explained the flowers. Then, even as the road was becoming bigger and busier, I found myself on the edge of a full-blown 'desert in bloom' phenomenon, the vivid, pink flowers suddenly everywhere, stretching out in thick stripes alongside the road, then completely carpeting the ground with magenta, so that the land stretched away pink under the blue sky all the way to the hills.

Soon, there was a scattering of bushes and a few rather straggly trees, as if someone had started to remake the world. The birds increased with the vegetation. There were hawk-like birds in the sky and small, scuffling bodies in the shrubs. On the hard shoulder, I began to see dead birds again, a handful each day, all shapes and sizes, plus the occasional sandy-coloured lizard. For days after I'd left the desert, re-encountering trees and birds nearly moved me to tears. It stayed amazing to see – and smell – green plants for a long time. *Water is life.*

Finally, I came close to Santiago. At dusk, nearly at the outskirts, I stopped at a small restaurant with a bit of land behind it. I leaned Woody against the white, wooden-slatted wall and found an elderly couple inside.

'Would it be possible, please, to camp behind your restaurant and then eat in it?' I asked, in my incompetent Spanish.

The woman had short brown hair in a hairnet. She wore a green sleeveless overall over a paisley patterned top and faded salmon trousers. She was even shorter than me and walked with a limp. Kind. Smiling.

'*Sí*', she said, and then checked with her husband. Her husband uttered that most welcome of phrases:

'*No hay problema.*'

He came with me round the back of the small building to a flat space with a scattering of straw under a tree. Perfect. Except for the grey and white puppy, tied up short to a digger. I put up my tent and gave the pup some water. He drank it, clearly thirsty. *How many pups live out their lives like this with otherwise compassionate people?* I wondered, again. The horses nearby seemed to have a much better life.

The next day, I woke early – useful, it turned out, as Woody had a flat. A spicule of wire was clearly visible, sticking out of the back tyre. I changed the inner tube and took the punctured tyre inside to patch over breakfast of eggs and coffee, before photographs and a kindly farewell.

Looking at the map in the tent the night before, I had come up with the brilliant idea of crossing Santiago on the motorway to save time. It was clearly the most direct route and would be easy to follow, unlike the cycle routes, which criss-crossed and meandered in a far less focused manner.

Crossing Santiago by bike on a motorway, it transpired, was not a good idea. It was midday by the time I reached the city proper. It was super-hot, and the road was very busy. But I was used to traffic – even fast traffic – and as usual with the bigger roads, there was a good hard shoulder. I got my head down and pedalled. After a few miles, the hard shoulder narrowed, then disappeared. Suddenly, the traffic was millimetres away. There followed some terrifying close shaves. This terror was then amplified beyond all previous measure when the road and everything on it, including me and Woody, dived down into a tunnel. The tunnel was dark. It turned out to be the first in a series. They all lacked a hard shoulder. The traffic was dense, loud and fast and, having had no time to switch on my lights, I was practically invisible. Not to mention on a slow, heavily laden touring bike that no one was expecting to encounter. At this point my optimism filter abruptly switched off, replaced by a single thought: *OK. Now you are going to die.* What a bloody stupid way to go. Suddenly, out of the corner of my eye, I saw a slip road joining the underground motorway from the world above and, realising I had to do something, however drastic, I veered off onto it. By almost unbelievably good fortune, the slip road I chose had been closed to traffic because of an accident just ahead of it, and Woody and I shot out of it unscathed, past some startled-looking traffic police and back into the light.

I stopped at a bakery and had a cold juice, an empanada and a sanity check. I had so much adrenaline surging through my body I could barely speak or eat. From the map, I figured out there was a main road that ran parallel to the still tunnel-based motorway. It looked as if I could use either this road, or a 'normal' above-ground motorway, for the rest of the ride across the city. The traffic was slower, so it felt a little less insane – though the hard shoulder still came and went. The scariest bits were bridges where the hard shoulder completely vanished, and junctions where I had to cross the path of traffic exiting at high speed. I was beyond grateful for the hi-vis gilet. But the thing that struck me most about the second half of that hair-raising, fast ride across Santiago – the adrenaline easily doubled my normal speed – was that there were onions on the road, presumably fallen from a truck, in excellent condition. Under normal circumstances, I would have stopped to pick them up for later. The onions were then joined by sweetcorn and a potato, a bizarrely reassuring trail of vegetables that I found myself looking for on this maddest of rides. I promised myself that, if I made it out alive, I would never, ever cycle across a major city on a motorway again.

Finally, on the far side of Santiago, I stopped under a bridge in the shade. After drinking an entire water bottle of warm water, I started eating a chunk of bread I'd been carrying for a while, until I realised the bread was full of ants. The adrenaline was gradually fading, though it was still powering me forward at an unnatural speed – by the end of that day I had ridden 89 miles. It left me feeling relieved but slightly sick.

For three days after that, I rode as hard as I could down Route 5. Leaving aside the Santiago stretch, I became weirdly fond of it. Route 5 is relatively flat, fast, has a huge hard shoulder and Copec service stations with real coffee in them. This struck me as almost as amazing as the salt flats, albeit in a very different way, obviously. There was a thirst I had in that heat, riding those miles, that could not be quenched by bottle-warm water, and I often arrived at a Copec station – which also featured friendly staff, food, Wi-Fi and spectacularly clean loos – fantasising about cold fruit juice as well as coffee. I was becoming a regular frequenter of these stations, greatly increasing the cost of life compared with my frugal days in the desert.

Beyond the road, it was getting more and more beautiful by the day. Now there were mountains, running alongside to the east, the highest and most distant glimmering with snow. Close up, everything was lush, bursting green with foliage. There were fruit stalls and flower stalls. Once, I was brought to a halt by a tree on the edge of a river that was full of egrets, the snowy mountains echoing their bright and delicate whiteness in the distance. And a vineyard with red dog roses all along the edge of it. What an astonishing transition, from the desert to this.

My last day alone was wonderful. There were roadside cherry and strawberry stalls. A Copec at just the right place. A long final push in lovely hot sun into the town of Chillán, a mere three or four days' ride from Pucón, where I arrived just before eight in the evening, cycling the wrong way up a one-way street to find Chris at an outside table with two cold beers, looking out for me coming from the other direction.

'Hello,' I said. 'Are you waiting for someone?' It was a very good moment

22. HAPPY NEW YEAR

[T]here is no more urgent intellectual task facing the human species ... than thoroughly to re-imagine its relationship with nature.

ROBERT MACFARLANE, 'TURNING POINTS'

Our economic system is incompatible with life on this planet.

JASON HICKEL

ARGENTINA

Happy Planet Index score: 50
Rank: 43rd out of 152

'I have a confession,' I said to Chris, as we walked steeply uphill. 'I've never much liked monkey puzzle trees.'

Chris threw me a quizzical glance but said nothing. It was possible he was out of breath.

'But I've only ever seen them in the UK,' I continued. 'They look out of place, somehow, almost artificial.'

Chris nodded.

'Very different here, though, eh?'

Another nod.

We were in the El Cani Nature Reserve not far from Pucón, walking up through an ancient southern beech forest into a forest of monkey puzzle, or Araucaria, trees. Here, the spectacularly spiky trees, with their dense branches of sharply pointed triangular leaves, felt wild and in place. Given the absence of monkeys in Chile, the name is a bit misleading, though, to be fair, nothing climbs these trees, and this would no doubt be true of monkeys, were there any around. The seeds are mainly dispersed by the most southerly parrot in the world, the cachaña or austral parakeet, and the forest was rich with birds and lichen and flowers and insects, including mosquitos and some low-flying creatures that looked like small dragonflies – one of which flew up my trouser leg and delivered a hefty sting, seconds after I'd said to Chris how lovely it was to travel in a country without too many tent-prowling predators or biting snakes and insects.

We'd spent Christmas Day in Pucón, more or less as planned, if you overlooked the five days we cycled together to get there. Chris, having travelled thousands of miles to join me, showed no resentment towards his unscheduled time on Route 5. He was riding strongly – ah, the joy of riding on someone's back wheel – and even agreed that Route 5 had exceptionally good hard shoulders, though he was a little bemused by my enthusiasm for Copec service stations, especially as a Christmas Eve treat. It seemed bizarre to be celebrating Christmas somewhere hot and sunny, though Pucón, an outdoor-sports town that looked like a cross between the mountain town of Aviemore in Scotland, an Alpine ski resort and Banff, Canada, was generously decorated with Christmas trees sporting fake snow. My presents from Chris included Twiglets and a new back tyre.

On Boxing Day, we met up with Chris' friend Rod and his partner Paula. Rod lived some miles outside Pucón, in a wooden house he'd built himself, on a steep slope at the edge of the forest. The walk up from the furthest point you could take a vehicle was rewarded by views of the snow-drenched Villarrica, one of Chile's most active volcanos.

'My daughter and I watched it erupt from here last spring,' Rod said, when we first saw it. 'Spectacular. Its other name is *Rucapillán*, Mapuche for "house of the great spirit".* In Mapuche mythology, *Pillán* is a very powerful being. Trust me, when you see it erupt, it fits.'

The Araucaria trees, he told me, were fire-adapted, with bark so thick they could survive the frequent, volcano-related fires.

We spent several days with Rod and Paula. I was tired, the relentless pushing of recent weeks having finally caught up with me, and the time with these generous new friends offered what I was most hungry for: rest, food, conversation. We talked and talked.

'After Pinochet stood down in 1990,' Paula told us one late afternoon, as we sat outside drinking beer, eating nachos and soaking up the last of the sun, 'the human rights violations fell away. But in terms of inequality and poverty, nothing much changed. He'd nailed the constitution, and the economic system remained the same.'

The economic system, an extreme form of free market economics, had been installed by the so-called 'Chicago Boys', Chilean economists who'd studied in the USA under Milton Friedman. In her disturbing book, *The Shock Doctrine*, Naomi Klein argues that Pinochet's CIA-backed coup was used as the 'shock' necessary to introduce these extreme economic policies while citizens were distracted by the coup, and by the disappearances, torture and murder that followed.

The sun gone, we moved inside, and Paula lit the fire against the suddenly cold night. 'The Chilean economy grew under Pinochet,' Rod said. 'Many people still think that's really all that matters.'

* The Mapuche are indigenous peoples of south-central Chile and south-west Argentina.

According to Rod, who was originally from the UK but had lived in Chile for 30 years, it was an article of faith across Chile that money and financial wealth were all-important.

'Language itself has been diminished and violated by consumerism here,' he said, grimacing. 'And if, for example, you try and argue for protecting nature as a priority instead, you are usually considered bonkers.'

Rod had been a lecturer, outdoor educator and environmentalist for decades. He'd worked with both Mapuche people and German settlers on small-scale, ecotourism projects and had been involved in various campaigns to protect the Araucaria forests.

'Here in Chile, nature is constantly under assault,' he said, taking a swig of the red wine we'd switched to. 'Agriculture is dominated by monocultures that require vast amounts of pesticides. The pesticides destroy biodiversity, including in the soil. Then we have mining, and mega-dams. All hugely damaging.'

In Rod's view, the root cause of all of it was that so many people felt disconnected from nature. The disconnection manifested as monocultures and destructive industries, which needed to be tackled. But ultimately, he argued, the solution required a profound consciousness shift, or the problems would just re-occur in another way. Experience was key: experience of nature.

'Nature herself is the most powerful change agent,' he said.

I learned from Chris later that Rod had pretty much introduced the notion of outdoor education to Chile, taking generations of Chilean students into their own mountains and forests. He was practical as well as philosophical, pointing me to a tool shed and helping me work on a slender crack that had appeared in Woody's frame.

It was hard to leave, though when we finally did cycle away, it was into rolling, rural country full of birdsong, flowers and lakes, with occasional heart-stoppingly beautiful views of the volcano.

'Don't write a word without being in touch with the earth,' Rod had said, when we hugged our goodbyes. 'Be a channel. Not a disconnected mind.'

It was advice both wise and daunting, especially when he threw, 'And be radical! Think of *The Life Cycle* as a palette, on which you can seriously challenge the colours of the status quo ...' at our receding backs.

We arrived at the small town of Paillaco just before dusk on New Year's Eve. There were blue mountains on the distant horizon and fields full of birds that looked like lapwings. I identified them as queltehue, thanks to a painting of them in a café. Outside the café, as in cafés the world over, (different) birds scrounged from table to table, hopping chaffinch-like across the chair backs; only here they were the size and shape of a small plump hawk, and had hooked beaks and big, lucent eyes.

On the road, at one point, hundreds of emerald-green beetles wandered into our path, their rainbow backs glinting and shimmering like a spill of oil on the tarmac. Beauty and surprise seemed everywhere. But the last few miles through the warm, sunny, windy evening were touch and go. Chris had spent the day suddenly dropping his bike and diving off into the bushes.

The town had a hotel that, luckily, had a room. In fact, it seemed that we were the only residents. The restaurant was beautifully laid out with wine glasses and candles on every table, though the kitchen, we were told, was closed. We were taken up wooden steps to a twin room with someone's work boots and protective eyewear still in it, and a wardrobe with a large plastic bag full of something that smelled strongly of fibreglass. Chris collapsed on the closest bed. I went to unpack the bikes. By the time I brought the first load up, he was under the lightly grimy duvet shaking and complaining of being icy-cold and achy. I put both sleeping bags over him and dug out the remaining monster-dose paracetamol tablets from my toothache days. And then I headed out in search of sustenance, especially beer.

Paillaco, or the part I saw, seemed poor. It was definitely not on the tourist belt. The place had a friendly feel, if a bit dishevelled – the small wooden houses were in various stages of disarray, lads played football in the street. I found a few shops still open, and bought beer, fruit juice, yoghurt, some fruit and veg and a large bag of crisps. I drew blanks on bread and aspirin.

Back at the hotel, the candles had been lit in the restaurant, despite there being no diners and, indeed, no food. I walked past a row of identical, camel-coloured work boots in the upstairs hallway to find Chris now unnaturally hot. He managed some fruit juice and a few crisps and talked about cycling a short day tomorrow. I was not convinced he'd be in any kind of shape to cycle at all. Meanwhile, there we were, on New Year's Eve, in a hotel not serving meals that couldn't sell us alcohol, with two bottles of beer and, for me, a solo dinner of crisps, cherries and pannier-melted chocolate. I ate it under a sign on the wall of our room that read: '***MANTENER EL ORDEN Y LA LIMPIEZA***' – '**MAINTAIN ORDER AND CLEANLINESS**' – in bold capitals. I was about to go to bed when I heard the New Year arrive, with sirens sounding and dogs barking across the whole town. I could hear people shouting and laughing and cheering in the streets. It made me grin, and I went to bed in good humour. Welcome, 2018.

The first of two standout highlights in the remaining time Chris and I had together involved penguins. Humboldt penguins, to be precise, and a detour off the bikes to visit a colony not far from the coastal town of Puerto Montt.

It had taken a while to get to Puerto Montt, between recovery days off – including, as predicted, 1 January – relatively short days and the first heavily wet-all-day day of the trip. In the rain, leaving town was the hardest bit, potholes hidden by the rapidly rising water and our faces constantly blasted with cold, gritty spray from the close-passing traffic. Once we were back on Route 5, life became easier. The hard shoulder was largely free of standing water and the traffic a little further away. Numerous bright orange flowers, like small, stripy lilies, were made even brighter by the rain, cheering us up.

Round lunchtime, we'd stopped at a most welcome café, the kind announced by flags and rows of parked trucks, and headed inside. There was a rush of fabulous warmth as we opened the door and dived for a table. We sat dripping, apologetically, while friendly staff waved away our concerns and brought me a disproportionately delicious scrambled egg and cheese sandwich – their suggestion as the vegetarian option – and Chris, still not entirely recovered, a bowl of plain rice, for which they wouldn't take any money.

Puerto Montt, when we reached it, was a mix of old, dilapidated wooden buildings and shiny new concrete and steel. There was a restaurant called Sherlock, where the man himself was reputed to have eaten. It took me a while to recall that Sherlock Holmes was fictional. Beyond Sherlock, a cruise ship lay in the bay and a long road led out of town along the waterfront. The road took us to our hostel, via a bus station, a ferry station, an artisan market and a fish market. Dark shapes flashed in the water nearby; the flippers of a group of frolicking sea lions, arching their huge bodies out of the water and returning with extravagant, drenching splashes.

We left the bikes at the hostel and took a bus, then a ferry, to the island of Chiloé, standing on deck on another grey, wet day, watching fulmar-like birds, gulls and terns swoop and squabble. From the town of Ancud, a minibus took us through streets of wooden buildings into rolling green hills before descending to a beach and driving straight across a shallow river. At the back of the beach, which arched around to a group of rocky islands at the far end, were a line of restaurants and penguin tour operators. Small boats sat in the waves near the shore. It was beautiful, despite the weather.

We were ushered through the drizzle and into one of the tour offices, the walls covered in colourful posters with information about the Humboldt and Magellanic penguins.

'You can tell them apart by the number of stripes,' our guide, a tall English-speaking German man, told us. 'Humboldt, one. Magellanic, two.'

Looking at the posters, we could see what he meant. The Humboldt penguins had a single, dark stripe curving above the white tummy and below the white throat. The Magellanic penguins had two, giving them more of a mint humbug look. Other birds we might see, according to the posters, included a blackish oystercatcher, brown as well as black-headed gulls and flightless steamer ducks – so bulky their small wings can't give them enough lift and are instead used like paddles to help them skim along the sea.

Our group of fifteen were shepherded onto one of the motorboats and pushed off the wooden boarding platform by men in thigh-length wellies and waterproofs, out into the waves. An otter, eating a crab while lying on its back, swam slowly away as we approached, then nonchalantly flipped back over and ate some more. We motored slowly towards the rocky islands, their tops in the low cloud and mist. A steamer duck, with a train of ducklings, surfed a wave onto a rock, the ducklings scurrying seamlessly up behind her, as if they had been doing this for much longer than their few days of life. As we got closer, we could see small groups of striped penguins, both kinds, hunched on the wet rocks. Humboldt penguins are super sensitive to the presence of people and the boat kept a respectful distance, although they displayed no response to us at all. Kitted out in identical bright yellow, generously hooded rain ponchos, we perhaps looked like harmless, if strange, sea creatures ourselves.

It was over too soon. Back on the beach, we headed for one of the cafés. It was wooden with gorgeous views from the ocean-side windows. We drank hot chocolate and ate empanadas, the best so far, and talked about penguins. Humboldts are named after Alexander von Humboldt of Chimborazo and other fame, as is the ocean current they swim in. Welling up from the deep ocean, the cold Humboldt Current runs from southern Chile to northern Peru, where it meets warm tropical waters. The effects of the current are many, varied and close to astonishing. It has a cooling influence on the climate of Ecuador, as well as Chile and Peru. It's largely responsible for the aridity of the Atacama Desert, because the marine air cooled by the current, while it generates cloud and fog, produces little rain. And, as I'd learned on the Peruvian El Ñuro visit, the current is fabulously rich in marine biodiversity. As it rises, it brings nutrients with it, which support the phytoplankton eaten by the krill eaten by the anchovies eaten by the penguins – among others – who dive underneath the fish to catch them, and who are able to hang on to them thanks to spines in their tongues. The more I read about the Humboldt current, the more the reality of the overused phrase 'everything is connected' came alive in my mind. As did the consequences of messing with it.[†] Ditto in relation to rainforests. Or páramo. Or indeed, any ecosystem I'd learned about. As the naturalist and passionate advocate for conservation John Muir put it: '[W]hen one

---

[†] One of the many predicted consequences of anthropogenic global heating (aka climate change) is an impact on global ocean currents, including the Gulf Stream (which could weaken) as well as the Humboldt (which is already warming).

tugs at a single thing in nature, he finds it attached to the rest of the world.'

Meanwhile, victims of the mad scramble for guano in the 19th century, the Humboldt penguins, who nest in holes they dig in the guano on rocky islands and coastlines, were now also threatened by rats and feral cats, dogs and goats; overfishing; oil spills and other pollution; and even, unexpectedly, mining.

That year, a Chilean company had been granted permission for the 'Dominga Complex', which included an open-pit copper mine, a processing facility, a desalination plant and a port. All were situated close to the Chilean Humboldt Penguin National Reserve – an area with so much wildlife it was called the Chilean Galápagos. The Dominga complex had originally been refused planning permission on the grounds that the environmental impacts would be simply unacceptable. The decision, which had been reversed and appealed and revoked on numerous occasions already, was back under review. The damage that Dominga would cause did not appear to be in dispute: the debate was more about whether the vast (if inequitably distributed) income it would bring justified it, or perhaps more realistically, whether there was any ethical or legal argument on earth that could halt the power of a highly lucrative project backed by wealthy corporates, in the context of an economic and political system driven by the value of profit above all else and structured in such a way it could not survive without endless growth. The basic story, it seemed, was the same everywhere, whether the main protagonists were involved with gold-mining, lead-mining, industrial fishing, palm oil, oil extraction or copper.

Here on the penguin beach, though, it was a different plotline altogether. The beach was busy, with its line of bobbing boats, a steady stream of minibuses bringing people to them and cafés and gift shops waiting for them afterwards. There was an upbeat, cheerful feel about the place. The local people, our waiter told us, protected the penguins, making sure no one landed on the islands or that any boats came too close. They were rewarded by the thriving ecotourism.

'Many local people work here,' she said. 'The penguins look after us, and we look after them.'

It was such an obvious win–win.

'This must be part of the way forward,' Chris said.

'Definitely,' I agreed, trying to resist the temptation to lick the plate the empanadas had arrived on. Lately I was finding I often had to remind myself I was in a public place, not in my tent.

The nights we were based in a town, Chris' preference was usually to eat out. This led to a bit of a trade-off situation. He'd come a long way to spend a significant chunk of his annual leave in Chile, with me. He was on holiday and wanting to make the most of it, whereas I was accustomed to spending most of the day cycling, and then using the evenings to tackle the journal, blogs, emails necessary to set up project visits, social media posts etc. I settled for tackling the most urgent things over breakfast, promising myself a work-binge after Chris had left. As it turned out, this was to be several days sooner than expected.

'There's been a landslide on the Carretera Austral,' the woman in the ferry terminal told us, while we were booking tickets for the penguin trip. 'South of Chaitén.'

The Carretera Austral, said to be one of the most beautiful roads in the world, was our route from there on. Work on it started under Pinochet's regime, connecting remote communities for the first time, though most people believed his motivation was more to do with creating access for the military than with rural regeneration.

'The road is completely closed,' she continued. 'You'll have to take a ferry around it if you insist on continuing south.'

She was rather stern and seemed disapproving of the whole bicycle idea, telling me in particular how dangerous it was. When we looked at the ferry timetable later, we realised that Chris would be cutting it very tight to get back to Puerto Montt and his airport bus if he came on the ferry with me. Chaitén would be as far as we'd travel together.

We rode away from Puerto Montt on the Carretera Austral, at that point a coast road that soon became gravel. For a while, wooden houses ran alongside it and the beach was busy with people having picnics.

That evening, after a short ferry crossing from sun into shade, the mountains right up against the road and the sea, stretching back green and forested to glimpses of ice, we found a camp spot at the back of a grey-sand beach. There was a snowy volcano visible in the direction we'd come from, and the outline of Chiloé on the horizon across the sound in front of us. Washing pots at the sea edge after dinner, I heard Chris shout, and looked up to see a graceful arching back and a tall fin. Dolphins. Then a dram, the first I'd had in over a year. Chris had brought out a miniature and we drank it between us, sitting outside the tent in the fast-gathering dark. I had forgotten how much I enjoyed that feeling of whisky-warmth hitting the back of your throat, then gently transforming your whole body. Even without the whisky, I could have sat on that nondescript beach for days, just watching and being there. I loved being on the move. But there were so many places that made me want to be still.

Alerce trees or Patagonian cypress – *Fitzroya cupressoides* – are among the tallest and longest-living trees in the world. Darwin reported finding one over 40 feet in diameter. Their Latin name honours Robert FitzRoy, the meteorologist captain of Darwin's ship HMS *Beagle* on her trip to Tierra del

Fuego.‡ Extensively logged in the 19th and 20th centuries, the largest alerces now are around fifteen feet in diameter. Those that were mature before the logging got underway, and survived it, are among the oldest trees in the world.

In Pumalín Park, one of several vast areas of land owned and managed for conservation purposes by Tompkins Conservation, we wandered among them, straining our necks to look up at their dizzying height and straining our arms to try and reach around the immense trunks with their thick stripes of orange-grey bark. Being among these giants was the second standout highlight of our time together, especially for me. I was still astounded by the contrast with the Atacama and marvelled in the lushness of the forests that the alerces were the heart of. We took time off the bikes and walked a trail to a waterfall, the trees heavy with ferns and epiphytes.

We'd landed in the park after a long ferry ride from Hornopirén. The chaffinch/hawk birds were also part-seagull and scrounged from the railings of the ferry as we sat in sheltered sunshine on deck and gazed at the snow-streaked, ragged range of peaks across the windswept water.

The jetty at the park end of the journey was immaculate. Everything was tasteful and beautifully done, the toilets, the cabins, the campsite, the café. 'Manicured' was the word Chris used, and it was a good one. It was all manicured.

In the campsite, we chose one of several carefully designated spots and drank Cup-a-Soup watching a crowd of smart, navy blue, black and white martins, arriving as the sun sank behind the heavily forested mountains. There was a haunting, almost tropical call from a thrush-size bird with a chest like a robin and a wren-like, up-tucked tail.

‡ FitzRoy pioneered the prediction of weather as a safety aid to sailors, referring to these predictions as 'forecasts' – the first time the word had been used.

The ranger who collected our money the next morning told us that the wren/robin was called a chucao.

'Neruda wrote about it,' he said.

Neruda, the Chilean poet and campaigner for the rights of the poorest workers – including miners[§] – had achieved something like rockstar status during his life, widely adored in a way that is hard to imagine for any poet in Europe. His life had been a high-octane one, including an escape on foot and horseback over the Andes into exile when his communist activities made him a target of official persecution.

From another campsite a day or so later we gazed out over delicately curled ferns and huge-leaved rhubarb-like plants to a gently steaming, double-coned volcano. On the gravel roads, pulses of ill-behaved traffic announced the earlier arrival of a ferry. Then the roads would fall silent, and we would ride on alone, bar the occasional cyclist.

At the far side of the park was an immaculate, wooden park building, and a sign that talked about the need to protect biodiversity and the redundancy of economics based on growth. Then we were back out into the scruffy world of people, with their houses and sheds and dilapidated old cars. I was surprised to find that I welcomed the scruffiness' return.

In Chaitén, bolstered by astonishing views of a white-flanked volcano, we checked into a hotel on the front as a treat and spent a final day together, walking the stony, driftwood-strewn beach until we were driven back by its dense population of biting black flies. Then we sat in the sun, watching the world go by from a bench in the small

[§] The history of mining in South America is one of abusive conditions for workers as much as degradation of the environment.

town, before eating fabulous wood-fired pizzas in a café, served with piscos flavoured by wild calafate berries.

The end of our time together seemed to arrive suddenly. Chris rode with me to the out-of-town terminal and saw me onto the night ferry, the car deck already lifting as I waved goodbye.

Inside, the lounge was packed with people fixing themselves up for the night, on and under the seating. I beat a hasty retreat and, collecting my sleeping bag from Woody, found an excellent spot on the top deck by the bridge with no one there at all.

It was seven when I woke. We were coming into a sheltered bay, with sand banks and forested mountains, a white glacier lying silent in the distance beyond.

Cycling down the ramp behind the sole motorbike on board, a giant kingfisher on the pier railings, I turned inland on a gravel road that disappeared into the forest. From the vantage point of the ferry, the land ahead had looked vast and wild, hundreds and hundreds of miles of green, densely forested mountains stretching to the far horizons. Hadn't I read somewhere that there were more trees on earth than stars in our galaxy? Here, I could believe it. Heading into it felt exhilarating, with just an undercurrent of daunting. Occasionally, the forest opened out enough to allow glimpses of distant white mountains, or the dark turquoise Río Palena, running more or less parallel to the road. As the day wore on, there were occasional, beautiful farms in spaces carved out between the thick stands of trees. The birdsong was extraordinary.

I rejoined the Carretera Austral at La Junta. It somehow took me eleven hours to get there, though the distance was a little under 50 miles. I was struggling to steer on the deeper gravel and going so slowly uphill it was hard to keep my balance. Then I rode south for days, sometimes camping, sometimes spending nights in hostels or bed and breakfasts. On inside nights, I felt increasingly cut off, missing my tent and the stars.

One day, the road swung away from the sea-lake I'd just cycled alongside, and up into a long climb in heavy rain. The climb was accompanied by a tantalising sense that there would be a stunning view at the top, if only I could see it, and increasingly crunchy noises from Woody's gears and chain. These were dramatically amplified as the front derailleur cage slipped down the frame – my first real mechanical issue. I sorted it and felt rather pleased with myself until the remaining crunchy noise turned out to be forewarning that the chain was about to break. I fixed that by the side of the road, too, an easy job with a spare chain link. But I was super sensitive to slight clicks and crunches from then on, with a suddenly enhanced awareness of what an extraordinary thing a chain is: how much strain it takes that it is possible for it to snap. There was some metaphor in there somewhere, I thought, about the drivetrain of capitalism and identifying the clichéd weak link.

Close to the city of Coyhaique, I spent a night in a campsite outside a beautiful wooden hostel. It was run by a man called Nacho, who was married to an Argentinian cowgirl and passionately interested in the gaucho culture she belonged to. There were several other camping–cyclists, including Beni from Canada, riding a fat bike with an envious capacity for gravel. Nacho treated us all to *mate*, the herbal infusion rich in caffeine and nutrients made from a type of holly that supplements the gauchos' almost entirely beef-based diet when they are on the move, drunk in ceremony.

'The gauchos are migratory. Or used to be. They drove cattle north to south, criss-crossing the Chile–Argentine border. On horses, of course,' Nacho told us. 'They would drink this to kickstart each day.'

Nacho passed the mate to Beni.

'It's a fascinating culture,' he continued. Tragically, it's almost certainly a dying one. The young now want to work in tourism, not with animals.'

Cycling up a long and lovely valley on a smooth road with a civilised gradient the next day, I was overtaken by a man on a mountain bike. His name was Mauricio, and he ran an independent radio station named Radio Genial, or 'Cool Radio'. Running Radio Genial meant constantly researching diverse topics, and Mauricio was a gold mine of information. We cycled side by side, swinging from topic to topic – including one I was encountering more and more frequently. The road. There were proposals to upgrade all the gravel sections of the Carretera Austral to asphalt. In some places, this had already happened – though there were plenty of gravel miles left. It was, of course, controversial. Providing improved access and infrastructure to the communities on the road was an obvious plus; the potential increase in tourism in an already busy area more of a mixed blessing.

As for the way the upgraded road would make logging, mineral extraction and other forms of extractivism easier, in Mauricio's view, that was clearly a downside.

'Development must be a balance between generating income and protecting the environment,' he said. 'Too often we don't get that balance right. And then it's too late.'

It was unsettling, as a cyclist, accustomed to riding on roads without much thought about them, to suddenly have them hove into view as a thing in their own right. To think about how much a road changes the region and communities it punches its way through, for good and ill. A road was not just a way of getting from A to B. It stood for development, in all its complex, contested senses.

We stopped at the top of a steeper section for a cold kiwi juice in a converted single-decker bus café, run by a woman called Nori. The juice was beyond delicious, and Nori, it turned out, was a big fan of Mauricio's radio station. It was a lovely exchange.

Reinvigorated, we rode the final section of the hill and stopped on the outskirts of the town. Mauricio invited me to camp in his garden should I struggle to find accommodation in town. Then he briefed me on the names, locations and relative superiority of all the bike shops and was gone.

23. PATAGONIA PARK

When industrial worldviews are applied to natural systems that support all life, we begin to treat the earth as a factory that produces all the things that we think that we need.
KRIS TOMPKINS

Coyhaique, surrounded by white-summitted mountains, is sometimes called 'the city of eternal snow'. It's a young city, home to about 50,000 people. It has an airport and nearby border crossings that allow access to Argentina, so it's also a tourist hub – with bike shops. One of these sold me a chain so that, if mine broke again, I'd have a backup.

'It'll be horrible to ride a new chain on the old block,' the young man in the shop told me. 'As you know.'

I did know.

'Properly crunchy. But better than no chain.'

Hard to disagree.

Coyhaique was busy. Finding somewhere to stay, places to buy food, and the usual laptop catch-ups all seemed to take more time than I'd left for them. Plus, in Coyhaique, I was joined by a former student, now friend, Neda from Switzerland, who was keen to travel with me for a while. It seemed only right to begin our time riding together with a pisco-enhanced reunion.

The next evening, we arrived at a packed campsite after a day's ride and a freezing ferry across a deep turquoise, white-flecked lake with views back to a fabulously spiky ridge, bristling with outrageous turrets and fingers of ice-covered rock. We ended up in a room in the house of the campsite owners, with a bathroom right next door and warmth of a literal and human kind. Best of all was a conversation with the owners' son, Victor. In between studying biology and training to be an eco-tour guide, he had installed what he thought was the first recycling centre on any campsite in Patagonia and was building beautifully designed sheltered areas for campers, using recycled wood and other materials.

All of 25, he was full of ideas and positive energy. Our 'how to save the world' conversation – as well as the tour of his campsite innovations – continued well into the next morning. When we eventually left, it was with big hugs and two bottles of his homebrew.

Neda and I rode south together for about five days. She was faster than me but had less stamina. Or so I thought until a relentlessly hard day with steep gravelly hill after steep gravelly hill, and one of the few times on the entire journey where water was hard to source. Neda dug in. A rumoured lake ahead turned out to be a small slime-covered, duck-inhabited pond that we reached at dusk. She gamely waded into the least slimy area to fetch water that, after being double-boiled, filtered and having purification tablets added to it, was just about tolerable to cook pasta in – or would have been had both of our stoves not gone on strike. We ate our half-cooked spaghetti with raw veg and cheese – better than it sounds, given how hungry we were – accompanied by outbursts of semi-hysterical laughter and Victor's beer, which could hardly have been more welcome.

We parted company about a day north of Cochrane so that Neda could travel at her own pace, get into town in time to find accommodation and – having realised she'd overdone it on the T-shirt front – find a post office to lighten her load the next day. I would be detouring into Patagonia Park and a meeting that was to be a high point of the entire journey.

I was cycling in mild weather into a different mountain range, the big white peaks falling away behind me and new ones hoving into view ahead. My main impression was of sheer vastness. The Southern Patagonia Icefield lay somewhere behind the mountains off to my right, one of the largest areas of permanent ice in the whole world, outside the polar regions. Vastness, and the ever-present, constantly changing turquoise rivers. I had it in my head that the last six miles from the road junction to the park office would be a lovely easy ride along one of these – the Río Chacabuco. It was not. It took another two hours and featured rough gravel, steep climbs, adverse camber and generous potholes, the river racing heedlessly along somewhere below me. When I finally dropped down into the park area, past a small herd of guanacos* towards the by-now familiar, dark, tasteful buildings, a man came out to greet me.

'Hello,' he said, in English. 'I am Paul. I work here. Because of the handover and all the security, the campsite is closed.'

He had clearly deduced, perhaps from the state of the bike, or perhaps simply by virtue of turning up on a bike, that I was unlikely to be a guest for the lodge.

'What will you do? Cry. As I thought. Since you are here, I suggest you carry on and camp anyway.'

'Thanks,' I said, 'that's really kind. And actually, I have a meeting here, tomorrow.'

* South America's other wild camelid, similar to the vicuña, only larger.

'A meeting? With whom?'

'With Kris Tompkins,' I said.

'Ah,' said Paul, temporarily lost for words.

At that moment, Kris and various other people emerged from the lodge restaurant. I recognised her from pictures online – pale blonde hair, compelling smile – and I guess I was pretty easy to identify, too. With a huge grin, she came over and gave me a hug. We posed for a photo. I was sweaty and trashed and probably not smelling too savoury, but Kris was full of warmth and welcome.

'Look at you!' she exclaimed. 'And the bike! And the ride! Amazing.'

It was a wonderful moment. We agreed to meet the next day at 10am in the restaurant. Then I cycled the final two kilometres – uphill and rough of course – before dropping down to the campsite on soft grass, cheered by the friendly encounter with Kris and looking forward to talking more. As the mosquitos found me, I retreated to my tent and read my notes.

Kris Tompkins was probably at the furthest end of the spectrum from the 'it's all about money' view than anyone I'd ever encountered. A former CEO of Patagonia, the company, and lifelong friend of its founder, Yvon Chouinard, she and her late husband Doug Tompkins had sunk millions of dollars of the money they'd both made from outdoor gear and clothing companies – North Face and Esprit in Doug Tompkins' case – into nature conservation. Into buying vast tracts of land and turning them into nature reserves, to be precise. Kris called it a 'capitalism jujitsu move' – taking money made from the capitalist system to reinvest in protecting nature from the capitalist system.

Doug had fallen in love with Patagonia on backpacking trips there in the 1960s. In 1968, he and a group of mates – including Yvon Chouinard – had climbed Mount Fitz Roy. They called themselves 'The Funhogs' and were only the third team ever to have reached Fitz Roy's fabulously jagged summit.[†] Then, at the age of 49, he had a revelation. As a climber, he was passionate about the hills. But as an entrepreneur he had, he wrote, 'climbed the wrong mountain', damaging the environments he loved via the pollution and resource use that goes along with the creation of millions of items of clothing and gear – items that most consumers of his products quite possibly didn't really need. He decided to dedicate his life to paying back his debt to nature and to encouraging others to do the same.

'Nature conservation,' he'd written, 'is simply paying rent for living on the planet.'

Kris and Doug got together not long after. She resigned from Patagonia, and they moved to southern Chile where they started buying huge tracts of land. Pumalín Park was their first major conservation project, with 400,000 hectares of exceptionally wet, botanically rich temperate rainforest running all the way to the ocean.

As wealthy incomers from North America, they were initially treated with suspicion, even hostility. Ranchers tended to hate them, for starters. Much of the ranchland they bought had been overstocked with sheep, the grasslands overgrazed and eroded. The removal of sheep had not gone down well. Others were suspicious of their motives. The large-scale purchase of land for nature conservation was then almost unheard of. Surely, people argued, their real motives were to do with the control of water, or of other resources, or even of Chile itself – at one point, the Tompkins owned so much land they cut the country in half. But over the years, attitudes had changed. People realised that they really were about nature conservation and, slowly, that nature conservation really mattered. The Tompkins were passionate about community engagement and created new and different local

---

[†] Lito Tejada-Flores also made the ascent. See his film *Mountain of Storms*.

employment. They joined with the locals in campaigning against the mega-dams that would brutally transform the beautiful Patagonia landscape – and the tourist revenues.

'People are generally positive about what is normally called "development"' Kris had said. 'But we saw the dark side of industrial growth.'

She and Doug saw themselves as being in resistance to that kind of development; the kind of development that the dams were part of; development driven primarily by profit.

By the time of Doug's death, he was lauded as a conservation hero. He had died, tragically, two years previously, after a kayaking accident in high winds on Lago General Carrera – the lake that Neda and I had not long crossed and then cycled three-quarters of the long length of. After a capsize in high waves, he had spent a long time in the desperately cold water and was eventually airlifted to hospital, where he died of severe hypothermia. Reading about it again brought back memories of Chris' kayaking near miss – a sort of link I felt with Kris I had no intention of sharing.

Kris and her team had carried on with the conservation work, creating millions of acres of privately owned reserve. In two days, 2 million of these were to be signed over to the Chilean government, as a national nature reserve. The president of Chile herself, Michelle Bachelet, was due to arrive for the ceremony. Hence the security. It was astonishing, under the circumstances, that Kris had agreed to make time to meet me.

In the end, the meeting was at around 11am, and at Butler House, a large, imposing building up the hill from the lodge. Inside, I was shown to a huge living room/kitchen area, with sofas in the window overlooking the grassy plain to the mountains beyond. There was a big table with various people working on it, all trying to sort the million things that needed writing or rewriting or sorted in other ways before Bachelet's arrival and the handover.

In between disruptions, Kris came and chatted, fully focused on what I was asking her, despite everything else going on around us. She was passionate about the value and role of the parks, not just in terms of their importance for nature conservation, but as a way of giving people access to nature.

'How can people be expected to care about something they've never experienced?' she asked.

Gifting them to the government as national reserves was a strategy not without risk, she acknowledged, and Tompkins Conservation would remain involved in the management of the reserves, even if they no longer owned them. But, she argued, the point was that the reserves should be managed democratically and that national parks should be accessible to all.

I asked her whether, as a response to our ecological crises, reserves and parks could ever be enough.

'No,' Kris said. 'No. In a truly sane society, maybe we wouldn't need them at all. We have to tackle the way we live,' she continued, 'and the way we view success. Our industrialised, globalised economy. They all need to change. But that will take a while.'

'And our philosophical outlook too, presumably,' I said. 'The worldview that underpins our approach to economics, and everything else.'

Forrest, a philanthropist from the USA who supported the foundation, came over with plates of rather disintegrated cake.

'Yes, definitely,' Kris agreed. 'Meanwhile,' she said, brushing off a cascade of crumbs, 'while we try and transform our unbelievably human-centred mindset, we're also transforming how we work on the ground.'

The revolution in their thinking about conservation strategy, she explained, had come when they started asking what else, beyond protecting habitats, they needed to do to create fully functional ecosystems.

'We began to ask ourselves, "Who's missing? What species had disappeared? Or whose numbers were low?"' she said.

This shift in focus led to painstaking work to bring back lost species and increase the numbers of threatened ones. Animals such as the huemul deer. Condors. Puma. There had been some hugely exciting results.

A couple of hours later, I was sitting on a log on the hillside with Tom. Tom – introduced as 'chief philosopher' – had suggested I join him for a hike up to the ridge behind the campsite. He planned to spend the night there, seeking a retreat from the intense work in the house below and a chance to reconnect with the wildness this work was all about protecting. A fast uphill hike on my day off the bike was probably not what I needed. But I loved the idea of this huge foundation having a chief philosopher and couldn't resist the chance to chat to him more.

We talked about deep ecology – a big influence on Tompkins Conservation's outlook – the need to reconnect people with nature; writers we both rated; and the problems of the 'techno-globalised industrial system'.

Tom's view was that we could tweak and dabble with 'greenness' all we liked. But what really had to change was the dominant, human-centred view that the earth is basically an immense supermarket full of stuff for us, and that everything else that lives here is either a resource or a pest. Until we saw a widespread shift away from that to a, humans-are-simply-citizens-of-the-ecological-community-not-masters-of-it sort of view, we would not avert the ecological crisis. We might postpone it, but it would keep coming back.

It was a revelation to hear Aldo Leopold's philosophical outlook endorsed by Tom and Kris: by people grappling with the best approach to protect millions of acres of wild land in practice, and with a multitude of specific questions, such as, how best to reintroduce giant anteaters to a formerly degraded grassland.

We walked a little further together before we parted company, Tom heading for the ridge that was now not far above us. His parting advice was brilliant.

'The more radical you are,' he said, 'and if you take this anti-industrialised capitalism, pro eco-centric combination of views you *are* radical, the more normal you need to appear.'

That he said, was his strength. He might look like some sort of eco-warrior/hippy/hiker at that point, but dressed in a suit, he could look like any average Joe (albeit a white, male one). It gave him a chance to meet people where they were; to find common ground and work from there. It seemed like something I could at least aspire to after a hot shower.

In the bar of the El Rincon Gaucho restaurant later, I thought about what made the Tompkins Conservation approach so powerful. It was something about the way their passion for conservation was informed and vitalised by lifetimes spent in the mountains, climbing, skiing, kayaking or just wandering. That personal experience of, and love for, wild nature was then combined with what you might call an ongoing philosophical enquiry: an enquiry that directly informed the practical one, constantly engaging with the worldview and values that lay behind both their immediate conservation aims, and the culture that made conservation necessary in the first place.

On top of this, of course, you could then add the money to do something spectacular on the conservation front. It was the first time in my life I'd ever wished I was seriously rich.

Forrest had invited me to join him and two friends at the restaurant for dinner. It was an excellent evening, beginning with pisco sours and progressing through fabulous food – especially fabulous in relation to the noodles and single remaining onion that would have been my meal at the campsite – and a generous quantity of red wine. The friends had a wicked sense of humour and there was much teasing, joking and chat. At some point, Forrest began to share the music he was lining up for the post-handover party and somehow we all ended up singing and dancing to Gloria Gaynor. It was rather late when I cycled back up the bumpy track to the campsite with the ridge of mountains off the left in sharp, moonlit silhouette. All so beautiful. And just a little piscoed.

The next morning, grateful that the pisco consumption had not continued further into the night, I packed slowly and, low on provisions, ate sultana butties. Finally, I headed back out on the terrible, washboardy, hilly, windy road. A fair amount of traffic was coming in – clearly things were gearing up for the president's visit. As I rejoined the Cochrane road, a drone buzzed overhead, presumably checking out security.

Before leaving, I'd had a coffee at the bar with one of the guides.

'Some people are worried, even scared, at the amount of land all around them that has become nature reserve,' he said. 'Actually, many people still see conservationists as hippies, and think it's crazy, using land for nature conservation rather than "cleaning" trees off it and growing sheep.'

I wondered how long it would take for this attitude to change.

The ride to Cochrane alongside the Río Baker, at that point a stunning bright turquoise, took me a lot longer than it should have done. Eventually, I passed a wooden sign that read: '*Bienvenido a Cochrane, Ultima Frontera*' – 'Welcome to Cochrane, the Ultimate Frontier' – and crawled into town, in search of something savoury, having existed on Forrest's cake for most of the day. I found a café and then a hostel, with a peaceful, outside camping space and a covered shelter. I fully intended to work in it.

'Hi!' said a very English-sounding voice.

Ben and Adam, two lads from the UK, emerged from a small tent already in situ. They were riding from the lowest point of the continent, somewhere in Argentina, to the highest, Aconcagua, which they hoped to climb. Their journey was fully human-powered and, until now, they had been cycling with pack rafts, which they were about to sell to some locals.

I'd just dug out my laptop and was looking at it, blearily, trying to figure out which backlog to start with, when there was another 'Hi!', this time from someone Canadian-sounding.

Two women, slim, athletic-looking, in their fifties perhaps, introduced themselves. Lisa and Shanne, from Canmore.

'Canmore, Canada? Do you know…?' I asked, and we were off.

They knew all the people I'd met when I'd been there. And they not only knew about the Y2Y project, but Lisa was very much involved with it.

I was delighted to chat, though the backlogs worried away at me, making me a touch distracted. On top of that, my words were coming out upside-down on a regular basis.

'To be quite honest, honey,' said Shanne, 'I think you probably need a beer more than you need to work.'

Lisa nodded in agreement. I closed the laptop. Several beers later, Ben and Adam returned. Several more, and we cooked dinner, all sharing stories as well as food. I finally made it to bed after eleven, having done no work at all but in much better spirits.

Next day, I woke to the realisation that hiking on a rest day had indeed not been a smart move. I was properly tired. I decided to tackle laptop work and leave late afternoon. My journey's end, I very much hoped, was Ushuaia: about as far south as you could cycle, the town whose nickname is *El Fin del Mundo*, 'The End of the World'. I wanted to get there with a couple of days in hand to arrange my travel back north to Valparaíso, where I was booked onto a cargo ship that would take me home. Ushuaia was still a long way ahead and I had less than three weeks left. Even a few hours cycling would help move me forward.

This plan was soon derailed too. Woody had developed a crack in his frame that, while at Rod's, I had filled with one of two sachets of the vegetable-based epoxy-resin equivalent I'd been carrying for that kind of scenario. But the crack had re-opened. In my inbox was a reply to a flagged email I'd sent to the Bamboo Bicycle Club asking for advice.

'Because the crack is lengthwise,' it read, 'the strength of the frame is not compromised.'

Phew.

'But you do need to seal it. Given the width,' (I'd sent a photo) 'epoxy alone won't work. Mix in some sawdust. Then tape it up and leave it overnight.'

So that settled the departure question. By great good fortune, the floor of the campsite kitchen had sawdust on it. I mixed some with the remaining epoxy on a plastic sultana packet and applied it with half a wooden peg I'd found while undertaking a similar procedure in Rod's shed. The result was a rather distinctive dark stripe along the down-tube, temporarily concealed by insulating tape.

The other reason to stay came in an email from Neda. Neda had had a flat tyre, been rained on and hitched forward. There were long delays ahead at O'Higgins, she wrote, the town that was the departure point for an unavoidable lake crossing. The normal ferry – a large one – had broken down. The replacement ferry – a small one – could only take a handful of bikes at a time. The word was that 50 cyclists were now in O'Higgins waiting to get on that boat. Some had been told they would have to wait a week.

A week! I didn't have a week. A three-day wait at O'Higgins would just about be compatible with making it to Ushuaia in time, if I only had tailwinds and could do silly mileage. But not a week. Neda's message said she herself was booked on to the ferry in a couple of days' time, and that she would try and book me on, too. Suddenly it looked as if whether I could make it to Ushuaia depended on Neda and on the ferry. Both were totally out of my hands.

By the time I woke the next day, feeling restored in a way that revealed how tired I must have been when I arrived, Lisa and Shanne had gone. They had left a small group of orangey-red plums from a tree in the campsite, carefully washed, in my stove pan. I was so glad to have stayed, reinvigorated by the time with them and with Ben and Adam.

Checking my emails before I left, I found a message from 'Portland to Penguins' – Tara and Aidan – who I'd last seen in Peru. They had announced their arrival in Ushuaia on Instagram not long before and would still be in Ushuaia around the time I was due to arrive. If I made it.

'Of course you will make it!' they'd written. 'You've got this. See you there for piscos.'

The message was so welcome it nearly made me cry.

24. PENGUINS AT THE END OF THE WORLD

Of all the paths you take in life, make sure a few of them are dirt.

JOHN MUIR

It took me four days to get to O'Higgins. Four days riding through vast landscapes, glacier-capped mountains always on the horizon, the immensity throwing the details of small-scale encounters into vibrant relief. A night camped under small fruit trees in the garden of an elderly couple who sold coffee with soft white rolls and sweet, dark-purple, homemade jam to passing tourists. They kept goats, a lone bay horse, some beige, disinterested sheep and a shy, grey dog. We shared the words for all these animals in both our languages.

Riding through gorgeous southern beech trees the next day, I passed a chucao, bathing in a stream and calling. Hours later, the sight of a blue-green hummingbird in a bright-red fuchsia bush lifted my energy and helped me on through the gathering tiredness.

On another day, I came to a cleared area of grassland with a white mountain-top vista. Wooden fencing, horses lying down, sheep, a distant dog; a beautiful farm in a scruffy, wildish sort of way. I stopped pedalling and stood there with the bike awhile, fantasising about ending my trip here, pushing Woody down the path, knocking on the door and asking if I could stay. Do some farm work. Help with the animals.

This was followed by a wet night in the tent at the back of a small beach, poised for a short ferry ride in the morning. Then more gravelly miles. Signs on the road about the protection of the endangered huemul deer. Two big hills. One a classic switchback, loops piled on loops. I took my cycling shoes off, put trainers on and pushed. At least with switchbacks you can see what you are up against. The next hill was somehow harder, a long, straight-line climb traversing the edge of the mountain. The turquoise river below transformed to cappuccino.

Two French cyclists went by.

'How are you?' I called to them.

'So-so,' was the reply.

So-so? How could anyone be here and only be 'so-so?' I realised how much I was relishing it all. The wildness and beauty was seeping into me, day by day, despite the toughness or perhaps because of it. I was beginning to feel that I had surrendered something, had given in to the place. And that that was just fine.

> *I love this road.* I wrote in the journal in my tent that night. *The country around it feels so vast and so untamed. Not mega-spectacular views like the photos of Torres del Paine but immense … I need more words for this kind of spaciousness. 'Immense' and 'vast' echo round my head all day. Boundless. Stupendous. Huge … The mountains come closer to the road, then move away as if in a gigantic, slow-motion dance.*

I was in a good mood despite the tiredness, which was like a background ache, a sort of fuzziness that dampened everything down and made me even slower than usual. The place still gave me energy. There was such lushness and variety and it was all so HUGE. I realised it was the hugeness, the gigantic, utterly un-British scale of things that was affecting me most. Millions and millions and millions of trees. I even enjoyed pushing. On foot, I could hear more wildlife and was better able to look around. On the bike, most of my attention was focused on potholes.

On the last of the four days, two more cyclists went by on recumbents. There were now at least eight people ahead of me, in addition to the 50 rumoured to be already there. Out of nowhere, I found I was feeling more at peace with the possibility of not making it. If I asked, 'Whose community am I cycling through today?' the answer would be, 'Thousands of us!' And that was just the ones I could imagine seeing. Throw in the invisible underground world and the answer would be 'Billions!' What an astonishing privilege it was to be here and how ridiculous to waste any mental energy on, *Will I / won't I get to Ushuaia?* As if it mattered in the bigger scheme of things one jot. And yet.

Mid-afternoon on the fourth day, there was a lovely, long, flat run into O'Higgins, after which I suddenly popped out of the trees into a cleared area – someone had told me that the non-native grasses arrived here on the soles of Brits accustomed to playing lawn tennis – with gardens, a street of small wooden houses, a tiny Copec, and a supermarket with outside speakers and the radio on. In the town square, I found various cyclists. They helped me to get on to the free Wi-Fi by the closed tourist office. And bingo, there was a message from Neda saying, go to Las Ruidos, a boat operator based in one of the hostels: you are on their wait list for Sunday/Monday.

At Las Ruidos I found I was indeed on the wait list.

'It could well be Monday,' the tired-looking man there told me when I reached the head of the queue of cyclists on a similar mission.

'Check in again tomorrow afternoon.'

And suddenly it was just about game-on again. *Neda, you are a bloody star.*

It was a Friday. The thought of two whole days ahead with nothing more urgent to do than catch up with the usual life admin and email backlogs felt unbelievably good. Later, camped outside El Mosco hostel, I had a beer, made some noodles, had another beer and tried to figure out how many miles per day I'd need to do, assuming I caught the ferry on Monday and was therefore able to leave the town of El Chaltén on the other side of the Argentinian border on 7 February. That would leave me nine days to reach Ushuaia by the 15th, which in turn would give me two days to sort bus logistics before I left again. But at 800 miles, that was still very tight. I whooped out loud when, putting distances into Google, I realised that it was actually only 640 miles. Then, trying to figure out what that meant in terms of miles per day, I realised how tired I really was. I was in the hostel, apparently functioning normally, albeit slowly, but I couldn't for the life of me recall the nine times table.

I did leave El Chaltén on the 7th, albeit in the afternoon. Being compelled to take those two days off in O'Higgins had made the likelihood of reaching Ushuaia both higher – I had clearly needed the rest – and lower. The time was now properly tight. And the time in O'Higgins had been of value in other ways, too.

I had been working on Woody on the deck outside the hostel when I'd met a man called Vincente. He'd stopped to say hello and to ask what I was up to. I showed him the crack in the frame, now unwrapped from its electrical-tape shield.

'Ah ha!' he said. 'Bamboo and epoxy. I can reassure you on that front.'

'How so?' I asked.

'I'm a boat builder,' he said. 'We're used to cracks in wood and cracks in bamboo and cracks in things that need both to stay flexible and cope with being wet. I can show you how to back up the repairs you've already done.'

'And then?' I asked.

'And then, it will be fine,' he said. 'The bike will outlast you, probably.'

I looked at him half-spooked. When I'd started working on the problem at Rod's, he'd suggested that if I had any more trouble with it, I should seek out a boat builder.

'A boat builder will know what to do,' he'd said, casually adding, 'if you need something, it will probably turn up. The universe is like that.'

By the time I joined the small crowd of cyclists scheduled for the Monday ferry, our bikes handed down to the crew on the boat below and stacked on the back deck in a chaos of handlebars and interlocking pedals, the new epoxy was set hard. This was to be a relief sooner rather than later – the official bike route from Chile into Argentina on the far side of the choppy water turned out to be technically the hardest thing I'd ridden, or pushed, in the whole of South America so far.

It had started well enough, with a beautiful ride through crinkly leaved southern beech trees along a stretch of 'Patagonia flat' road, then a shortish climb up to the highest point where I camped off the track among the trees.

Early the next day, I'd reached a sign that read: '*Bienvenido a Argentina. Verduras y frutas prohibidas*' – 'Welcome to Argentina. Vegetables and Fruit Forbidden'.

After that, the road became single track, which I managed to cycle for almost three feet. The road, generally speaking, was heading downwards. But it went down in an up-and-down sort of way, some of the ups on the limit of pushing and some of the downs plunging into small streams and then sharply up again. There were tree roots across the trail and sometimes trees. There were big rocks, loose rocks and occasional deep channels only just wide enough for the bike so that I had to push Woody through them while balancing on a slippery mud ledge a foot or so above. The panniers ripped and snagged. The streams became easier when I gave up on the ambition of keeping my feet dry and walked straight through them. Then I did the same with the mud puddles. And the bogs.

An American cyclist, who had been packing up his tent when I cycled by in the morning, caught up with me as I reached a section I simply could not push the bike up and helped unload and carry the panniers. Who knew that a bike ride could be so good for your biceps. The exact same thing happened at the next un-pushable spot, with a group of hikers. I found I was grinning almost all the time, despite it being one of the toughest few miles ever. At times it was so tough it was funny. Ridiculous, even. Miles are so very relative.

Suddenly, I could see buildings and a jetty – the lake had been visible for a while. I stopped and ate my remaining O'Higgins-purchased apple on a rock, in case the fruit and vegetable prohibition was policed, and it was confiscated. Then the final descent.

I popped out over a wooden bridge and into a world of short-mown lawns and compact buildings. The immigration building had a lovely, big, obvious sign outside it. The American cyclist was in there already. When he came out, I was ushered in. My passport was checked and stamped.

'Welcome to Argentina,' said one of the two immigration officers.

'Thank you,' I said. And that was it.

I joined the cluster of cyclists down by the boat shed, waiting for the next ferry. Not long after, Beni turned up. His fat tyre bike was much better suited for the terrain we'd all just crossed. But he was towing a trailer. How he had made it along that track I would never know. Then, even more amazingly, an elderly French couple arrived, pedalling recumbents. I felt as if I had just survived some sort of bizarre, gruelling survival challenge and was visibly battered and grungy. They seemed as unruffled as if they'd just been cycling along a seaside promenade. Cyclists are astonishing, I thought, as I watched the ferry appear in the distance out of a cloud of spume and wash.

Underway, white-capped waves switched between a milky, dove grey and turquoise. Glaciers shimmered at the top of the dark peaks that framed the lake. I couldn't bring myself to stay inside and sat on deck with the bikes. For an exhilarating, slightly crazy while, after we'd motored out of the bay's shelter and back into the wind, the captain surfed down the by-now steep waves, then turned the boat back into them and surfed again. It felt like my grin was being stretched backwards to my ears by the motion and the wind.

On the far side, the other cyclists decided to camp together on a site close to the jetty. There was a shared sense of achievement and camaraderie, and the evening had the clear makings of a party. I wanted to stay, but my deadline was making itself felt. There were still a couple of daylight hours, and I could use them to gain a few miles. When I told the others why I wasn't going to join them and when I hoped to reached Ushuaia, they all laughed. Except Beni.

'I think you will do it,' he said. 'It will be tough. But you will do it.'

That evening's ride was not exactly a hardship. The road was wide and flat. Cloud and sun played across the spiky ridges. I camped late outside a hotel that I discovered, after they let me in for breakfast next morning, had fabulous views through a picture window of the extraordinary shape of Mount Fitz Roy coming and going in the morning mist.

I was in El Chaltén by 1pm, stocking up on food and cash. Then I sat in a café and tried again to figure out what I needed to do to make it to Ushuaia in time. The answer was a good 70 miles a day, every day, for eight and a half days, that day being the half. Another 70 miles would make today's total 85 – I'd just ridden fifteen miles to get to El Chaltén.

If I had known how hard those days were going to be, would I have stayed with the other cyclists, joined the party and jacked it in? I guess I'll never know. But probably, no. I could feel that old madness stirring, that unlikely combination of utter, irrational, bloody-minded determination and a sort of calm; the combination that had seen me out of Cartagena and kept me cycling in those initial crazy days and weeks, and that had seen me through countless challenges since, from lung-emptying mountain passes to the tear-reducing corrugated gravel of the Eduardo Avaroa National Reserve. *OK Kate. Game on. Let's do this.*

DAY ½

Distance to Ushuaia: 647 miles

It is 3pm when I leave El Chaltén. The road sweeps out of town past the bus station, over the Río Fitz Roy and away. The sun is out, the wind blowing strong. For a while, it is gusty in new ways, and I have to learn how to ride it. More than once, I am slam-swept off the road into the hard shoulder – away from the traffic at least. The wind scares me. Its force and power are of a different magnitude. Other cyclists joke about suddenly finding themselves on the far side of the road, but I find that thought unnerving in the extreme. Today, though, the wind becomes an ally, settling after an hour or so into that entity I'd begun to assume mythical ... a tailwind!

I am heading south-east on a long straight stretch. Behind me, the Fitz Roy range, dark and grey and still in cloud. Alongside, though at a distance, Lago Viedma, huge, vividly turquoise, flecked with white. Chunks of glacier? Yes. Later, the glacier itself comes into view, a huge, chiselled tongue of grey/white/cobalt ice reaching right down to the water's edge. Above, sky-blue sky, white clouds. At one point the colours line up like an abstract painting: gold grass, turquoise lake, an almost lilac line of low hills, blue sky. Beautiful. Slowly the mountains and then the hills fall away, and I am into hugeness of another kind. It is stunning.

The tailwind ride lasts for a fabulous 55 miles. Sometimes I simply sit there, doing nothing but hang on as I am blown forwards at 15 mph. Sometimes, temporarily lacking top gears, I pedal like crazy. My shadow, thrown right in front of me, shows my head and body bobbing madly from side to side. I stop for a late lunch – toast, (hoarded from the hotel breakfast), cheese and cherries, hiding from the wind behind the road barrier. It is only cold when I stop.

At the junction with Route 40, I put on gloves and jacket, eat the other bit of toast and steel myself for a strong side-wind/headwind. As soon as I turn at the junction, I stop flying and start grinding. It is a reminder of how hard it could be from then on. Soon I start seeing signs for a hotel at La Leona, at an almost perfect distance, only two miles short of my 85-mile aim. Psychologically, the hotel is a good goal. It's also helpful since, for some reason, my milometer is intermittently quitting, and I can't really tell the day's mileage from it. Above the road, small hawks. On the road, a dead armadillo, upside down, its guts half-eaten, all red above the shell. Nearby, three vultures, all hopping on the spot.

At last, the hotel comes into view across a river. It looks shut, no cars outside. But it is down a track with trees. Trees mean shelter. I ride down. Lights! The door is locked but someone comes and opens it.

'Yes, we have a room. The price is $75'

'$75?!'

'Camping? Yes, that too.'

A kindly man shows me round the back.

There is shelter from the wind, and I can buy breakfast in the hotel in the morning. Perfect. By now it is well gone 9.30, and the light is fading fast. I put the tent up by a group of willows, and then chat to a voluble, German-speaking woman, cycling north.

'Won't you get more headwinds going that way?' I ask.

'Winds are everywhere,' she says.

I fall asleep thinking, *what a wonderful day*. I had ridden it often laughing aloud, sometimes almost crying with gratitude. Of course, I know it won't always be like this to Ushuaia. But if nothing else, the day gives me hope. I could still make it. I could.

DAY 1½

Distance to Ushuaia: 579 miles

Good sleep. Crazy southern hemisphere stars when I crawl out for a pee in the night. In the morning, I go for breakfast and find the place heaving with tour-bus tourists. La Leona Hotel has a history. A famous Argentinian scientist and explorer, Perito Moreno, had been attacked here by a female puma (La Leona). Butch Cassidy had stopped by.

I leave La Leona at 11ish, up the gravel track and jump back out into the tailwind. A late start but, with this wind, it really doesn't matter. Ten miles later, I stop to take a photo of the river. In the length of time it takes to eat a biscuit, the wind switches. It is as if the god of the headwinds has just noticed me.

'Ha! that's where you are! And with a tailwind. We can't have that.'

Booff. The north-westerly is now a south-easterly and I am riding straight into it.

Or maybe the road has changed direction. It is a weird section of road, climbing up and away from the river with lots of wriggles not marked on the map. Ominous clouds. I put on rain gear, but the rain never actually arrives. It just stays dull and cloudy. And windy. I am constantly recalculating. At 5 mph, with X hours before darkness I can do Y miles ... It is like being back in northern Colombia at the turtle school, doing turtle maths. *At this speed for this many hours, the turtle can go this far…* The calculations are not encouraging. It is never enough. And what do I learn from this? That if you wake to blue skies and a tailwind, you jump straight into it. No matter that the journal is not up to date or that the hotel you are camped behind does omelette breakfasts. Go!

On the map, it looks like a relatively short section from La Leona hotel to the junction with the El Calafate road. It is not. It goes on, and on, and on.

Finally, a long 48 miles later, I reach the junction and turn hard left back into a tailwind. For a while I fly again, shouting 'good luck!' to hitchhikers, my bad humour at the endless road evaporated. Then the road starts to climb. It climbs steadily for fifteen miles up the side of a hill. A pair of small, white wing-tipped eagles float alongside. Other than the hill I am on, the land stretches out flat for hundreds of miles.

The milometer quits working again. I keep trying different alignments of the magnet on the spoke, but nothing works. Yesterday, the solar charger quit functioning too, quite suddenly, when I went to charge my phone. I am very tired. At one point on the flat I feel I could fall asleep on the bike. And this is day 1½ of 8½!

About halfway up the hill I pull out my tiny music player. I haven't listened to music in ages. I am suddenly flooded with beautiful sounds. Wonderful! I haven't heard much music at all of late. So different from Colombia. The player cuts through a song to announce a low battery. I plead with it. *Just one more song? Please?* It plays the whole album – twice. Through many false summits.

The light is fading fast down across the long, flat valley floor below. At about 9.30, it looks as if the summit really is close. A campervan has pulled off the road just ahead, I think at a viewpoint. I resist the urge to search for a scenic spot – it's about to be dark! – and push the bike into a more-or-less-OK-as-long-as-the-wind-doesn't-really-get-up spot just off the road.

Tent up. I make myself cook noodles and veg. The gas is very low. I could easily have bought some in El Chaltén. Stupid. There is not much ahead that will help. Esperanza is unlikely to have a camping shop, though the town of Río Gallegos in a couple of days might. I am low on food in general, not just food that needs cooking. *It will be OK. Keep it together.*

DAY 2½

Distance to Ushuaia: 516 miles

I am now about twelve miles down on the miles I need to get to Ushuaia by the 15th. But a chunk of today will be a descent. If there isn't a headwind, I should be able to catch up...

The gas lets me make almost-boiling water for coffee. Then marinated porridge. It's a cold morning, dry and overcast. Not much of a view from my camp spot.

The solar charger is working again – until it starts to hail. I get going. The top of the hill is very close. On the far side, it is much windier. I have a monster sidewind most of the rest of the day.

I eat lunch off the road, hunkered down behind slight shelter in brief sun, feeling happy. I will miss these set-aside moments, the bike on its side on the verge.

The last six miles to Esperanza are fast. I had been going to pull out the treats/emergences credit card and sleep in a bed, but the hotel is full and they send me to the petrol station. The petrol station has rooms but is also full. I put up my tent on a scratty bit of grass hidden behind a bush on the forecourt. The petrol station café is open 24/7 and is warm, has toilets and Wi-Fi. Eating egg and chips at 10.45pm. This is a sanctuary.

DAY 3½

Distance to Ushuaia: 451 miles

I wake snug in the tent, feeling conscious of being warm. I had heard rain earlier. I feel very unmotivated to get up. It is cold packing the tent. I put gloves on.

Back on the road, a cloudy sky, like the soft, ruffled underbelly of a huge, grey dove. Ostriches running panicky along a fence line. Ostriches?! Wrong continent, surely. Rheas. *Get a grip. This is a biodiversity bike ride. You are supposed to know where ostriches do and don't live.*

In the petrol station café, I'd picked up an email from Lisa and Shanne. At the airport on their way home they'd seen Yvon Chouinard and couldn't resist introducing themselves.

'And you'll never guess what he said,' Shanne had written. 'He said: "Have you heard about the woman who is cycling the length of South America on a bamboo bike? She wants to use the journey to raise awareness of biodiversity loss." How cool is that? Yvon Chouinard is talking about you!'

I end up pushing on past a lovely potential camp spot – sheltered, with views – chasing those last few miles before sundown and the promises dangled by a 'Hotel and Restaurant' sign. Ahead in Güer Aike. It is almost dark when I arrive. There is no trace of a hotel. Güer Aike seems to be a tiny place off the road. I take a wrong turn trying to get to it and end up at a large, red-painted shrine to the Virgin Mary. The people gathered there point me to a campsite down a gravel track by the river. This proves to be some sort of fishing camp. Two large men emerge from a hangar-like building where meat is being roasted to say, I can camp, but they are having a music event later, and if I want any sleep I should camp as far away as possible. Or I could join them. It will be rock music tonight. Folk music tomorrow. I pitch the tent in a tiny space between the BBQ oven and the table. I down a beer and sleep just fine, despite cheerfully awake neighbours and the distant music.

DAY 4½

Distance to Ushuaia: 380 miles

I wake to find myself listening for the wind. It dropped late in the evening yesterday. I have been heading south-east across the tail of the continent and am nearly at the Atlantic Ocean. Where I should have had the wind behind me, there was nothing. But it will be worth it if it stays nothing tomorrow. At Río Gallegos, the road swings south, then south-west and straight into the wind. Or at least, into what has been the main direction of the wind so far.

It is hard to get out of the tent. I find a hot shower, which is disproportionately reviving. An easy ride, if not super-fast, to Río Gallegos. A service station. Wi-Fi! Coffee! Veggie rolls! Fruit salad! I send messages to Chris and Neda. The cash machine declines to give me cash. Then, realising I have spent too much time not cycling, I decide to leg it with just the food I can buy from the service station, rather than go hunting for something more substantial.

It is a mood-swing day. I turn into what should have been headwind after cycling out of town past the Malvinas military base. There is no headwind. I ride fast, at least initially, full of coffee and glee. Road signs announce Ushuaia to be 573 kilometres away. I do the maths and work it out at 60-something miles per day now, not 70! More glee until I realise I've got the number of cycling days wrong and it is definitely still 70-something, only a higher 70-something than before. I find myself laughing out loud at the lack of headwind and at the fast, forward motion. Unperturbed by rain showers.

The hours slip by. Carrying on cycling long after I want to quit is now the main game and I develop a strategy. I ride for an hour. Stop, hop off, jump about, eat something, then get straight back on the bike and go. The breaks are super-short, but knowing I only have to cycle for an hour and then I can have one seems to help, at least mentally. Physically, I have been working so hard for so long that my head hair has become thin and my fingernails, toenails and leg hair have all stopped growing. Intriguing.

I realise gradually that I am not going to make it by the 15th. This is probably the easier section of the ride, and I am already slipping behind. Tiredness is sinking back in, and soon I am feeling low and flat. This whole race to the finish thing is such a cliché. It's a cliché entirely of my own making, yet I have trapped myself in it. At some point, I decided that the end of my journey was Ushuaia and, worse, announced it on Instagram. Now I can't find a way to rethink the destination without feeling like I've failed.

It is ridiculous. But there it is. I still can't bring myself to give up. I will just keep cycling stubbornly forward, as many hours as I can, however impossible the task has become. And on the other side of this feeling, something stranger. A regular flicker of almost-comprehension, as if the exhaustion is allowing me to tune into a different, parallel reality, and on some level I know that I just have to trust. It will be OK, even if you can't for the life of you see how it possibly can be, with your rational mind and your turtle maths.

I finally reach the border at 6.30pm. I am about to cross back into Chile again. It takes an hour to get through, but the room is warm, and I needed a rest. It is only 7.30pm when I emerge, the toilets locked up and the café closed. I cycle away.

At 9ish, I start looking for somewhere to camp. I am on a long, straight, increasingly headwindy road. I pitch the tent in a marginal spot, the other side of a wire fence, in view of the road and only partially wind-protected.

I study the map. I eat bread and cheese. I have nowhere near enough food and no gas to cook noodles. The next big town is three days' away. There's sufficient signal in the tent to search: there's another lake crossing ahead and Maps.me shows a café at the jetty. Then, yes! A small town 53 miles ahead has a supermarket, a hotel and a Copec petrol station! I must try and get there tomorrow whatever the wind is doing (and ideally well beyond it).

Hopefully sleep will help. An earlier night than usual. It's 11ish now. Leaving the tent for a pee in the dark, I see two eyes looking straight at me from the other side of the wire fence. They startle me, and my first thought is, *Gollum!* Then, *no, the eyes are the wrong shape*. The eyes move off to the left, but slowly and while still looking at me. The body of whoever's eyes they were must have been slinking sideways, quite close to the ground. *Probably a fox*, I think later. *Gollum. How ridiculous.*

DAY 5½

Distance to Ushuaia: 314 miles

Raw porridge, plus nuts and raisins, plus powdered milk, plus water = muesli!

My camp spot was better chosen than I'd realised. I only feel the full force of the wind after I've lifted Woody and all the stuff back over the fence and am on the road. What follows is the hardest four hours ever. I am going straight into a wind with a force and power that is barely comprehensible. I am at various limits. Of strength just to keep going into this ferocity. Of ability to keep the bike going straight and not lunge into the traffic or off onto the hard shoulder and down the verge as the wind gusts and slams around me. Of mental ability to deal with it. It is just so bloody knackering. I cry at one point, huge sobs of exhaustion and frustration. I am never going to make it. The goal of reaching Ushuaia by bike from Colombia is slipping away in this maelstrom, despite all this effort. I hate this bloody road.

The big, left-hand bend that I know is ahead keeps not materialising. By now I am tuned into all the deviations in the road, and precisely what they mean. Bends to the right take me more fully into a headwind. I am talking out loud, my words shredded and unheard. *Just go left. Please, JUST GO LEFT.* More tears when an unmistakably right-hand bend appears.

At long last, I turn left onto the road to the ferry crossing. And I fly! Ten miles, effortless. Whereas the previous fourteen near killed me.

I eat fried eggs and drink two mugs of coffee in the ferry-side café. I am super hungry. I stow away some bread for later. The high-sided ferry manoeuvres in despite humongous winds. Through the wet windows I can see cormorant-like birds lunging in the wind. Earlier, enormous geese, grey and pale.

On the other side, I am now on a gigantic island. Isla Grande de Tierra Del Fuego, half in Chile, half Argentina. It owes its name to Ferdinand Magellan, the Portuguese explorer who visited in the early 16th century, en route from the Philippines to Spain. The fires he saw – and named the place after – were lit by the indigenous Yámana who lived there. Magellan believed them to be waiting in the forests to ambush his armada. More likely they were simply warding off the cold.

On the Big Island of the Land of Fire it is, for me, a mixed ride. Stretches of flying, longer stretches of straining. It is grey and cloudy, with a dark, wild sky. The wind is against me for the last many hours. I reach the town. It is about 8.30pm. I have only done 53 miles. But what do I gain by battering on into the wind for half an hour, then searching in the near dark for somewhere to camp?

There is a hotel on the edge of town. Two friendly, interested women take me to an upstairs room in an annexe. They are excited by my journey and that I am travelling alone.

DAY 6½

Distance to Ushuaia: 261 miles

I wake at 5am and check the weather. It is still blowing a hooley. I decide to sleep some more. I get up again at 7 to gain an extra hour.

Message from Tara and Aidan. Their plane is not until 8pm on the 16th! So, I may see them yet. To arrive in Ushuaia on the 15th is now 87 miles per day but 16th is 65 miles per day and almost still feasible. Today and tomorrow will be tougher, as I cycle cross-country back to the main road. Then it should be easier. The main road more or less follows the Atlantic coast before it sways off south and then bends back around the end of a long lake and heads west to Ushuaia.

I head out onto the road that yesterday would have been blasted with a screaming headwind. It is absolutely fine. The wind has dropped. To my delight, I do about 50 miles of 'normal' cycling, at speeds of 10–15 mph. The land around is khaki, mostly flat.

The rain starts somewhere before the junction. I come to a white building that is on a bit of the new road that is still closed off and push across through the by now blistering wetness to investigate. It is open, and clearly used as a shelter. Messages from other cyclists are scrawled in multiple languages on the plastered walls. I take out another layer of fleece, take off the soaked-through rain jacket, put the fleece on, put the jacket back on. I eat a mouthful of something and set off again on a section I am anticipating to be a flier. The wind is still wonderfully low, but the asphalt abruptly ends, leaving a rough, corrugated, loose and increasingly muddy surface. *Ripio*, gravel. My speed plummets.

There is something scraping. Hang in there, Woody. The new road is now running alongside the dirt track I am on. It is nearly finished but blocked off. I heave the bike over the barricade at a low point and ride on smooth asphalt with no other traffic at all. My speed picks up again.

At about 7.30pm the rain finally eases and weak sun appears.

At 8ish I arrive at the Chilean part of the border. There's now about ten miles of no-man's land to the Argentinian border. I am soggy from the ride and chilled from standing around in the immigration queue. I could push on. But there is a hostel right here on the Chilean side. I turn back and check in.

There are 186 miles between here and Ushuaia. I've given up on the idea of making it by the 15th and am now focused on getting there on the 16th, in time to catch Tara and Aidan before their flight home. That's in three days' time. If the headwind returns, there is no way the maths will add up.

DAY 7½

Distance to Ushuaia: 185 miles

Away early and across the no-man's land – dirt road but fast – and through customs and border control.

The section of the road that headed east towards the coast, also fast. I am waiting for the right-hander that will put me on the south-easterly road, parallel to the coast. I will be on that road all day, and I am tensing for a strong sidewind. After about an hour I realise I am already on that road. With a tailwind!

I ride the tailwind all day. I ride it for 99 miles, laughing out loud, sometimes crying out loud. There is only one tough, sidewind section, when the road swings off through the town of Río Grande. I watch a beetle half-running, half-blown across the tarmac. Then a full-on headwind as the road curves back around, revealing how very little I have in reserve and how very little appetite I have left for winds. Then it's back to tailwind.

I ride on and on, into the beautiful evening, watching cloud shadows race away instead of toward me.

It's almost 9pm when I pull off the road down a track to a flat spot among trees, near a small river. It's clearly been well-used as a camp spot/BBQ site, but there are no other people here now, only a hawk sitting in one of the larger trees. I walk up the slope a little, admiring the trees. I have spent eleven and a half hours on the bike and am a little dizzy.

I pitch the tent near a patch of purple-headed grasses and crawl inside. I love my tent.

I will miss being in these huge spaces, day after day, such vast distances behind and ahead.

I am ready for this to stop.

DAY 8½

Distance to Ushuaia: 86 miles

I'm back on the road by 8am. This is perilously early for me. I don't dare get up earlier – I am sure my body is only coping as well as it is because I am allowing it to sleep.

I'm going slowly, though there is now no wind. This section I'm heading into would be tougher than anything if there was. I'm aiming for a café I know is about 24 miles away. It takes me a good three hours. By the time I get there, it is raining again.

La Union café is clearly an institution. It is large and busy. Sitting with a mound of food and coffee, I begin to realise how tired I am. Bleak tired. Incompetent tired.

There is a small group of people around Woody when I go to leave. Then, 'Wait a minute!'

A man goes into the café to fetch someone. A young woman called Iria.

Iria is a cyclist, from Ushuaia. She works there, in a cyclist-friendly hostel.

'If you arrive by bike, you get a free night!' she says. 'And I will buy you a beer.'

It is a lovely encounter and I leave cheered, into easing rain. The road keeps swinging around, now running high above a long lake.

At around 6pm, I pass the turn-off to a camp spot that Iria had suggested. I keep cycling. I head up towards the Garibaldi Pass. The lake falls further away below. I feel better when I'm obviously climbing than when I'm just going very slowly up a road that appears to be flat but isn't. I climb for a good hour. I am expecting another hour, but suddenly I'm on the descent.

A car pulls over. Iria gets out, flags me down, hands me some bread. I'd been berating myself for not having bought any.

'See you tomorrow!' she says and is gone.

On the descent, I am talking out loud, laughing, crying, close to delirious with that strange and potent mix of joy and exhaustion.

However, it is not quite in the bag yet as Woody's chain is making all kinds of noises, most of them ominous.

I turn off just before dark down another track, this time steep, and camp close enough to the river to hear it all night.

DAY 9½: 16 FEBRUARY

Distance to Ushuaia: 20 miles

I wake to blue sky and coldness. The tent is in shade. I move everything into the sun. It's still cold. The parakeets I heard when I arrived have gone; birds of prey are now calling from the higher branches. I lube and clean Woody's chain as best I can with icy, un-cooperative fingers. Then I pack and haul Woody and all the stuff back up the steep slope and out onto the road. The first sign I see says, '*Ushuaia*'. The chain sounds less crunchy, and soon we are bowling along.

I reached Ushuaia four hours before Tara and Aidan's flight home. I have never been so exhausted nor so exhilarated in my life. Already at the airport, they found a way to leave their bikes and luggage and come back into town to join me. I was so very happy to see them: the only people I knew in the whole world who truly understood what it meant to have reached Ushuaia by bike. And what it had taken to get there. You can drink a lot of pisco sours in four hours.

Later, I met with Iria and had that beer, too, before sending lots of celebratory messages I had no recollection of in the morning, and collapsing into bed.

The next day, I grappled, somewhat fuzzily, with logistics. As an end-of-ride treat – a big one – for part of the journey back up to Valparaíso to catch the cargo ship, I had booked onto a ferry from Punta Arenas through the tangle of islands and straits that constitute Patagonia's western shores. It was a four-day ride and I promised myself that, once on it, I would do nothing at all bar eat, sleep, watch the mountains slide by and reread *One Hundred Years of Solitude*. So many aspects of that seemed like almost unbelievable luxury.

But first, I had to get to Punta Arenas. This, I thought, would be straightforward – there were regular buses. But the buses were all booked up, not a single seat on any of them. Hitching seemed a bit too uncertain. After a short period of panic, I found a tour operator who could book me a taxi. A very expensive taxi. It was roughly 400 miles and two border crossings.

It took me most of that journey to adjust to the new speed. Retracing my ride, unravelling all those hard-fought-for miles without effort. Speed is one of the things that ancient sunlight has bought us. In this case, a speed roughly ten times faster than the one I was accustomed to; than the speed you can achieve on two wheels, powered by a biscuit. We shot by my last campsite, then the one before that, back over the mountain and into the forest of two-tone trees, dark olive festooned with pale-grey/green lichens, then the dead trees, then the long flat, treeless pampas. I was a bit stunned by how far I had come; by how long the 100-mile day was, even in a car. I had no idea which way the wind was blowing, or even whether it was blowing at all.

We passed a spot by a blue shed where I had sat on a burned-out log and eaten bread and cheese. I pressed my face against the glass and tried to make sense of what it meant to have cycled this; and what it meant to be reversing the miles so fast.

We drove through the town of Porvenir and right to the ferry terminal, where a regular crossing would deliver me directly to Punta Arenas.

It was the slowest-loading ferry ever. A single person checked each car's passengers and tickets. Finally on board, Woody stowed below with the motorbikes, I headed into the warmth. The boat's small lounge was packed. A kindly family shuffled closer and made space, squeezing me into a window seat. As the ferry pulled away, there was a shout. Dolphins! A small pod of them, leaping high out of the water and racing towards us. Then they swam alongside the boat, leaping in great curves. People were laughing and pointing. It was joyfully, straightforwardly wonderful.

I would have plenty of time on the cargo ship to think about the journey as a whole; what it meant, where it had left me. But this would do just fine as a summary. The arching, leaping, wet backs of the dolphins, surely one of the most moving examples of the sheer exuberance of life; and the people on the boat, cheered by these beautiful, energetic animals, all laughing. Vital. Vitality. How utterly, indescribably precious.

This is what it has all been about. Exactly this.

EPILOGUE

When you say that you are urgently looking for climate solutions, yet continue to build a world economy based on extraction and pollution, we know you are lying because we are the closest to the land, and the first to hear her cries.

NEMONTE NENQUIMO

All the problems we have are symptoms of stories that were wrong.

CARL SAFINA, *ARTIFISHAL*

Can biodiversity loss really be as great a threat as climate change? The question that kick-started *The Life Cycle* journey is almost its ending, too. The answer that I encountered again and again during those 8,288 miles was a resounding yes. Biodiversity is crucial for functioning ecosystems, and ecosystems are the foundation of the living world, providing 'services' vital for all life. Fresh water. Clean air. Fertile soil. Pollination. Waste disposal. Food. The dramatic loss of biodiversity, the collapse of ecosystems: these things are potentially catastrophic. Already catastrophic in many parts of the world. And of course, 'biodiversity loss' is not just a threat to us. It also entails the tragic loss of other-than-human beings every bit as entitled to be here as we are.

So, yes, biodiversity loss really is as great a threat as climate change – terrifying as that is. But it is a 'yes/and' because these are not the only issues we face. Rockström and his team identified nine 'processes that regulate the stability and resilience of the earth system' and the planetary boundaries associated with them.[1] Of these, six are now considered to have been crossed. We are in the danger zone in relation to biodiversity loss (now referred to as 'biosphere integrity'); climate change; the nitrogen and phosphorous cycles; 'land-system change', e.g. deforestation and desertification; plastic and other forms of synthetic, chemical pollution – the 'novel entities' boundary – and freshwater.* And all of these issues are interconnected, both with each other, the astonishing catalogue of human impacts on the living earth in the Anthropocene and with poverty, social injustice and the inequitable distribution of land, power and wealth.

The apparent detour to visit anti-gold-mining activists – then anti-copper, lead and oil activists – turned out not to be a detour at all, but a direct route to the dark and glittering heart of it all. Was the behaviour and the values of the companies I encountered an aberration? A case, or several cases, of stray bad apples among an otherwise fresh, green and shiny bunch? This was the other question that haunted me throughout the journey.

That terrible, deadly pit in the centre of Cerro de Pasco was probably the most powerful visual answer I encountered, an appalling summary of the rest. Of mercury seeping into water in the search for gold; of gas flares on the Amazon's oil-polluted rivers; of beautiful and biologically rich cloud forests at risk from copper-mining. As my journey unfolded, these were joined by the image of vast factory ships decimating ocean biodiversity; of industrial-scale beef farming destroying rainforests; of plastic-wrapped, pesticide-drenched bananas on those endless acres of biodiversity deserts known as plantations.

* In relation to 'green water' – water available to plants.

The companies – often multinational corporations – behind these travesties are not aberrations. They are the inevitable result of the economic and political systems that currently dominate our planet; the systems that we've devised to 'meet our needs' but without meeting the needs of so many, and at such appalling cost. Of course, not all companies and corporations are bad: and of course, business has to be part of the solution. But even the good guys are still operating within systems that demand that they grow. And in the absence of absolute decoupling of growth from environmental impact, aspects of that growth will always, inevitably, be problematic.

These systems are themselves the result of the mindsets, worldviews and values that underpin and legitimise them. They tell us that gold is of higher value than water; that profit is more important than life. They tell us that nature is essentially a vast warehouse of commodities we – or some of us – are entitled to exploit without limit; that any costs in terms of human and ecological well-being are justified by the all-important pursuit of economic and material wealth; that the industrialised world is still entitled to exploit the rest of the world, too; that our knowledge is, after all superior; that we are separate from and superior to the rest of nature; that quality of life is primarily about money and possessions.

Witnessing all this brought me back to a conclusion I'd known all along but that was made vivid through this journey. We must arrest and reverse biodiversity loss as a matter of urgency. But we cannot do this in isolation from tackling the other environmental and social issues we face. And nature conservation, as carried out in conventional reserves and protected areas, like reducing our carbon footprints, while crucial, will never be enough. We cannot tweak our way, piecemeal, to environmental safety, to a truly sustainable and just society. We need to look outside the reserves as well. And when we do, we see that we need a radical overhaul, with systems driven by different values. Systems whose primary aims are the well-being of people and planet, not whose overriding value or goal is profit, with human and ecological well-being as begrudging constraints, an afterthought, side-reins on a malevolent dragon. We have to tackle the deep roots – economic, political and philosophical. We have to change the system.

UPDATES ON PROJECTS/PEOPLE I VISITED

While I was still cycling, AngloGold Ashanti announced that they would accept the results of the *consulta popular* and leave the Cajamarca area. Overjoyed to hear this, I learned later that Cosajuca had gone on to offer advice and support to other Colombian communities who wanted to hold a similar referendum as part of their resistance to extractivist industries.

But the situation has since changed. In part due to the Colombian government's increasing investment in extractivism as a driver for post-Covid economic recovery, the threat of open-pit gold-mining has returned to Cajamarca, and AngloGold Ashanti is pushing hard for La Colosa to go ahead.

Doubts have been cast on the legitimacy of the referendum, and the individuals and organisations who led it have been stigmatised as being both anti-development and opposed to the national government. There have been further murders of community leaders and several have had to leave Cajamarca to seek safety elsewhere.[2]

A group of the Cajamarca activists are planning to visit London in the spring of 2023 as part of an international tour to raise awareness of and support for their situation. They are supported in this by the London Mining Network.[3]

Walking down that small, muddy track into the cloud forest in Ecuador and listening to Carlos tell me about the astonishing variety of plants and animals that live there is one of the most powerful memories I have of the journey. But the threat of large-scale copper-mining still hangs heavily over the Intag region, and Carlos is still fighting.

In 2019, there was a win in a lower court, with legal action based 'on the inevitable violation of Ecuador's Constitutional Rights of Nature' that copper-mining in one of the 'biological jewels of the world' would constitute. Some exceedingly rare frogs played a key role – two that were thought to have become extinct but recently rediscovered in the area that mining would devastate, the confusing rocket frog and the long nose harlequin, and one new species, recently discovered and named, after a public vote, as the Intag resistance rocket frog. These frogs are only the tip of the biodiversity iceberg – hundreds of endangered species are among those who will be threatened if the mining goes ahead.

This win was appealed by the mining companies and the judge's decision to uphold the Constitutional Rights of Nature was reversed. A team of lawyers and members of the local community, including Carlos, is now trying to reverse the court decision that gave permission in 2022 for the Llurimagua copper open-cast mine to proceed.†

Later in 2022, there were exciting positive developments as a result of three scientific expeditions to gain a clearer picture of the biodiversity at the Junín site. They catalogued many more endangered species, counteracting the environmental impact reports carried out by Ecuador's ENAMI and the Chilean company, Codelco, the two companies behind the Lulurimagua mine proposal. In relation to the results of the mammal survey, Carlos said, 'The results of the study will be really helpful for the court case. Another monkey now is on the critically endangered list, and it is believed there should be 70 species of mammals just within the community reserve; fifteen of which are in danger of extinction. This is a huge difference from what the company's environmental impact studies were reporting.'

---

† As this book went to press, we heard the news that the provincial court had revoked the decision for the Llurimagua copper-mining project to go ahead.

There is, as Carlos highlights, a deeply uncomfortable issue underlying all this. As part of the drive to tackle climate change, the rapidly accelerating shift to electric vehicles is driving the demand for copper – and the appalling impacts on human and ecological communities that come with copper-mining. The so-called clean energy transformation is vital. But Carlos asks, 'Shouldn't the question be, how can we contain runaway climate crisis without being complicit in human rights violations, the devastation of communities and the decimation of forests harbouring threatened species?'[4]

Bethan spent months in remote parts of the Peruvian Amazon with fellow conservationist Eilidh Munro filming their documentary about the impacts that building the road through the Manú Biosphere Reserve will have. *Voices on the Road* includes interviews with local people on both sides of the debate and went on to win numerous awards.

Kris Tompkins successfully handed over millions of acres to the Chilean government as national nature reserves. Tompkins Conservation has increased its emphasis on the reintroduction of missing species to strengthen and complete local ecosystems, which led to the creation of Rewilding Chile and Rewilding Argentina. Among the many species they are working to bring back is the jaguar.

POLITICAL UPDATES

Gustavo Petro, an economist, long-time senator and former guerrilla was elected as president of Colombia in August 2022. He is considered Colombia's first left-leaning president and his running mate, Francia Márquez, the first female, Afro-Colombian vice-president. Petro pledged to reform the economic system in Colombia, promising 'to expand social programmes, increase taxes on the wealthy, and transition the country away from extractive industries and towards a greener economy'. He is pushing for a total rethink of Colombia's conservation strategy, seeking a huge increase in funding for

Amazon rainforest defence, support for farmers to transition to sustainable practices and strong climate change policies.

In June 2022, a series of protests began in Ecuador. Rooted in 'structural social exclusion of indigenous peoples in Ecuador' against rising food and fuel prices, they were driven primarily by indigenous leaders and student and worker collectives. A state of emergency was declared in response and the Ecuadorian president, Guillermo Lasso, who succeeded Lenín Moreno in May 2021, was criticised for allowing violent and deadly responses to the protest, narrowly escaping impeachment. Ecuador, like Bolivia, Brazil and other South American countries was particularly hard hit by Covid, with the city of Guayaquil so overwhelmed at one point that bodies were left in the street.

Peru has emerged from a period of political turmoil with its first female president, lawyer and former vice-president Dina Boluarte. Boluarte, considered by some to be 'far right', was sworn in after President Pedro Castillo was impeached after only sixteen months in office. Violent protests continue in Peru.

The Bolivian president, Evo Morales, resigned after protests over the new road into the Bolivian Amazon; his response to wildfires that decimated millions of hectares of forest and grassland; the loss of a case against Chile in the international courts claiming land-locked Bolivia's rights of access to the sea; and finally, his disputed re-election for a fourth term in 2019. Withdrawal of support by the police and the army was a key factor in his resignation: allies argue that he was forced out by right-wing political interests. Opposition senator Jeanine Áñez became the interim president before Louis Arce, a former colleague of Evo Morales, won the presidential election in 2020.

Chile joined the South American countries taking a swing to the left with the election of its youngest ever leader, Gabriel Boric, aged 35, in 2021.

Following the run-off between President Jair Bolsonaro, credited with being leader of Brazil during the worst ever period of Amazon deforestation, and former president

Luiz Inaćio Lula da Silva, the world now holds it breath to see whether Lula can provide protection for indigenous communities, tackle illegal farming and logging and halt deforestation as promised. Many commentators considered the election to be the single most important ever held, in terms of the planet's future.

OTHER DEVELOPMENTS

The UN Biodiversity COP (COP 15), postponed due to the Covid pandemic and moved from China to Canada, finally took place in 2022, with governments signing up to the overall goal of halting and reversing biodiversity loss by 2030. The headline target agreed was to protect at least 30 per cent of land, fresh water and oceans by 2030; and there was explicit recognition of indigenous peoples' rights, roles, territories and knowledge as the most effective biodiversity protection. While the headlines focused on the '30 by 30' target, there was also a significant agreement 'outside the reserves and protected areas', for example, stating that 'countries will reduce the global footprint of consumption "in an equitable manner" and "significantly reduce overconsumption"'. Debate continues about how this will be funded and about the need for more precise targets on overconsumption, harmful subsidies, the restoration of degraded ecosystems and the halting of human-caused extinctions.

A UN report recognised that the indigenous peoples of Latin America are the best guardians of biodiversity and have a key role to play in the climate emergency, with deforestation up to 50 per cent lower in their territories. But the ever-increasing demand for beef, soy, timber, oil and minerals means they are also increasingly under threat.

The WWF's 2020 'Living Planet Report' used 'ecological footprinting' to estimate the pressure on the earth resulting from humanity's consumption of natural resources. The report concluded that humanity as a whole exceeded the regenerative capacity of the earth in the 1970s, and that this 'is the ultimate cause of the decline in the natural wealth of the world's forest, freshwater, and marine ecosystems'. It also showed that, while the steepest declines in biodiversity have been greatest in tropical ecosystems over the past 30 years, 'the loss of natural wealth in northern temperate ecosystems largely took place more than 30 years ago ... in 1996, the Ecological Footprint of an average consumer in the industrialized world was four times that of an average consumer in the lower income countries. This implies that rich nations (located mainly in northern temperate zones) are primarily responsible for the ongoing loss of natural wealth in the southern temperate and tropical regions of the world.'

In addition, it was among many analyses that linked the destruction of nature to the emergence of global pandemics. And it argued that 'bending the curve on biodiversity loss' will require changes to both production and consumption in relation to farming and, especially, to the production of and consumption of meat.

The 2022 'Living Planet Report' shows that there has now been a 69 per cent decline in the relative abundance of wildlife populations worldwide since 1970.

A slew of reports found that just four commodities – beef, soy, palm oil and wood products – are the biggest drivers of deforestation, with beef the worst offender. These can in turn be linked to major financiers and various household-name businesses. The vital work of tracing backwords through supply chains has linked UK supermarkets, such as Morrison's Sainsbury's, Iceland and Asda, via JBS – one of the largest food companies on earth – to beef cattle that have been raised on land cleared of rainforest.[5]

A report by Portfolio Earth found that 'in 2019, the world's largest banks invested more than USD 2.6 trillion ... in sectors which governments and scientists agree are the primary drivers of biodiversity destruction.' It concluded that the financial sector is bankrolling the mass extinction crisis while undermining human rights and indigenous sovereignty. These banks include HSBC and Barclays.[6]

Other links between commodities consumed in the UK and Europe and industries operating in South America with devastating consequences can readily be made in relation to gold, lead, oil and of course copper and lithium – driven in large part by the rapid rise in demand for electric vehicles.

In 2019, the Huaorani people won a major legal battle to have half a million acres of their territory in the Ecuadorian Amazon protected from oil-drilling. Nemonte Nenquimo, a Huaorani leader and co-founder of the indigenous-led non-profit organisation Ceibo Alliance was a winner of the Goldman Environmental Prize in 2020 and one of *TIME*'s 100 most influential people. In a stark echo of the Kogi's message to Younger Brother, Nenquimo wrote: 'To all the world leaders that share responsibility for the plundering of our rainforest ... this is my message to the Western world – your civilisation is killing life on earth.'[7]

In 2020, there was another major oil spill in the Ecuadorian Amazon. 672,000 gallons of crude oil and fuel spilled from two pipelines into the Coca and Napo rivers. A year later, 27,000 Kichwa people were still without safe water and food.

There were further Amazon oil spills in 2022 in both Ecuador and Peru (where tourists were taken hostage in a desperate attempt by indigenous groups to get medical help, clean water and food). And a devastating oil spill off the coast of Peru near Lima, wreaking havoc on one of the world's richest areas of marine diversity.

In 2022, oil extraction began at the Ishpingo oil field that lies partly within Ecuador's Yasuní National Park. Extraction at the Tiputini and Tambococha oil fields, also within Yasuní, has been underway since 2016. President Guillermo Lasso is said to have 'plans to double Ecuador's oil production in spite of opposition from indigenous communities and environmentalists'.[8]

The true cost of banana consumption continues to be revealed. In 2022, a court in the United States ruled that families suing the banana company Chiquita Brands International – the successor to United Fruit – for its role in funding paramilitary death squads in Colombia can proceed toward a jury trial.

The 2021 documentary *Seaspiracy* exposed the immense impacts of industrial fishing on ocean life and ecosystems. Among other things, it pointed out that $35 billion in subsidies is given to the fishing industry each year – and that plastic straws comprise 0.03 per cent of plastic entering the oceans each year. The campaign against plastic straws is an example of what George Monbiot refers to as 'micro-consumerist bollocks' that he sees as 'a displacement activity: a safe substitute for confronting economic power. Far from saving the planet, it distracts us from systemic problems and undermines effective action'.[9]

The 2021 Global Witness report revealed that 227 'land and environmental activists' were murdered in 2020, the worst figures on record. Colombia is now considered the most dangerous country in the world to be an environmental defender, with 65 (recorded) killings. In 2022, it reported that 1,700 had been murdered in the past decade, with the majority of those murders taking place in Latin America.[10]

Global Witness argues that there is an imperative need to hold corporations to account in relation to these attacks – on people and nature: 'Too many of the world's biggest and most powerful corporations are profiting from practices that are destroying our planet, driving people from their homes and, at times, fuelling violence against environmental defenders standing up against unfettered resource extraction in their communities.' What we need, they argue, 'are new global standards to stop companies being able to operate in a way that drives environmental and human rights abuses'.[11]

To end with the need for system change. According to the UK-based Colombian academic and activist, Oscar Guardiola-Rivera, my journey through South America put me in exactly the right place to seek for clues as to what a changed system might look like; to what 'different ways of inhabiting the earth' might entail.[12] Clearly, it includes a system with different models of development that makes a radical shift away from its current dependency on relentless, resource-dependent economic growth. A system with different values. Different power structures. Different ideas about what quality of life entails and about our relationship with nature. A system in service of human and environmental well-being, and not the other way around.

Guardiola-Rivera's point is that Latin America is rich with detail about these 'alternative' visions – many of them much older and deep-rooted than those based in consumerist capitalism. And it has a long and vital history of rebellion and resistance to colonialism in its many forms.

Among the clues I found on my journey, three crystalised into what might be thought of as guiding lights.[‡]

- *Buen vivir* – or at least the 'subversive and rebellious' versions – and other accounts of what it means to live well as a human on earth that are less focused on financial wealth and material possessions and more on positive and peaceful relations between people and each other, people and the rest of nature. Many of these insights are captured in the Happy Planet Index, and its regular assessments of how different countries are doing in terms of 'delivering long and meaningful lives within the earth's environmental limits'.[13] These accounts of what

‡ A fully articulated, comprehensive account of alternative systems is beyond the scope of this book (and probably its author). If my Spanish had progressed better, I'm sure that even more of these clues would be South American in origin.

it means to live well – with growth only in some areas, only when needed and as a means to an end rather than an end in itself – clearly connect and overlap with:

- Doughnut Economics – on Kate Raworth's brilliant model, the inner ring of the doughnut is a social foundation, ensuring that 'no one is left falling short on life's essentials' while the outer ring is an ecological ceiling, ensuring that 'humanity does not collectively overshoot the planetary boundaries'. There are many ways to live within the two – within the doughnut – but live within it we must. There is now a Doughnut Economics Action Lab, figuring out ways that communities, from local groups to cities to nations, can turn 'a radical idea into transformative action'.[14] Related approaches are taken by the Wellbeing Economy Alliance, whose basic premise is that the economic system must be transformed so that economics supports the well-being of people and planet, rather than vice versa.
- Aldo Leopold's environmental ethics – thinking of ourselves as fellow citizens of an ecological community on much the same terms as any other being, rather than as the managers of nature understood solely as a sort of warehouse of resources – profoundly alters the way we think about ourselves as well as nature. It displaces the eye-wateringly arrogant human-centrism and sense of entitlement that lies at the roots of so many of our environmental challenges. It, and other environmental ethics like it, provides the philosophical foundations for vastly better ways of treating other living beings; better, fairer ways of inhabiting the earth. As do the worldviews and ways of living of numerous indigenous peoples, of course.[15]

These vital ideas, visions and models are already out there. What an immense source of hope! It is a catastrophic failure of imagination to think that there is no alternative. And it isn't true. We like to think that we are the most intelligent species on earth: we have everything we need to figure out how to achieve that most important of ambitions – *all* of us living well on our single, beautiful, vibrant and flourishing planet.

There are plenty of other reasons for hope, too. The win-wins of 'natural solutions', such as regenerating ecosystems proving to be the most powerful way of removing carbon from the atmosphere; the way nature and biodiversity bounce back from even extreme assaults much faster than we thought; the brilliance, bravery, creativity and tenacity of activists all over the word. In sum, *we have the solutions*. We just need to implement them. And that of course is where it becomes about power and vested interests. Those who benefit most from the system as it is currently structured are also those with the most power and the most to lose. There are a million grounds for hope. Nevertheless, this is, undoubtedly, a fight.

CODA

Back in the UK after a month on the cargo ship, Woody had a thorough service at my local bike shop, the ever-supportive Ride Bikes in Ulverston. He is currently in my living room, though he and I will be back on the road giving *Life Cycle* talks soon.

A report from the Weizmann Institute of Science showed that 'the mass of all our stuff – buildings, roads, cars and everything else we manufacture – now exceeds the weight of all living things on the planet'. The mass of plastic alone is double that of all animals.

I remain amazed at the way in which touring on a bike while hauling your own stuff offers a fast-track understanding of the key to happiness and sustainability both. Have less, appreciate more, celebrate often.

For regular updates, further reading and thoughts about positive actions we can all take, please see The Life Cycle *page of my website Outdoor Philosophy: www.outdoorphilosophy.co.uk.*

ACKNOWLEDGEMENTS

Heartfelt thanks to all the activists I met, working to protect our ecological and human communities in so many different, brilliant, creative and often brave ways. I haven't been able to include all your stories, but you are all remembered. Thanks also to everyone who offered friendship and hospitality on the road.

The activist/hospitality list of names overlap, of course, and include: the crew of *Fort St Pierre*, Alicia, Olivia, Manuela, Alan, Sandra and Wilmer, Rosamira, Ybeth, aka Chile, Calu, Gabriel, Francy, Felix, Elizabeth, Vivien, Tinka, Diane, Paula, Francis and David, César, Claudia, Cedric, Sara, Maffi, Nhora, Christian, Oscar, Rosa and Daniella, Alejandro, Jennifer, Bart the Belgian, Costanza and Anna Marie, Javier, Jonnathan, Milton, Kendra, Eliana, Adriana, Juan, Dorys, Carlos, Ivonne, Javier, Andreas, Gabrielle, Alex, Tania, Eric, Lorenzo, Armando, Cindy, Christopher, Cynthia, Edilberto, Sol, Margarita, Guido, Lucho, Wilmer, Elli, Congressman Arana, Margarita, Carmen, Jorge, Diego, Bethan, Eilidh, Bryn, Chris, Christina, Adam, Kike, Nasario, Lilian, Ely, Valentina, José, Carmen, Okiram, Nohely, Ronal, Dani, Rod, Paula, Nacho, Mauricio, Kris, Tom, Forrest, Vincente, Iria and the crew of *Cap San Sounio*.

Thanks to the various Casa de Ciclistas I stayed at, and to all the cyclists I met, or who joined me, especially Tara and Aidan, Lisa and Shanne, the Bulgarians, Lee, Ferga, Chris and Neda. Your friendship is so very valued.

Thanks to Lizzie for filming with such tenacity and for the great footage that resulted; to the Pikes on Bikes for numerous briefings, brilliant routes and wise advice about cycling in South America; to Tim for fabulous spreadsheets detailing the best stoves and other camping kit; to Andy for invaluable help with project research and translation and to Jacs, 'thank you for the music'!

To everyone who contributed to the Crowdfunder that paid for Woody – including the Bamboo Bicycle Club course – and much of the kit I used to write and communicate while I was away, thank you. Especial thanks to the anonymous donor who contributed most of it, aka my brother, Bill, and to the Eden Project who donated the bamboo.

Thanks to Graham for suggesting *The Life Cycle* journey could be done as a John Muir Award and for helping make sure I completed the paperwork as well as the ride.

Special thanks to Shane and Nigel for the wonderful welcome back party at the RGS – the promise of that kept me going through some tough old miles! And to Richard at the Lake District National Park for agreeing to host the Cumbrian book-launch party – another great incentive to keep going.

Writing the book has been a much greater challenge than riding the Andes. Thanks to fellow cyclists and writers who helped keep me sane while engaged with similar challenges themselves, including Julian, who's been encouraging me for years, and to Dougie for all those mutual writing-challenge chats. Especial thanks to Bex – your emails never failed to make me laugh and sharing the information that no less a writer/traveller than Freya Stark grappled with self-loathing and despair cheered me up no end; to Jenny, I don't think I would've made it without all those wonderful messages and chats; and to Lee for the inspired idea of a writers' retreat that mostly consisted of dog walks and bike rides.

Thanks to many for useful conversations, especially Matt at WWF, Ian at the University of Cumbria and Tony at Natural England; and to everyone who has read and commented on drafts, including Lucy, for reading the early Colombia chapters; Laura and Tim for reading the later versions of those chapters and for being my 'ideal readers'; to Jonathon, Robert and Chris for heroically reading all of it and for giving such useful feedback and even more useful encouragement.

Special thanks to James, my agent, who took me and *The Life Cycle* book on, found me the best possible publisher, was endlessly encouraging and helpful and always emailed a 'So, how is it going?' message at just the right time. Emily, I can't thank you enough for connecting us – and for your own trail-breaking books and support, not least when I was applying to Banff.

I am very grateful to the Banff Centre for the Mountain and Wilderness writing programme, and the (generously subsidised) chance to attend it – a dream come true. Thanks to Jo, for the invitation to speak at the festival; to ace tutors Marni, Anthony and Harley; to all the other writers on the course; and, especially to the 'Banffies' that continued with Zoom meetings after the course, especially Maria, Gloria, Brian, Martina, Marni, Anthony, Michael and Louise. Your feedback, friendship and support has truly been a lifeline.

I have been incredibly fortunate to be able to do the ride and then write this book while earning practically no money. This would not have been possible without the endless generosity – financial and in so many other ways – of my partner, Chris. I owe you, big time. Much of the finance initially came in the form of a legacy from Chris' father, Danny, who was always generous in his belief in *The Life Cycle* project. Thank you, in memory. Thanks also to my dad, John, for a lifetime of support – that has only increased of late; and to the Authors' Licensing and Collecting Society for a generous hardship grant when I lost the little income I had from giving talks during Covid. When *The Life Cycle* is a best-seller, I promise I will repay you all.

To everyone at Icon Books and especially James, whose rigorous editing, help with identifying the many words that needed to be lost, fact-checking and attention to detail has always been so positive and supportive: thank you! Thanks also to Kiera, Duncan and Ruth; and to Fritha, who helped so much in an earlier round of cutting when the word count was near overwhelming.

Thank you to Endura, Lyon Equipment, Heart of the Lakes, Ride Bikes Ulverston – especially Jen, Dan and Jack – and Small World Consulting for sponsorship and support of various kinds – all very much appreciated. Thanks to Mihela at Patagonia Europe for the jacket and to Lisa for involving me in Patagonia UK's brilliant Adventure Activism tour when I got back.

Becky, thanks so much for the term 'bio-diversion', for great chats and, above all, for calling the coastguard.

Polly, I am so very sorry that our postponed reunion after I got back from South America turned out to mean that we never saw each other again. I think of you with such love, and so often.

To everyone at Holme House Farm, the small eco-community where I will soon live, thank you for putting up with endless whinges about deadlines, for all your support and for being my neighbours, friends and allies in figuring out how to live well by different values.

To all the other-than-human beings whose communities I travelled through and live in now, my biggest hope is that *The Life Cycle* will in some small way contribute to your future.

To Carter, whose transformation from stressed and difficult rescue dog to joyful, stick-chasing hill companion never fails to cheer.

And finally, to Chris, again, without whose support and love none of this would have been possible. Thank you. You're still a keeper.

NOTES

Introduction

1. More info about the CO2e savings – about fifteen kilograms compared with an average steel frame, according to Mike Berners-Lee – and building the bike can be found in these blog-posts on my Outdoor Philosophy website: https://www.outdoorphilosophy.co.uk/2016/10/05/how-to-build-a-bamboo-bicycle/ https://www.outdoorphilosophy.co.uk/2016/11/29/why-to-build-a-bamboo-bicycle/.

Chapter 1

1. Edward O. Wilson, recorded in the BioMuseo, Panama City.
2. Tony Juniper, *What has Nature Ever Done for Us? How Money Really Does Grow on Trees* (London: Profile Books, 2013), p. 31.
3. Alan Ereira, *The Elder Brothers' Warning* (London: Tairona Heritage Trust, 2008), pp. 126–7.
4. Ibid., p. 5.

Chapter 2

1. 'Living Planet Report', World Wildlife Fund (WWF), 2016, p. 39.
2. Michael Jacobs, *The Robber of Memories: A River Journey Through Colombia* (London: Granta Publications, 2012), p. 29.

Chapter 4

1. Francisco Sánchez-Bayo and Kris Wyckhuys, 'Worldwide decline of the etomofauna', *Biological Conservation*, vol. 232, April 2019, pp. 8–27.
2. Tony Juniper, *What has Nature Ever Done for Us? How Money Really Does Grow on Trees* (London: Profile Books, 2013)
3. 'Insect declines are a stark warning to reality', UN, 20 March 2019, https://www.unep.org/news-and-stories/story/insect-declines-are-stark-warning-humanity.
4. You can read all about it in *The Carbon Cycle: Crossing the Great Divide* (Uig: Two Ravens Press, 2012 and Edinburgh: Word Power Books, 2016 (revised 2nd edition)).

Chapter 5

1. *From the Heart of the World: The Elder Brothers' Warning*, first released in 1992.
2. Alan Ereira, *The Heart of the World* (London: Cape, 1990), p. 201.
3. Ibid., *The Elder Brothers' Warning*, pp. 156–7.
4. Ibid., pp. 156–7.
5. 'The Curse of Gold', Human Rights Watch, 1 June 2015, https://www.hrw.org/report/2005/06/01/curse-gold.
6. Ibid., *The Heart of the World*, p. 11.

Chapter 7

1. Carlos Zorrilla, 'Letter from Ecuador – where defending nature and community is a crime', The Ecologist, 15 March 2015, https://theecologist.org/2015/mar/25/letter-ecuador-where-defending-nature-and-community-crime.
2. Ibid.
3. Ibid., *Birds, Butterflies and Orchids: A Life in the Cloud Forest* (Carlos Zorrilla: 2008).
4. Ibid., 'Current Work', DECOIN, January 2010, https://www.decoin.org/currentwork/.

5. Ibid., 'Protecting Your Community Against Mining Companies and Other Extractive Industries', Global Response, 2009, https://www.culturalsurvival.org/sites/default/files/guide_for_communities_0.pdf.
6. Stewart M. Patrick, 'Why Natural Resources Are a Curse on Developing Countries and How to Fix It', *The Atlantic*, 30 April 2012.

Chapter 8

1. Henry Beston, *The Outermost House: A Year of Life on the Great Beach of Cape Cod* (London: Selwyn & Blount, 1928), p. 24.
2. Joe Kane, *Savages* (London: Pan, 1997), p. 70.
3. Aldo Leopold, *A Sound County Almanac: And Sketches Here and There* (Oxford: Oxford University Press, 1949).
4. Joe Kane, *Savages*, p. 4.
5. Sue Branford, 'Indigenous best Amazon stewards, but only when property rights assured: Study', Mongabay, 17 August 2020, https://news.mongabay.com/2020/08/indigenous-best-amazon-stewards-but-only-when-property-rights-assured-study/.

Chapter 9

1. Andrea Wulf, *The Invention of Nature: The Adventures of Alexander von Humboldt, the Lost Hero of Science* (London: Hodder & Stoughton, 2015), p. 5.
2. Aurore Chaillou, Louise Roblin and Malcolm Ferdinand, 'Why We need a Decolonial Ecology', *Green European Journal*, 4 June 2020.

Chapter 10

1. Karsten Heur, *Walking the Big Wild: From Yellowstone to the Yukon on the Grizzly Bears' Trail* (Toronto: McClelland and Stewart, 2002).

Chapter 11

1. A Canadian-based research group recently reached the following conclusion: 'In many cases, a lack of information is not the major barrier to biodiversity conservation; instead, mechanisms to translate information into action are most urgently needed.' See Rachel T. Bucton, et al., 'Key information needs to move from knowledge to action for biodiversity conservation in Canada', *Biological Conservation*, vol. 256 (April 2021).

Chapter 12

1. Neil Pike and Harriet Pike, *Peru's Cordilleras Blanca & Huayhuash: The Hiking and Biking Guide* (Hindhead: Trailblazer Publications, 2015).

Chapter 13

1. Dervla Murphy, *Eight Feet in the Andes: Travels with a Mule in Unknown Peru* (London: John Murray, 2003), p. 148.

Chapter 14

1. Tony Dajer, 'High in the Andes, A Mine Eats a 400-Year-Old City', *National Geographic*, 2 December 2015.
2. The film can be watched on YouTube: 'Video cortometraje mensaje de conservación en bici de bambú "Kate Rawles y su paso por Pasco"', Centro Labor, YouTube, https://www.youtube.com/watch?v=c5V-tJHMgIQ.
3. 'Yanacocha Mine, Peru', Environmental Justice Atlas, https://ejatlas.org/conflict/yanacocha-mine-peru.
4. Stephanie Boyd, *The Devil Operation* (Cusco: Quisca Productions, 2010).
5. Anna Leka Miller, 'Meet the Badass Grandma Standing Up To Big Mining', Daily Beast, 13 April 2017, https://www.thedailybeast.com/meet-the-badass-grandma-standing-up-to-big-mining.

Chapter 15

1. Davi Kopenawa, *The Falling Sky: Words of a Yanomami Shaman* (Cambridge, MA: Harvard University Press, 2013).

Chapter 16

1. Andrew Whitworth, et al., 'How much potential biodiversity and conservation value can a regenerating rainforest provide? A "best-case scenario" approach from the Peruvian Amazon', *Tropical Conservation Science*, vol. 9, 1 (2016), pp. 224–5.

Chapter 17

1. Thora Amend, et al., *Protected Landscapes and Seascapes* (Heidelberg: IUCN & GTZ, 2008).
2. 'Global Witness reports 227 land and environmental activists murdered in a single year, the worst figure on record', Global Witness, 13 September 2021, https://www.globalwitness.org/en/press-releases/global-witness-reports-227-land-and-environmental-activists-murdered-single-year-worst-figure-record/.
3. 'Machu Picchu', *National Geographic*, 15 November 2010.
4. Darrell La Lone, 'The Inca as a nonmarket economy: supply on command versus supply and demand', *Contexts for Prehistoric Exchange* (Academic Press Inc.,1982), pp. 292–316, p. 292.

Chapter 19

1. 'Pablo Solón Targeted: The Bolivian Government Must Stop Persecuting Those Defending Nature and Rights and Address the Real Problems', Committee for the Abolition of Illegitimate Debt, 10 July 2017, https://www.cadtm.org/Pablo-Solon-targeted-The-Bolivian.
2. Jeffrey Webber, 'Bolivian Horizons: An Interview with Pablo Solón', Political Economy Research Centre, 22

October 2019, https://www.perc.org.uk/project_posts/bolivian-horizons-interview-pablo-solon/.
3. 'Cargill: The Worst Company in the World', Mighty Earth, https://stories.mightyearth.org/cargill-worst-company-in-the-world/.
4. 'Pablo Solón targeted: The Bolivian Government Must Stop Persecuting Those Defending Nature and Rights and Address the Real Problems', Committee for the Abolition of Illegitimate Debt, https://www.cadtm.org/Pablo-Solon-targeted-The-Bolivian.

Epilogue

1. 'Planetary boundaries', Stockholm Resilience Centre, https://www.stockholmresilience.org/research/planetary-boundaries.html.
2. 'Communities resist the La Colosa Mine open-pit gold mine proposed in the Paramos of Tolima', ABColombia, https://www.abcolombia.org.uk/communities-resist-the-la-colosa-mine-open-pit-gold-mine-proposed-in-the-paramos-of-tolima/.
3. https://londonminingnetwork.org/.
4. Carlos Zorrilla, 'If the frogs should win', The Ecologist, 11 January 2021, https://theecologist.org/2021/jan/11/if-frogs-should-win.
5. Chris Moye, 'Cash Cow', Global Witness, 23 June 2022, https://www.globalwitness.org/en/campaigns/forests/cash-cow/.
6. 'Bankrolling Extinction', Portfolio Earth, https://portfolio.earth/campaigns/bankrolling-extinction/.
7. Nemonte Nenquimo, 'This is my message to the western world – your civilisation is killing life on Earth', *Guardian*, 12 October 2020.
8. 'Ecuador expands oil extraction from Amazon reserve', 14 April 2022, https://phys.org/news/2022-04-ecuador-oil-amazon-reserve.html.

9. George Monbiot, 'Sea Change', 9 April 2021, Monbiot.com, https://www.monbiot.com/2021/04/09/sea-change/.
10. 'Land and environmental defenders', Global Witness, https://www.globalwitness.org/en/campaigns/environmental-activists/.
11. 'Holding corporates to account', Global Witness, https://www.globalwitness.org/en/campaigns/holding-corporates-account/.
12. Oscar Guardiola-Rivera, *What if Latin America Ruled the World? How the South will take the North into the 22nd Century* (London: Bloomsbury, 2011).
13. Happy Planet Index, https://happyplanetindex.org/wp-content/themes/hpi/public/downloads/happy-planet-index-briefing-paper.pdf.
14. https://doughnuteconomics.org/.
15. For example, see Wade Davis, *The Wayfinders: Why Ancient Wisdom Matters in the Modern World*, (Toronto: House of Anansi Press, 2009).